FINITE ORTHOGONAL SERIES IN THE DESIGN OF DIGITAL DEVICES

(Analysis, Synthesis and Optimization)

M. G. KARPOVSKY

A HALSTED PRESS BOOK

JOHN WILEY & SONS, New York · Toronto
ISRAEL UNIVERSITIES PRESS, Jerusalem

ISRAEL UNIVERSITIES PRESS
is a publishing division of
KETER PUBLISHING HOUSE JERUSALEM LTD.
P.O.Box 7145, Jerusalem, Israel

Published in the Western Hemisphere by
HALSTED PRESS, a division of
JOHN WILEY & SONS, INC., NEW YORK

Library of Congress Cataloging in Publication Data

Karpovsky, Mark Girshevich.
Finite orthogonal series in the design of digital devices.

"A Halsted Press book."
Bibliography: p.
Includes index.
1. Digital electronics. 2. Digital control systems. 3. Series, Orthogonal. I. Title.
TK7869.D5K37 621.3815 75-44018
ISBN 0-470-15015-7

Distributors for the U.K., Europe, Africa and the Middle East
JOHN WILEY & SONS, LTD., CHICHESTER

Distributors for Japan, Southeast Asia and India
TOPPAN COMPANY, LTD., TOKYO AND SINGAPORE

Distributed in the rest of the world by
KETER PUBLISHING HOUSE JERUSALEM LTD.
IUP Cat No. 24200 4
ISBN 0 7065 1521 8

Set, printed and bound by Keterpress Enterprises, Jerusalem
PRINTED IN ISRAEL

CONTENTS

PREFACE

This book is concerned with methods for the analysis, synthesis and optimization of digital control devices, based on the use of finite orthogonal series. We develop a unified approach to these problems which is largely analogous to the Laplace-transform and Z-transform methods of automatic control theory. The functions that describe the behavior of the devices are expanded primarily in series of Walsh and Haar functions and their various generalizations.

There is today a fairly large literature on the use of Walsh and Haar functions in such fields as television and picture processing, electromagnetic radiation, radar, multiplexing, data processing, voice communication, pattern recognition, random access communication, and so on. These topics have been discussed, in particular, in the monograph of Harmuth [63] and in several recent symposia. The present book is apparently the first monograph in English on the application of these functions to analysis, synthesis and optimization of digital devices (of the many articles on the subject, special mention should be made of Lechner [100]).

The book contains results of research begun by the author and Dr. E. S. Moskalev in 1966. Some of them are accessible to the English-speaking reader (see [78, 80, 83–85, 88]), others have been published only in Russian; most of them are published here for the first time. The book was originally planned as a joint publication of the author and Dr. Moskalev as far back as 1971, but later circumstances prevented further collaboration. The author continued his work on the subject; in the process, new results were obtained and new methods evolved, so that the amount of material now presented is considerably more than in the preliminary version. The author is deeply indebted to Dr. Moskalev, whose fruitful collaboration was of essential importance for this book.

Almost all the results and methods described in the book are original, a major exception being the bulk of Chapter I, which is in the nature of a mathematical introduction.

The foundation of all the methods of analysis, synthesis and optimization considered in the book is expansion of the functions describing the operation of the devices as finite Fourier series in Walsh and Haar functions or their generalizations. It thus becomes possible to replace the original functions by their sequences of Fourier coefficients (spectra), which possess several remarkable properties, many of which are analogous to the properties of Laplace transforms and Z-transforms. This is true in particular of the Walsh functions and their generalizations (Chrestenson functions). For this reason, Walsh and Chrestenson expansions (Walsh and Chrestenson spectra) furnish our main tools in the solution of problems of analysis and optimization.

Besides Walsh and Chrestenson spectra, considerable use is made of various correlation functions derived from the original functions by iterated application of Walsh and Chrestenson transformations (i.e., the transformations associating with any function its Walsh or Chrestenson spectrum). These correlation functions recall the classical correlation functions used in communication theory, the theory of stochastic processes, etc., and have recently begun to receive wide application in the synthesis of digital filters, pattern recognition, and so on.

Unlike Walsh (Chrestenson) expansions and correlation functions, Haar expansions (spectra) and their generalizations are used primarily in the solution of synthesis problems. Underlying this situation is the fact that the Fourier coefficients with respect to Haar-type bases depend only on the "local" behavior of the original function, i.e., on its behavior at several "neighboring" points, whereas each of the Fourier coefficients in a Walsh-type expansion depends on the "global" behavior of the function, i.e., its behavior at all points of definition.

All of the synthesis methods discussed may be used both to design networks implementing digital devices (network interpretation of synthesis methods) and to set up algorithms simulating the behavior of the devices on a computer (program interpretation). Most of our attention will be taken up with the network interpretation.

The special languages introduced here—Walsh- and Haar-type spectra, correlation functions—are eminently suited for the solution of many problems which are quite difficult to tackle in terms of the original functions (a number of problems of this type will be considered in Chaps. II and V).

The proposed methods are analytical in the sense that, as a rule, they do not require examination of alternatives (the "brute-force" approach). It thus becomes possible to cope with problems of comparatively high dimensions (unlike, say, most of the methods used in applied switching theory and applied automata theory). In addition, the availability of fairly simple analytical procedures opens the way to efficient use of computers in the problems under consideration.

All of the methods examined are intended for devices based on elements with two or p ($p > 2$) stable states. Indeed, functions describing the operation of such devices may be interpreted as functions defined on the groups of binary or p-ary vectors with respect to componentwise addition modulo 2 or p, respectively. The Walsh and Chrestenson functions are the characters of these groups, so that methods for solving problems of analysis and optimization based on their use fall into the framework of abstract harmonic analysis on finite commutative groups. Thus, in problems connected with the design of digital devices from bistable and multistable elements the Walsh and Chrestenson functions seem to play a role analogous to that of the sine and cosine functions in automatic control theory.

On the other hand, our analysis and optimization methods may be generalized without difficulty to devices whose behavior is described by functions defined on arbitrary finite commutative groups (e.g., devices built from elements with different numbers of stable states, devices operating in systems of residue classes, etc.). The role of the Walsh and Chrestenson functions will then be taken over by the characters of the appropriate groups (this approach is outlined in §1.6). The only reason that most of our attention is devoted to devices whose elements have the same number of stable states is that such devices are the most widely employed in engineering practice.

A more challenging and interesting problem would be to generalize the methods and results of the book to functions defined on finite noncommutative groups; in this context the Walsh and Chrestenson functions would probably be replaced by irreducible representations of the groups in question. This question is not considered in the present book.

A topic of particular importance in engineering practice, which receives special attention in the book, is the analysis of reliable (fault-tolerant) digital devices and the synthesis of reliable systems for the transmission and processing of information. We shall also consider questions involved in the technical realization of the synthesis methods. Our discussion of the reliability of digital devices (Chap. V) makes essential use of correlation functions, while the main tools used in connection with technical realization problems (Chap. VI) are the Haar functions.

The network solutions considered in the technical realization section are essentially of a schematic nature. As it turns out, these solutions are practically independent of the specific features of the functions implemented by the device. These features affect only the content of the storage unit holding the expansion coefficients, so that it is relatively easy to adapt the functions to changes in the medium in which they are to operate.

Moreover, the networks proposed contain only such standard computer modules as registers, counters, decoders, memory and adders; this yields a high degree of technological feasibility and permits the use of highly integrated circuitry.

We have arranged the book in such a way that the various sections are comparatively independent. The interdependence of the sections is shown in Figure 0.1.

The mathematical tools employed in the book comprise parts of group theory and the theory of characters, switching function theory, automata theory and the theory of error-correcting codes. Although we present all necessary definitions in the text, a more thorough familiarity of the reader with these disciplines would be highly desirable.

The essential contribution of Dr. E. S. Moskalev should be re-emphasized here.

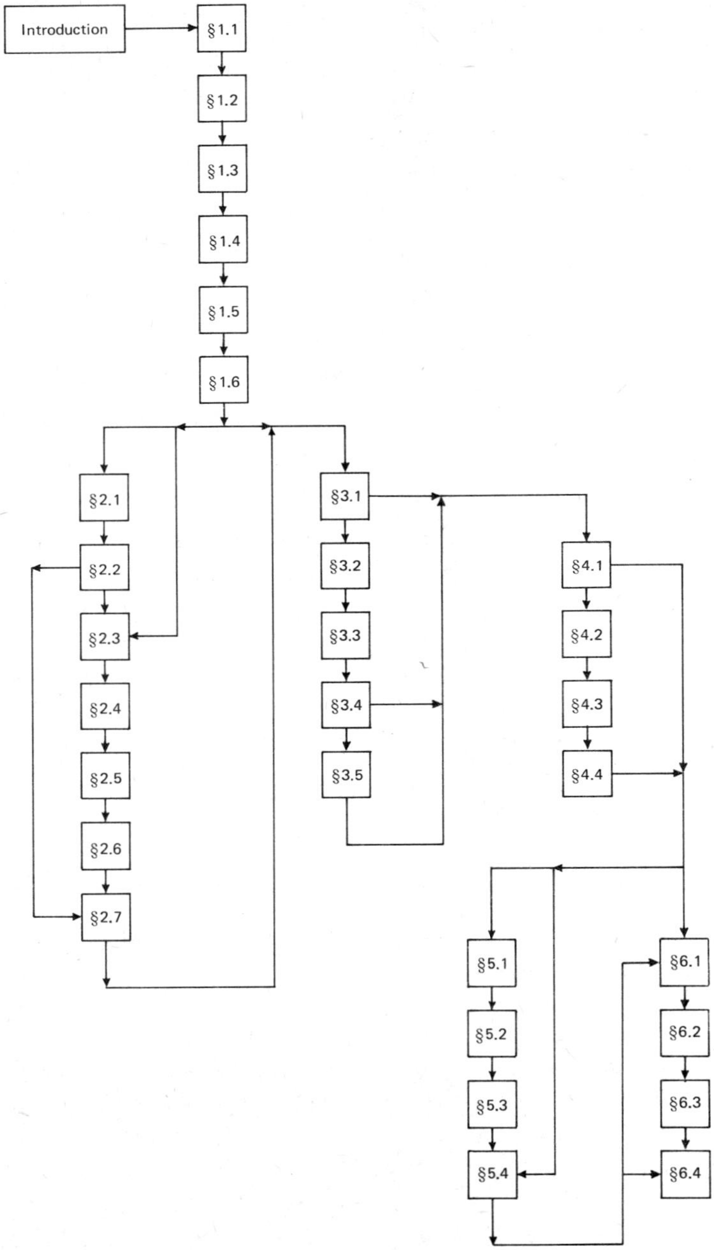
Introduction
§1.1
§1.2
§1.3
§1.4
§1.5
§1.6
§2.1
§2.2
§2.3
§2.4
§2.5
§2.6
§2.7
§3.1
§3.2
§3.3
§3.4
§3.5
§4.1
§4.2
§4.3
§4.4
§5.1
§5.2
§5.3
§5.4
§6.1
§6.2
§6.3
§6.4

FIGURE 0.1.

Of the many persons who made the publication of this book possible, I am particularly indebted to Mrs. M. Podolsky and Mr. A. Trachtenberg of the Technion (Israel Institute of Technology), Haifa. Special thanks are due to the translator and editor of the manuscript, Mr. D. Louvish, whose skill contributed in no small degree to the quality of the final version.

Tel-Aviv University,
Tel-Aviv, May 1975

Mark Karpovsky

Chapter One

SPECTRAL TRANSFORMS OF LOGICAL FUNCTIONS*

The subject of this chapter is the representation of systems of logical functions by orthogonal series. We describe the basic properties of various discrete functional transforms relating the original systems of functions to the series coefficients, introduce spectral and correlation characteristics of systems of logical functions, and demonstrate their use for the analysis and synthesis of digital devices. In short, the chapter sets forth the main mathematical tools to be used in the sequel. A thorough grasp of the results presented here is a prerequisite for an understanding of the subsequent chapters.

1.1. ORTHOGONAL-SERIES REPRESENTATION OF A SYSTEM OF LOGICAL FUNCTIONS

1.1.1. Devices intended for the processing of discrete information may be divided into two classes: digital devices without memory (*combinational networks*), in which the output depends only on the present input, and digital devices with *memory*, whose output is determined by the input plus the internal state of the device. In devices with memory one can always distinguish two blocks, one a combinational network and the other containing only memory elements [70, 92]. Our first topic will therefore be systems of functions describing the operation of combinational networks.

The operation of a combinational network built from elements with p ($p \geqq 2$) stable states, m p-ary inputs and k p-ary outputs is described by a *system of p-valued logical functions*:

$$y^{(s)} = f^{(s)}(z^{(0)}, z^{(1)}, \ldots, z^{(m-1)}) \qquad (s = 0, 1, \ldots, k-1;\, y^{(s)}, z^{(s)} \in \{0, 1, \ldots, p-1\}). \tag{1.1.1}$$

In the particular case $p = 2$, the functions in system (1.1.1) are known as *Boolean functions* (*BF*); this is the case most frequently used in engineering practice.

* Some authors use the term "switching functions" for both Boolean and many-valued logical functions, but since "switching logic" is generally understood to refer to Boolean (two-valued) logic we prefer the less ambiguous "logical function."

System (1.1.1) may also be specified by a table (Table 1.1.1), known as a *truth table*. This table has p^m rows, each corresponding to a p-ary argument vector of length m and an appropriate p-ary vector of length k specifying the values of the functions.

The design of a digital device (DD) involves two basic problems: *synthesis* and *analysis*. The synthesis problem is: given a system of type (1.1.1) or a truth table (Table 1.1.1), construct an optimal DD (in a sense to be specified) realizing the system; that is to say, a DD whose input signals $z^{(0)}, z^{(1)}, \ldots, z^{(m-1)}$ and output signals $y^{(0)}, y^{(1)}, \ldots, y^{(k-1)}$ satisfy equations (1.1.1). The analysis problem is to describe the operation of a given DD in terms of a system (1.1.1) and to determine the properties of this system.

TABLE 1.1.1.

$z^{(0)}$	$z^{(1)}$	$\ldots$	z^{m-1}	$y^{(0)}$	$y^{(1)}$	$\ldots$	$y^{(k-1)}$

The analysis and synthesis problems may be solved with the help of Boolean algebra or many-valued logic [70, 92, 156]. The main shortcoming of these tools, however, is that in the basic problems of the above type they require an exhaustive examination of all alternatives (the so-called *brute-force method*). Since the number of alternatives may increase exponentially with the number m of arguments, the actual use of these tools in engineering practice is already well-nigh impossible (even when computers are employed) for problems involving comparatively few arguments.

In this book we shall examine another approach to analysis and synthesis, based on the use of spectral expansions of systems of logical functions (LF): the systems will be represented by functions of a continuous argument, and the latter expanded in finite orthogonal series.

The pros and cons of this "spectral" approach will be considered in §6.4, where it will also be compared with Boolean algebra and many-valued logic and recommendations will be made as to its use.

1.1.2. Consider an LF described by (1.1.1) or Table 1.1.1. We shall represent this function by a step function $\Phi(z)$ of a real argument, defined on a half-open interval $[0, p^m)$, as follows.

Set

$$z = \sum_{s=0}^{m-1} z^{(s)} p^{m-1-s}, \tag{1.1.2}$$

$$y = \sum_{s=0}^{k-1} y^{(s)} p^{k-1-s}. \tag{1.1.3}$$

Using this notation, we can represent system (1.1.1) by a discrete function $y = f(z)$, defined at the points $0, 1, \ldots, p^{m-1}$ of the interval $[0, p^m)$. Finally, we complete $y = f(x)$ to a step function $\Phi(z)$ by defining

$$\Phi(z) = f(\delta) \quad \text{when} \quad z \in [\delta, \delta + 1). \tag{1.1.4}$$

We shall say that a *step function* $\Phi(z)$ *represents the original system of LF* if $\Phi(z)$ satisfies (1.1.4) for the function $f(z)$ defined as in (1.1.1)–(1.1.3).

The solution of all analysis and synthesis problems for systems of LF will be based on their step-function representations. Throughout the sequel we shall use the same notation for the arguments of the functions and the vectors of their p-ary expansions, provided no confusion can arise.

EXAMPLE 1.1.1. Let $p = 2$, and consider the system of BF (Boolean functions) defined by Table 1.1.2. This table describes the operation of a one-digit adder, where $y^{(0)}$ is the sum output and $y^{(1)}$ the carry output.

TABLE 1.1.2.

$z^{(0)}$	$z^{(1)}$	$z^{(2)}$	$y^{(0)}$	$y^{(1)}$
0	0	0	0	0
0	0	1	1	0
0	1	0	1	0
0	1	1	0	1
1	0	0	1	0
1	0	1	0	1
1	1	0	0	1
1	1	1	1	1

TABLE 1.1.3.

z	$f(z)$
0	0
1	2
2	2
3	1
4	2
5	1
6	1
7	3

The corresponding function $f(z)$ is defined by Table 1.1.3, and the step function $\Phi(z)$ representing the system is shown in Figure 1.1.1.

1.1.3. The next step in solution of the analysis and synthesis problems is to expand the step function $\Phi(z)$ representing the system of LF as an *orthogonal series* or *Fourier series*:

$$\Phi(z) = \sum_{\omega=0}^{\infty} S(\omega)\Psi_\omega(z), \tag{1.1.5}$$

where $\{\Psi_\omega(z)\}$ is a complete system of orthogonal step functions (*basis*) defined on $[0, p^m)$, i.e.,

$$\left(\int_0^{p^m} \Psi_\omega(z)\overline{\Psi_\omega(z)}\,dz\right)^{-1} \int_0^{p^m} \Psi_\omega(z)\overline{\Psi_r(z)}\,dz = \begin{cases} 1 & \text{if} \quad \omega = r, \\ 0 & \text{if} \quad \omega \neq r. \end{cases} \tag{1.1.6}$$

Here and below $\overline{\Psi(z)}$ is the function complex conjugate to $\Psi(z)$.

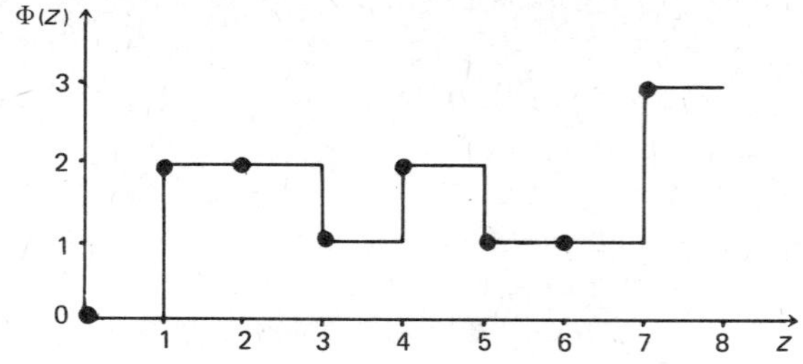

FIGURE 1.1.1.

The coefficients $S(\omega)$ of the series (1.1.5) are the Fourier coefficients, defined by

$$S(\omega) = \left(\int_0^{p^m} \Psi_\omega(z) \overline{\Psi_\omega(z)} \, dz \right)^{-1} \int_0^{p^m} \Phi(z) \overline{\Psi_\omega(z)} \, dz. \tag{1.1.7}$$

The sequence of coefficients $S(0), S(1), \ldots$ of the step function $\Phi(z)$ representing the system of LF, relative to the basis $\{\Psi_\omega(z)\}$, will be called the *spectrum* of the system relative to $\{\Psi_\omega(z)\}$.

Thus, formulas (1.1.2) through (1.1.7) define a chain of one-to-one transformations:

$$\{f^{(s)}(z^{(0)}, \ldots, z^{(m-1)})\} \leftrightarrow f(z) \leftrightarrow \Phi(z) \leftrightarrow S(\omega).$$

Since there is a one-to-one correspondence between the original system of LF $\{f^{(s)}(z^{(0)}, \ldots, z^{(m-1)})\}$ and its spectrum $S(\omega)$, the problems of analysis and synthesis for such systems will generally be formulated and solved in the language of spectra (or in the language of correlation functions (see §1.5), which are double spectral transforms of the representing step function).

The solution of our problems clearly depends in an essential manner on the choice of basis. The relevant questions will be considered below: for two-valued (Boolean) logic in §1.2, and for many-valued logic in §1.3.

1.2. BASES FOR SYSTEMS OF BOOLEAN FUNCTIONS

1.2.1. As mentioned above, the choice of basis is a decisive factor in the solution of analysis and synthesis problems using spectral representations of systems of LF. In particular, it will determine the number of nonzero values of spectrum, and this number in turn (see §§6.1–6.3) will determine the complexity of the realization of a given system of LF. By a *realization* of a system we mean a device whose input and output signals stand in the relation defined by the functions of the system.

Since we are representing the initial system of p-ary LF of m arguments by a step function $\Phi(z)$ on $[0, p^m)$ of *span* 1 (i.e., each "step" is of length 1) it is natural to stipulate that the basis consist of step functions with the same span on $[0, p^m)$. As we shall show below, one can always choose a basis such that the number of nonzero terms in the series (1.1.5) representing $\Phi(z)$ is finite, in fact, at most p^m.

Another natural requirement is that each of the basis functions take on a relatively small number of values, thus simplifying physical realization of the system of LF. For example, if $p = 2$, it will be convenient to have basis functions with values 0, ± 1. If this is the case, in fact, there is no need for the relatively difficult multiplication operation in calculation of $\Phi(z)$ from (1.1.5).

Finally, it will be convenient if the values of the basis functions $\Psi(z)$ can be calculated by simple, standard, analytical procedures applied to the values of z.

We now exhibit a few bases which meet these requirements.

We first consider two systems of functions (Walsh functions and Haar functions) which are admirably suited for spectral representations of systems of Boolean functions (BF). Systems of these types arise quite frequently in present-day design of digital devices.

1.2.2. The *Walsh functions* are step functions on $[0, 2^m)$, defined for any natural ω and m, $0 \leqq \omega \leqq 2^m - 1$, by

$$W_\omega(z) = (-1)^{\sum_{s=0}^{m-1} \omega^{(m-1-s)} z^{(s)}} = \exp\left(\frac{2\pi}{2} i \sum_{s=0}^{m-1} \omega^{(m-1-s)} z^{(s)}\right), \tag{1.2.1}$$

where $\omega^{(s)}$, $z^{(s)}$ are determined by the binary expansions of ω and z:

$$\omega = \sum_{s=0}^{m-1} \omega^{(s)} 2^{m-1-s}, \tag{1.2.2}$$

$$z = \sum_{s=0}^{m-1} z^{(s)} 2^{m-1-s} \quad \text{and} \quad i = \sqrt{-1}. \tag{1.2.3}$$

The Walsh functions defined by (1.2.1)–(1.2.3) are discrete functions, defined at the points $0, 1, \ldots, 2^{m-1}$ of the interval $[0, 2^m)$; they may now be completed to step functions, as done above for $f(z)$ in (1.1.4). This completion to a step function will be adhered to throughout the sequel for all discrete functions that arise.

The Walsh functions are an important tool in the approximation of continuous functions. In such problems one usually normalizes them to the interval [0, 1) and considers the set of Walsh functions for $m \to \infty$. Recently, Walsh functions have also found application in such fields as radar, television, electromagnetic radiation, data processing and pattern recognition [63, 76, 103, 120, 123]. There are good grounds for the assertion that Walsh functions and their generalizations (§1.3) should play a role in the analysis, synthesis and optimization of digital devices analogous to that of the exponential (or trigonometric) functions in analysis and synthesis of continuous-time (analog) devices.

The Walsh functions for $m = 3$ are illustrated in Figure 1.2.1.

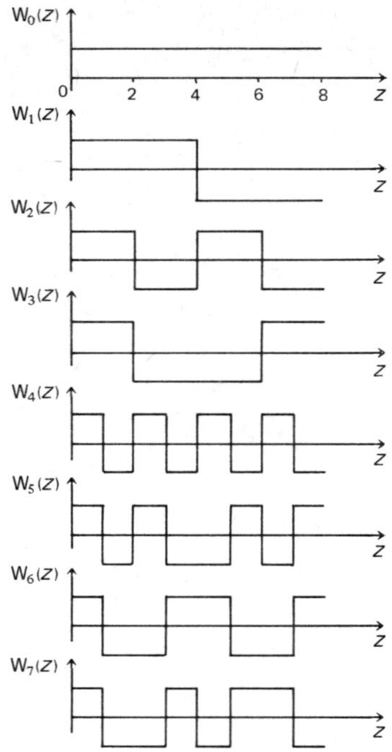

FIGURE 1.2.1.

The subscript ω of the function $W_\omega(z)$ will be called its *index*, and the number of ones in the binary expansion of ω (i.e., the quantity $\|\omega\| = \sum_{s=0}^{m-1} \omega^{(s)}$) will be called the *weight* of the index.

Consider the subset of Walsh functions with indices of weight 1. Denote these functions by $R_s(z)$ ($s = 1, \ldots, m$). Then by (1.2.1), (1.2.2),

$$R_s(z) = W_{2^{s-1}}(z) = (-1)^{z^{(s-1)}} = \exp\left(\frac{2\pi}{2} iz^{(s-1)}\right), \qquad i = \sqrt{-1}. \qquad (1.2.4)$$

The functions $R_s(z)$ thus defined are known as the *Rademacher functions*. The first four Rademacher functions are shown in Figure 1.2.2. Note that if we change -1 to $+1$ and $+1$ to 0, the function $R_s(z)$ is simply $z^{(s-1)}$.*

* The following equivalent definition of the Rademacher functions is sometimes employed:

$$R_s(z) = \operatorname{sgn} \sin 2^{s-m}\pi z,$$

where

$$\operatorname{sgn} a = \begin{cases} 1 & \text{if } a > 0, \\ -1 & \text{if } a \leq 0. \end{cases}$$

This formula illustrates the relationship of the Rademacher functions (hence also the Walsh functions) to the ordinary harmonic functions.

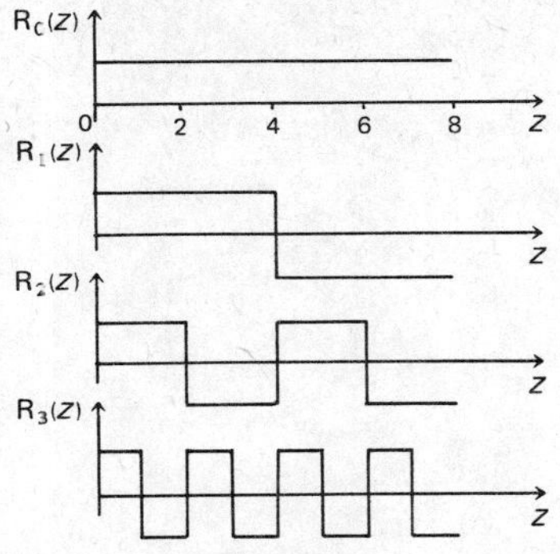

FIGURE 1.2.2.

Any Walsh function may be expressed as a product of Rademacher functions. Indeed, by (1.2.1) and (1.2.4),

$$W_\omega(z) = \prod_{s=0}^{m-1} (R_{s+1}(z))^{\omega(m-1-s)}. \tag{1.2.5}$$

Thus the set of Walsh functions is the multiplicative closure of the set of Rademacher functions.

We now consider a few properties of Walsh functions which will be used constantly in what follows.

I. *Completeness and orthogonality.*

THEOREM 1.2.1. *The Walsh functions form a complete orthogonal system. In other words*:

$$\sum_{z=0}^{2^m-1} W_t(z)\, W_\omega(z) = \begin{cases} 2^m & \text{if} \quad t = \omega, \\ 0 & \text{if} \quad t \neq \omega, \end{cases} \tag{1.2.6}$$

and if $\Phi(z)$ *is a step function representing some system of BF of m arguments and*

$$\sum_{z=0}^{2^m-1} \Phi(z)\, W_\omega(z) = 0 \qquad (\omega = 0, 1, \ldots, 2^m - 1),$$

then $\Phi(z) \equiv 0$.

Theorem 1.2.1 shows that we can use the Walsh functions as a basis for orthogonal expansion of step functions representing systems of BF.

COROLLARY 1.2.1. *For any natural* $\omega, 0 \leqq \omega \leqq 2^m - 1$,

$$\sum_{z=0}^{2^m-1} W_\omega(z) = \begin{cases} 2^m & \text{if} \quad \omega = 0, \\ 0 & \text{if} \quad \omega \neq 0. \end{cases} \tag{1.2.7}$$

This follows from (1.2.6) by setting $t = 0$, since $W_0(z) \equiv 1$.

II. *Finiteness of representing series.*

THEOREM 1.2.2. *Let $\Phi(z)$ be a step function representing a system of BF of m arguments. Then*

$$\Phi(z) = \sum_{\omega=0}^{2^m-1} S(\omega)\, W_\omega(z), \tag{1.2.8}$$

where

$$S(\omega) = 2^{-m} \sum_{z=0}^{2^m-1} \Phi(z)\, W_\omega(z). \tag{1.2.9}$$

Theorem 1.2.2 is proved by inserting (1.2.9) into (1.2.8) and using Theorem 1.2.1.

The expansion (1.2.9) is the representation of $\Phi(z)$ as a finite Fourier series, and the numbers $S(\omega)$ are its Fourier coefficients.

It follows from Theorem 1.2.2 that the Walsh expansion of any step function representing a system of BF of m arguments contains at most 2^m terms with Walsh functions with index weight $\leqq m$. Thus, if the original system of BF is defined by a sequence of 2^m values, its Walsh spectrum will also contain at most 2^m nonzero terms.

EXAMPLE 1.2.1. Consider a system of BF describing the operation of an adder, as in Table 1.1.2 (see Example 1.1.1). Let us expand the corresponding step function (see Table 1.1.3, Figure 1.1.1) in a Walsh series. The result is

$$\begin{aligned}\Phi(z) &= 2^{-3}(12 - 2W_1(z) - 2W_2(z) - 2W_4(z) - 6W_7(z)) \\ &= 2^{-2}(6 - R_1(z) - R_2(z) - R_3(z) - 3R_1(z)\, R_2(z)\, R_3(z)).\end{aligned}$$

III. *Symmetry of index and argument.*

THEOREM 1.2.3. *For any $\omega, z \in \{0, 1, \ldots, 2^m - 1\}$,*

$$W_\omega(z) = W_z(\omega). \tag{1.2.10}$$

This follows directly from the definition of Walsh functions (see formulas (1.2.1)–(1.2.3)).

IV. *Translation of argument.*

THEOREM 1.2.4. *For any $\omega, z, \tau \in \{0, 1, \ldots, 2^m - 1\}$,*

$$W_\omega(z \oplus \tau) = W_\omega(z)\, W_\omega(\tau) \pmod 2. \tag{1.2.11}$$

Here and below the notation $a \oplus b \pmod p$ denotes the number whose p-ary expansion is the sum modulo p of the corresponding p-ary components of a and b; similarly, $a \ominus b \pmod p$ denotes the difference modulo p.

The proof of Theorem 1.2.4 follows readily from (1.2.1)–(1.2.3).

It follows from Theorems 1.2.3 and 1.2.4 that the set of Walsh functions is closed under multiplication.

V. *Isomorphism of set of linear BF and set of Walsh functions.*

A Boolean function $f(z^{(0)}, \ldots, z^{(m-1)})$ is said to be *linear* if it may be expressed as

$$f(z^{(0)}, \ldots, z^{(m-1)}) = c_0 z^{(0)} \oplus \cdots \oplus c_{m-1} z^{(m-1)} \pmod 2 \tag{1.2.12}$$

where $c_s, z^{(s)} \in \{0, 1\}$ $(s = 0, 1, \ldots, m-1)$.

The number of linear BF of m arguments is 2^m. The set of linear BF is a commutative group,* with $(f \oplus g)(z) = f(z) \oplus g(z) \pmod 2$. In other words, the group operation is addition modulo 2, the identity e is 0, and $f^{-1} = f$ for any f.

The Walsh functions also form a group with respect to multiplication, with $e = W_0(z) \equiv 1$, $W_\omega^{-1}(z) = W_\omega(z)$ and $(W_\omega \cdot W_q)(z) = W_\omega(z)\, W_q(z)$.

Two groups G_1 and G_2 are said to be *isomorphic* if there exists a one-to-one correspondence h: $a \leftrightarrow h(a)$, $a \in G_1$, $h(a) \in G_2$, such that $h(a \circ b) = h(a) \circ h(b)$, where $\circ$ denotes the group operation in G_1 and G_2, as the case may be (these operations may of course be quite different). The correspondence h is assumed to exhaust the entire group G_2.

THEOREM 1.2.5. *The group of Walsh functions is isomorphic to the group of linear BF.*

Indeed, we shall show that the one-to-one correspondence

$$h(W_\omega(z)) = \bigoplus_{s=0}^{m-1} \omega^{(m-1-s)} z^{(s)} \pmod 2$$

* Recall that a *commutative group* is a set G with a binary operation $\circ$ such that:

1) for any $a, b \in G$, $a \circ b \in G$;
2) there exists $e \in G$ such that for any $a \in G$

$$a \circ e = e \circ a = a;$$

3) for any $a \in G$, there exists $a^{-1} \in G$ such that $a \circ a^{-1} = e$;
4) for any $a, b, c \in G$,

$$a \circ (b \circ c) = (a \circ b) \circ c;$$

5) for any $a, b \in G$,

$$a \circ b = b \circ a.$$

is an isomorphism. Using the symmetry and translation properties, we have

$$h(W_t(z)\,W_q(z)) = h(W_{t\oplus q}(z)) = \bigoplus_{s=0}^{m-1} (t^{(m-1-s)} \oplus q^{(m-1-s)})\, z^{(s)}$$

$$= \bigoplus_{s=0}^{m-1} t^{(m-1-s)} z^{(s)} \oplus \bigoplus_{s=0}^{m-1} q^{(m-1-s)} z^{(s)} = h(W_t(z)) \oplus h(W_q(z)) \qquad (\text{mod } 2). \quad \blacksquare$$

By virtue of the isomorphism between the group of Walsh functions and the group of linear BF, one can solve various problems involving linear BF and linearization of BF (see Chapter II) in the language of Walsh expansions, and, conversely, Walsh expansions may be treated as expansions in linear BF.

1.2.3. Henceforth, Walsh spectra will be used extensively as a working tool in solution of analysis and synthesis problems for network realizations of BF. We shall therefore devote some attention to methods for their construction and computation.

We first consider an effective algorithm for construction of the Walsh spectrum and estimate its complexity. This algorithm is similar to the analogous algorithm used for the basis of trigonometric functions ("fast Fourier transform" [27, 114]).

THEOREM 1.2.6. [11]. *Let $\Phi(z)$ be a step function representing a system of BF of m arguments, and $S(\omega)$ $(\omega = 0, 1, \ldots, 2^m - 1)$ its Walsh spectrum. Set*

$$\begin{cases} a_0(\omega) = \Phi(\omega), & \\ a_0(2^{m-1} + \omega) = \Phi(2^{m-1} + \omega) & (\omega = 0, 1, \ldots, 2^{m-1} - 1), \\ a_q(\omega) = a_{q-1}(2\omega) + a_{q-1}(2\omega + 1), & \\ a_q(2^{m-1} + \omega) = a_{q-1}(2\omega) - a_{q-1}(2\omega + 1) & (q = 1, \ldots, m). \end{cases} \tag{1.2.13}$$

Then

$$S(\omega) = 2^{-m} a_m(\overleftarrow{\omega}),$$

where $\overleftarrow{\omega} = \sum_{s=0}^{m-1} \omega^{(s)} 2^s$ *if* $\omega = \sum_{s=0}^{m-1} \omega^{(s)} 2^{m-1-s}$.

Theorem 1.2.6 describes a simple algorithm for calculating the Walsh spectrum, in terms of a recursive procedure yielding the sequence $a_q(\omega)$.

The number of operations N_W required to calculate $S(\omega)$ with this algorithm is $N_W = m \cdot 2^m$, where m is the number of binary digits in the argument. A sample calculation of the spectrum is shown in Table 1.2.1.

Another method for construction of the Walsh spectrum is very convenient if m is not too large. Let W_m be the matrix whose ω-th row is $(W_\omega(0), W_\omega(1), \ldots, W_\omega(2^m - 1))$

TABLE 1.2.1.

z, ω	$\Phi(z)$	$a_1(\omega)$	$a_2(\omega)$	$a_3(\omega)$	$\overleftarrow{\omega}$	$8S(\omega)$
0	1	3	7	27	0	27
1	2	4	20	−5	4	−13
2	0	4	−5	−13	2	−13
3	4	16	0	7	6	11
4	3	−1	−1	−13	1	−5
5	1	−4	−12	−5	5	−5
6	7	2	3	11	3	7
7	9	−2	4	−1	7	−1

($\omega = 0, 1, \ldots, 2^{m-1}$). For example, for $m = 2$ and $m = 3$ the matrices W_2 and W_3 are

$$W_2 = \begin{pmatrix} 1 & 1 & 1 & 1 \\ 1 & 1 & -1 & -1 \\ 1 & -1 & 1 & -1 \\ 1 & -1 & -1 & 1 \end{pmatrix}, \qquad W_3 = \begin{bmatrix} 1 & 1 & 1 & 1 & 1 & 1 & 1 & 1 \\ 1 & 1 & 1 & 1 & -1 & -1 & -1 & -1 \\ 1 & 1 & -1 & -1 & 1 & 1 & -1 & -1 \\ 1 & 1 & -1 & -1 & -1 & -1 & 1 & 1 \\ 1 & -1 & 1 & -1 & 1 & -1 & 1 & -1 \\ 1 & -1 & 1 & -1 & -1 & 1 & -1 & 1 \\ 1 & -1 & -1 & 1 & 1 & -1 & -1 & 1 \\ 1 & -1 & -1 & 1 & -1 & 1 & 1 & -1 \end{bmatrix}.$$

Matrices of this type are *Hadamard matrices* [11, 126], and so all the properties of Walsh functions may be phrased in terms of Hadamard matrices. For this reason, the Walsh transform is sometimes called the *Hadamard-Walsh transform* [126]. The translation of the properties of Walsh functions is as follows:

a) Formula (1.2.6): the scalar product of any two rows of W_m vanishes.

b) Formula (1.2.7): the sum of elements of any row except the first vanishes.

c) Formula (1.2.10): W_m is a symmetric matrix.

d) Formula (1.2.11): the element-by-element product of the z-th and τ-th rows (columns) of W_m is its $(z \oplus \tau)$-th row (column) (mod 2).

If we express the spectrum and the original function as vectors,

$$\mathbf{S} = (S(0), S(1), \ldots, S(2^m - 1)) \quad \text{and} \quad \mathbf{\Phi} = (\Phi(0), \ldots, \Phi(2^m - 1)),$$

then, as is evident from (1.2.9),

$$\mathbf{S} = 2^{-m} W_m \mathbf{\Phi}. \tag{1.2.14}$$

For example, for the system of BF defined by Table 1.1.2 (Example 1.1.1), we have

$$S = 2^{-3}W_3 \begin{pmatrix} 0 \\ 2 \\ 2 \\ 1 \\ 2 \\ 1 \\ 1 \\ 3 \end{pmatrix} = \begin{pmatrix} 1.5 \\ -0.25 \\ -0.25 \\ 0 \\ -0.25 \\ 0 \\ 0 \\ -0.75 \end{pmatrix}.$$

It is clear from (1.2.14) that we could have defined the Walsh functions in terms of Hadamard matrices W_m, without troubling to complete the functions $f(z)$ and $W_\omega(z)$ to step functions. Nevertheless, our procedure is technically very convenient, since it allows us to utilize the classical theory of orthogonal series (see, e.g., §3.5).

Calculation of the Walsh spectrum by Hadamard matrices requires 2^{2m} operations of addition or subtraction for a function of m variables (2^m operations per coefficient). We now describe a matrix method for calculating the Walsh spectrum which requires only $m2^m$ operations; it is the matrix analog of Theorem 1.2.6 and defines the so-called "*fast Hadamard–Walsh transform*" [126].

Let A be a symmetric $2^m \times 2^m$ matrix. For any $\omega = \sum_{s=0}^{m-1} \omega^{(s)} 2^{m-1-s}$, we put $\overleftarrow{\omega} = \sum_{s=0}^{m-1} \omega^{(s)} 2^s$. Now define $\overleftarrow{A}$ to be the matrix whose ω-th row (column) is the $\overleftarrow{\omega}$-th row (column) of A.

Let

$$A_m = \begin{pmatrix} 1 & 1 & 0 & 0 & 0 & \cdots & 0 & 0 & 0 \\ 0 & 0 & 1 & 1 & 0 & \cdots & 0 & 0 & 0 \\ & \cdot & & \cdot & & \vdots & \cdot & \cdot & \\ 0 & 0 & 0 & 0 & 0 & \cdots & 0 & 1 & 1 \\ 1 & -1 & 0 & 0 & 0 & \cdots & 0 & 0 & 0 \\ 0 & 0 & 1 & -1 & 0 & \cdots & 0 & 0 & 0 \\ & \cdot & & \cdot & & \vdots & \cdot & \cdot & \\ 0 & 0 & 0 & 0 & 0 & \cdots & 0 & 1 & -1 \end{pmatrix}.$$

Then by Theorem 1.2.6,

$$W_m = (\overleftarrow{A_m})^m. \tag{1.2.16}$$

Together with (1.2.14), formula (1.2.16) provides us with a matrix method of calculating the Walsh spectrum, involving m iterated multiplications of the vector $\mathbf{\Phi}$ by the matrix A_m, each multiplication requiring just 2^m operations of addition or subtraction.

The representation of the Hadamard matrix W_m defined by the above formula is known as matrix factorization. It can be shown that the fast Hadamard–Walsh transform defined by (1.2.16) computes the Walsh spectrum with the minimum number of addition or subtraction operations [11, 126].

To end this subsection, we present another m-recursive construction of the Hadamard matrix W_m. It is based on the factorization (1.2.16) and described by

$$\overleftarrow{W}_m = \begin{pmatrix} \overleftarrow{W}_{m-1} & \overleftarrow{W}_{m-1} \\ \overleftarrow{W}_{m-1} & -\overleftarrow{W}_{m-1} \end{pmatrix}. \tag{1.2.17}$$

The recursive procedure defined by (1.2.17) for calculation of the spectrum is essentially the same as the fast Hadamard–Walsh transform.

1.2.4. Each of the expansion coefficients of a function relative to the Walsh basis depends on the behavior of the function at all its points of definition. In some cases it is more convenient to use bases in which the expansion coefficients depend only on the "local" behavior of the function, that is, on its behavior at a few points which are in some sense "close together." An example is the system of Haar functions.

The unnormalized *Haar functions* $H_l^{(q)}(z)$ are defined as follows:

$$H_0^{(0)}(z) \equiv 1$$

$$H_l^{(q)}(z) = \begin{cases} 1 & \text{if } \ z \in [(2q-2)\,2^{m-l-1}, (2q-1)\,2^{m-l-1}), \\ -1 & \text{if } \ z \in [(2q-1)\,2^{m-l-1}, 2q \cdot 2^{m-l-1}), \\ 0 & \text{at all other points of } [0, 2^m) \end{cases} \tag{1.2.18}$$

$$(l = 0, 1, \ldots, m-1; q = 1, 2, \ldots, 2^l).$$

The Haar functions for $m = 3$ are illustrated in Figure 1.2.3. The Haar functions are step functions of span 1, taking the values 0, ± 1.

We now consider the properties of the Haar functions.

I. *Completeness and orthogonality.*

THEOREM 1.2.7. *The set of Haar functions is a complete orthogonal system:*

$$\sum_{z=0}^{2^m-1} H_l^{(q)}(z)\, H_t^{(r)}(z) = \begin{cases} 2^{m-l} & \text{if } \ t = l, q = r, \\ 0 & \text{otherwise} \end{cases} \tag{1.2.19}$$

$$(l = 0, 1, \ldots, m-1; q = 1, 2, \ldots, 2^l),$$

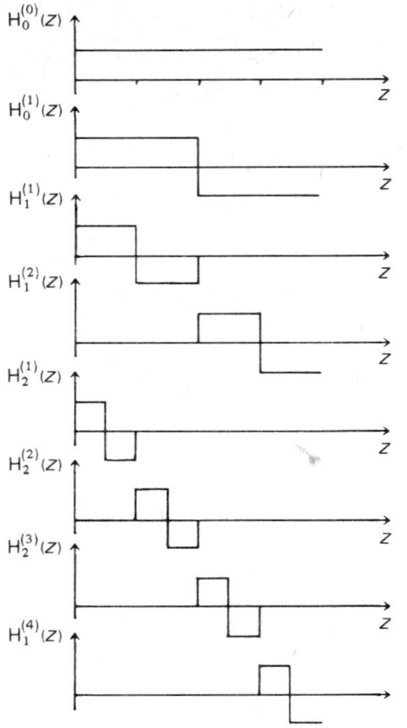

FIGURE 1.2.3.

and if $\Phi(z)$ *is a step function such that*

$$\sum_{z=0}^{2^m-1} \Phi(z) H_l^{(q)}(z) = 0 \qquad (l = 0, \ldots, m-1; q = 0, 1, \ldots, 2^l),$$

then $\Phi(z) \equiv 0$.

Theorem 1.2.7 shows that the Haar system is a suitable basis for expansions of step functions representing systems of BF.

Like Walsh expansions, Fourier expansions in Haar functions are widely applied in numerical computation, since they yield a uniform approximation of $\Phi(z)$ [59, 146, 158]. In such applications the interval of definition is usually normalized to [0, 1), the integrals of the squared Haar functions are normalized to unity, and as the basis system one takes the set of Haar functions for $m \to \infty$.

II. *Finiteness of representing series.*

THEOREM 1.2.8. *For any step function* $\Phi(z)$ *representing a system of BF of m arguments,*

$$\Phi(z) = c_0^{(0)} H_0^{(0)}(z) + \sum_{l=0}^{m-1} \sum_{q=1}^{2^l} c_l^{(q)} H_l^{(q)}(z), \tag{1.2.20}$$

where

$$c_l^{(q)} = 2^{-m+l} \sum_{z=0}^{2^m-1} \Phi(z) H_l^{(q)}(z). \tag{1.2.21}$$

The proof follows from Theorem 1.2.7 by substitution of (1.2.21) into (1.2.20).

It follows from Theorem 1.2.8 that a step function representing a system of BF of m variables may be expanded in a series of Haar functions containing at most 2^m terms.

By (1.2.21) and (1.2.18),

$$c_{m-1}^{(q)} = 2^{-1}(\Phi(2q-2) - \Phi(2q-1)) \qquad (q = 1, 2, \ldots, 2^{m-1}), \tag{1.2.22}$$

$$c_{m-2}^{(q)} = 2^{-2}(\Phi(4q-4) + \Phi(4q-3) - \Phi(4q-2) - \Phi(4q-1)) \tag{1.2.23}$$
$$(q = 1, 2, \ldots, 2^{m-2}),$$

. . . .

$$c_{m-l}^{(q)} = 2^{-l}\left(\sum_{s=2^{l-1}+1}^{2^l} \Phi(2^l q - s) - \sum_{s=1}^{2^{l-1}} \Phi(2^l q - s) \right) \tag{1.2.24}$$
$$(l = 1, 2, \ldots, m-1; q = 1, 2, \ldots, 2^{m-l}).$$

Thus, $c_{m-1}^{(q)}$ depends only on two consecutive values $\Phi(2q-2)$, $\Phi(2q-1)$ of $\Phi(z)$, $c_{m-2}^{(q)}$ on four consecutive values $\Phi(4q-4)$, $\Phi(4q-3)$, $\Phi(4q-2)$, $\Phi(4q-1)$, and so on. In general, each of the coefficients $c_{m-l}^{(q)}$ depends on the behavior of $\Phi(z)$ only on the interval $[2^l q - 2^l, 2^l q - 1)$. Thus, the expansion coefficients of $\Phi(z)$ in Haar series depend only on the local behavior of the function, over an interval which is shorter, the higher the subscript of the coefficient. This points to the essential difference between the Walsh and Haar expansions: every coefficient of the Walsh expansion depends on the function values over the entire interval $[0, 2^m)$. It will be shown later that this feature of the Haar expansions considerably simplifies the selection of functions with prescribed Haar spectra out of a given class of functions.

We now propose an algorithm for calculation of the Haar spectrum and determine its complexity.

THEOREM 1.2.9 [11]. *Let $\Phi(z)$ be a step function representing a system of BF of m arguments and $c_l^{(q)}$ ($l = 0, 1, \ldots, m-1; q = 1, 2, \ldots, 2^l$) its Haar spectrum. Set*

$$a_0(t) = \Phi(t) \qquad (t = 0, 1, \ldots, 2^m - 1),$$

$$\begin{cases} a_s(t) = a_{s-1}(2t) + a_{s-1}(2t+1) & (t = 0, 1, \ldots, 2^{m-s} - 1), \\ a_s(2^{m-s} + t) = a_{s-1}(2t) - a_{s-1}(2t+1) & (s = 1, 2, \ldots, m). \end{cases}$$

Then

$$c_{m-s}^{(q)} = 2^{-s} a_s(2^{m-s} - 1 + q) \tag{1.2.25}$$

Theorem 1.2.9 yields a simple algorithm calculating the Haar spectrum by s-recursive construction of the sequence $a_s(t)$ ($s = 1, 2, \ldots, m; t = 0, 1, \ldots, 2^{m-s+1}$). It is apparent from a comparison of Theorems 1.2.6 and 1.2.9 that the Walsh and Haar spectra may be calculated simultaneously using the sequence $a_s(t)$.

The number of addition or subtraction operations required by this algorithm to calculate the 2^m coefficients of the Haar series is $2 \cdot 2^m - 2$. It can be shown that this is the minimum complexity for an algorithm computing the Haar spectrum.

For example, consider the system of BF defined by Table 1.1.2. The computation of the Haar spectrum is shown in Table 1.2.2, where we have put $c_{m-s}^{(q)} = c(t)$ for $t = 2^{m-s} - 1 + q$. (Note that for given m the values of s and q are uniquely determined for each t, in view of the fact that $q \in \{1, 2, \ldots, 2^{m-s}\}$, where $s \in \{1, 2, \ldots, m\}$.) The Haar expansion of our system is

$$\Phi(z) = \tfrac{3}{2} - \tfrac{1}{4}H_0^{(1)}(z) - \tfrac{1}{4}H_1^{(1)}(z) - \tfrac{1}{4}H_1^{(2)}(z) - H_2^{(1)}(z)$$
$$+ \tfrac{1}{2}H_2^{(2)}(z) + \tfrac{1}{2}H_2^{(3)}(z) - H_2^{(4)}(z).$$

There is an interesting relationship between the Haar functions and the Rademacher functions $R_s(z)$ ($s = 1, 2, \ldots, m$). By (1.2.4) and (1.2.17), we have

$$H_l^{(q)}(z) = \begin{cases} R_{l+1}(z) & \text{if } z \in [(q-1)\,2^{m-l}, q2^{m-l}), \\ 0 & \text{otherwise.} \end{cases} \tag{1.2.26}$$

In view of this formula, it is a fairly easy matter to calculate a Haar function $H_l^{(q)}(z)$ given the binary code of its argument $(z^{(0)}, \ldots, z^{(m-1)})$, since (except for notation) the value of the Rademacher function $R_{l+1}(z)$ is the l-th digit $z^{(l)}$ of the code z (see §1.2.2).

Unlike the Walsh system, the Haar system is not closed under multiplication: the product of two Haar functions need not be a Haar function. However, the Haar

TABLE 1.2.2.

z, t	$\Phi(z)$	$a_1(t)$	$a_2(t)$	$a_3(t)$	$c(t)$
0	0	2	5	12	1.5
1	2	3	7	−12	−0.25
2	2	3	−1		−0.25
3	1	4	−1		−0.25
4	2	−2			−1
5	1	1			0.5
6	1	1			0.5
7	3	−2			−1

basis possesses a highly important advantage over the Walsh basis as regards computation of $\Phi(z)$ by serial summation of the terms of the series. This follows from

THEOREM 1.2.10. *Let $\Phi(z)$ be a step function representing a system of BF of m arguments. Then for any fixed $z = z^*$ ($z^* \in \{0, 1, \ldots, 2^m - 1\}$) the number of nonvanishing terms of the series* (1.2.20) *is at most $m + 1$.*

PROOF. To prove Theorem 1.2.10, one need only observe that for any fixed $z^* \in [0, 2^m)$ there are exactly $m + 1$ distinct pairs (q, l) such that $z^* \in [(q - 1)2^{m-l}, q \cdot 2^{m-l})$. The assertion now follows in view of (1.2.26). ∎

Note that the number of nonvanishing terms depends on the choice of z^*. For example, for $m = 3$, $z^* = 5$, we have

$$\Phi(z^*) = \Phi(5) = c_0^{(0)} - c_0^{(1)} + c_1^{(2)} - c_2^{(3)},$$

since $H_1^{(1)}(5) = H_2^{(1)}(5) = H_2^{(2)}(5) = H_2^{(4)}(5) = 0$ (see Figure 1.2.3).

Thus, for any fixed $z = z^*$, we can express $\Phi(z^*)$ as an algebraic sum of coefficients $c_l^{(q)}$, containing exactly one coefficient with each subscript $l \neq 0$ and two with subscript $l = 0$.

To end this section, we reemphasize the three most important distinguishing features of the Walsh and Haar bases, thanks to which they find wide application in analysis and synthesis of networks realizing systems of BF.

1. The expansion of any step function representing a system of BF of m arguments contains at most 2^m nonvanishing terms.
2. The basis functions assume the values ± 1 or $0, \pm 1$.
3. The values of any basis function are calculated with extreme ease from the binary code of its argument, the calculation involving at most m elementary operations over single-digit binary numbers.

More information about the properties of Walsh and Haar functions may be found, e.g., in [59, 77, 146, 157].

REMARK. Since the basis functions we have been considering take the values $\{1, -1\}$ or $\{0, \pm 1\}$, it is sometimes convenient, when calculating the values of systems of BF from (1.1.5), to let Boolean functions take the values 1 and -1 instead of 0 and 1, respectively, i.e., to replace $f^{(s)}(z^{(0)}, \ldots, z^{(m-1)})$ by $\varphi^{(s)}(z^{(0)}, \ldots, z^{(m-1)})$, where

$$\varphi^{(s)}(z^{(0)}, \ldots, z^{(m-1)}) = 1 - 2f^{(s)}(z^{(0)}, \ldots, z^{(m-1)}).$$

It should be clear that this transformation has no effect on the validity of our results.

1.3. BASES FOR SYSTEMS OF p-VALUED LOGICAL FUNCTIONS

1.3.1. In this section we go over to complete systems of orthogonal functions suitable for the analysis and synthesis of systems of p-valued logical functions. These orthogonal systems generalize the Walsh and Haar functions to the p-valued case. The new basis functions will assume not two or three real values, as was the case previously, but p or $p + 1$ complex values.

1.3.2. The generalization of Walsh functions to the p-valued case, where p is a prime, is the system of Chrestenson functions [24, 154]. The *Chrestenson functions* $\chi_\omega^{(p)}(z)$ are step functions defined on the interval $[0, p^m)$, defined for any natural numbers ω, m $(0 \leqq \omega \leqq p^m - 1)$ by

$$\chi_\omega^{(p)}(z) = \exp\left(\frac{2\pi}{p} i \sum_{s=0}^{m-1} \omega^{(m-1-s)} z^{(s)}\right) \tag{1.3.1}$$

where $i = \sqrt{-1}$, $\omega^{(s)}, z^{(s)} \in \{0, 1, \ldots, p-1\}$, and

$$\omega = \sum_{s=0}^{m-1} \omega^{(s)} p^{m-1-s}, \qquad z = \sum_{s=0}^{m-1} z^{(s)} p^{m-1-s}. \tag{1.3.2}$$

A comparison of formulas (1.2.1) and (1.3.1) reveals that

$$\chi_\omega^{(2)}(z) = W_\omega(z); \tag{1.3.3}$$

in other words, the Chrestenson system for $p = 2$ is simply the Walsh system.

Table 1.3.1 lists the Chrestenson functions for $p = 3$, $m = 2$. Both here and in Table 1.3.2 below, $e_1 = -\frac{1}{2}(1 - i\sqrt{3}) = \exp(2\pi i/3)$, $e_2 = \bar{e}_1 = -\frac{1}{2}(1 + i\sqrt{3}) = \exp(4\pi i/3)$.

TABLE 1.3.1.

$\chi_\omega^{(3)}(z)$ \ z	0	1	2	3	4	5	6	7	8
$\chi_0^{(3)}(z)$	1	1	1	1	1	1	1	1	1
$\chi_1^{(3)}(z)$	1	1	1	e_1	e_1	e_1	e_2	e_2	e_2
$\chi_2^{(3)}(z)$	1	1	1	e_2	e_2	e_2	e_1	e_1	e_1
$\chi_3^{(3)}(z)$	1	e_1	e_2	1	e_1	e_2	1	e_1	e_2
$\chi_4^{(3)}(z)$	1	e_1	e_2	e_1	e_2	1	e_2	1	e_1
$\chi_5^{(3)}(z)$	1	e_1	e_2	e_2	1	e_1	e_1	e_2	1
$\chi_6^{(3)}(z)$	1	e_2	e_1	1	e_2	e_1	1	e_2	e_1
$\chi_7^{(3)}(z)$	1	e_2	e_1	e_1	1	e_2	e_2	e_1	1
$\chi_8^{(3)}(z)$	1	e_2	e_1	e_2	e_1	1	e_1	1	e_2

Consider the subset $\{K_{r,s}^{(p)}(z)\}$ of the Chrestenson system $\chi_{\omega}^{(p)}(z)$ consisting of all those functions whose subscript ω is divisible by a power of p. These functions generalize the Rademacher functions:

$$K_{r,s}^{(p)}(z) = \chi_{r \cdot p^s}^{(p)}(z) = \exp\left(\frac{2\pi}{p} i \cdot r \cdot z^{(s)}\right). \tag{1.3.4}$$

A comparison of formulas (1.2.4) and (1.3.4) shows that

$$K_{1,s}^{(2)}(z) = R_{s+1}(z), \tag{1.3.5}$$

and so the functions $K_{r,s}^{(p)}(z)$ are known as *generalized Rademacher functions.* The generalized Rademacher functions for $p = 3$, $m = 2$ are listed in Table 1.3.2.

TABLE 1.3.2.

$K_{r,i}^{(3)}(z)$ \ z	0	1	2	3	4	5	6	7	8
$K_{0,0}^{(3)}(z)$	1	1	1	1	1	1	1	1	1
$K_{1,0}^{(3)}(z)$	1	1	1	e_1	e_1	e_1	e_2	e_2	e_2
$K_{2,0}^{(3)}(z)$	1	1	1	e_2	e_2	e_2	e_1	e_1	e_1
$K_{1,1}^{(3)}(z)$	1	e_1	e_2	1	e_1	e_2	1	e_1	e_2
$K_{2,1}^{(3)}(z)$	1	e_2	e_1	1	e_2	e_1	1	e_2	e_1

There is a one-to-one correspondence h between the set of values assumed by the Chrestenson functions and the set $\{0, 1, \ldots, p-1\}$. Indeed, set

$$h\left(\exp\left(\frac{2\pi}{p} iq\right)\right) = q. \tag{1.3.6}$$

Then, as is evident from (1.3.4), the mapping h takes the value of $K_{1,s}^{(p)}(z)$ onto the corresponding digit $z^{(s)}$ of the p-ary code of z.

The Chrestenson functions may be expressed in terms of the generalized Rademacher functions. Indeed, by (1.3.4) and (1.3.1),

$$\chi_{\omega}^{(p)}(z) = \prod_{s=0}^{m-1} (K_{1,s}(z))^{\omega(m-1-s)} \tag{1.3.7}$$

so that the set of Chrestenson functions is the multiplicative closure of the set of generalized Rademacher functions.

Formula (1.3.7) generalizes the formula (1.2.5) relating the Rademacher and Walsh functions.

We now turn to the main properties of the Chrestenson functions. (The proofs are analogous to those of the properties of the Walsh functions.)

I. *Completeness and orthogonality.*

THEOREM 1.3.1. *The set of Chrestenson functions is a complete orthogonal system:*

$$\sum_{z=0}^{p^m-1} \chi_t^{(p)}(z)\,\overline{\chi_q^{(p)}(z)} = \begin{cases} p^m & \text{if } t = q, \\ 0 & \text{if } t \neq q, \end{cases} \tag{1.3.8}$$

where $\overline{\chi_q^{(p)}(z)}$ is the complex conjugate of $\chi_q^{(p)}(z)$. If $\Phi(z)$ is a step function representing a system of p-valued logical functions of m arguments, such that

$$\sum_{z=0}^{p^m-1} \Phi(z)\,\overline{\chi_\omega^{(p)}(z)} = 0 \qquad (\omega = 0, 1, \ldots, p^m - 1),$$

then $\Phi(z) \equiv 0$.

It follows from (1.3.8) for $t = 0$ that

$$\sum_{z=0}^{p^m-1} \chi_\omega(z) = 0 \qquad (\omega \neq 0). \tag{1.3.9}$$

II. *Finiteness of representing series.*

THEOREM 1.3.2. *Let $\Phi(z)$ be a step function representing a system of p-valued logical functions of m arguments. Then*

$$\Phi(z) = \sum_{\omega=0}^{p^m-1} S(\omega)\,\chi_\omega^{(p)}(z), \tag{1.3.10}$$

where

$$S(\omega) = p^{-m} \sum_{z=0}^{p^m-1} \Phi(z)\,\overline{\chi_\omega^{(p)}(z)}. \tag{1.3.11}$$

It follows from Theorem 1.3.2. that if the original system assumes nonzero values at $\leqq p^m$ points, the same is true of its Chrestenson spectrum $S(\omega)$. This observation makes it possible to reconstruct any system of p-ary functions from its Chrestenson spectrum.

EXAMPLE 1.3.1. Consider the system of two ternary functions of two arguments defined by Table 1.3.3 ($p = 3, m = 2$). The corresponding step function $\Phi(z)$ and its Chrestenson spectrum are also given in Table 1.3.3. We have

$$\Phi(z) = 3^{-2}(36 + \tfrac{3}{2}(1 - i\sqrt{3})\,\chi_4^{(3)}(z) + (-\tfrac{21}{2} - \tfrac{15}{2}i\sqrt{3})\,\chi_5^{(3)}(z)$$
$$+ (-\tfrac{21}{2} + \tfrac{15}{2}i\sqrt{3})\,\chi_7^{(3)}(x) + (\tfrac{3}{2} + \tfrac{3}{2}i\sqrt{3})\,\chi_8^{(3)}(z).$$

TABLE 1.3.3.

z, ω	$z^{(0)}$	$z^{(1)}$	$f^{(0)}(z)$	$f^{(1)}(z)$	$\Phi(z)$	$8S(\omega)$
0	0	0	0	2	2	36
1	0	1	2	2	8	0
2	0	2	0	2	2	0
3	1	0	1	0	3	0
4	1	1	0	1	1	$1.5 - 1.5\sqrt{3}i$
5	1	2	2	2	8	$-10.5 - 7.5\sqrt{3}i$
6	2	0	2	1	7	0
7	2	1	1	0	3	$-10.5 + 7.5\sqrt{3}i$
8	2	2	0	2	2	$1.5 + 1.5\sqrt{3}i$

III. *Symmetry of index and argument.*

THEOREM 1.3.3. *For any* $\omega, z \in \{0, 1, \ldots, p^m - 1\}$,

$$\chi_\omega^{(p)}(z) = \chi_z^{(p)}(\omega). \tag{1.3.12}$$

IV. *Translation of argument.*

THEOREM 1.3.4. *For any* $\omega, z, \tau \in \{0, 1, \ldots, p^m - 1\}$,

$$\chi_\omega^{(p)}(z \oplus \tau) = \chi_\omega^{(p)}(z)\,\chi_\omega^{(p)}(\tau). \tag{1.3.13}$$

It follows from Theorems 1.3.3 and 1.3.4 that the set of Chrestenson functions is closed under multiplication.

V. *Isomorphism of set of linear p-valued logical functions and Chrestenson system.*

A p-valued logical function $f(z^{(0)}, \ldots, z^{(m-1)})$ is said to be *linear* if it may be expressed as

$$f(z^{(0)}, \ldots, z^{(m-1)}) = \bigoplus_{s=0}^{m-1} c_s z^{(s)} \quad (\text{mod } p). \tag{1.3.14}$$

The linear functions form a commutative group of order p^m with respect to addition modulo p; the inverse of an element $\bigoplus_{s=0}^{m-1} c_s z^{(s)}$ (mod p) ($c_s \in \{0, 1, \ldots, p-1\}$) is $\bigoplus_{s=0}^{m-1} (p - c_s)\, z^{(s)}$ (mod p).

The Chrestenson functions $\chi_\omega^{(p)}(z)$ also form a multiplicative group.

THEOREM 1.3.5. *The group of Chrestenson functions is isomorphic to the group of linear p-valued logical functions. The isomorphism h is defined by*

$$h(\chi_\omega^{(p)}(z)) = \bigoplus_{s=0}^{m-1} \omega^{(m-1-s)} z^{(s)} \quad (\text{mod } p). \tag{1.3.15}$$

This isomorphism is an extension of the isomorphism between the generalized Rademacher functions and the components of the p-ary expansion of the argument; it will be used in the sequel in analysis and synthesis of networks realizing systems of p-valued logical functions.

1.3.3. We now examine some methods for calculating the Chrestenson spectrum. A natural method is provided by formulas (1.1.1) through (1.1.7) for $\Psi_\omega(z) = \chi_\omega^{(p)}(z)$. We proceed to a different method, generalizing the method of §1.2.3 calculating the Walsh spectrum through Hadamard matrices.

Let $\chi_m^{(p)}$ be the matrix whose ω-th row is $(\chi_\omega^{(p)}(0), \chi_\omega^{(p)}(1), \ldots, \chi_\omega^{(p)}(p^m - 1))$ $(\omega = 0, 1, \ldots, p^m - 1)$. The matrices $\chi_1^{(3)}$ and $\chi_2^{(3)}$ $(m = 1, m = 2, p = 3)$ are shown below:

$$\chi_1^{(3)} = \begin{pmatrix} 1 & 1 & 1 \\ 1 & e_1 & e_2 \\ 1 & e_2 & e_1 \end{pmatrix}, \qquad \chi_2^{(3)} = \begin{pmatrix} 1 & 1 & 1 & 1 & 1 & 1 & 1 & 1 & 1 \\ 1 & 1 & 1 & e_1 & e_1 & e_1 & e_2 & e_2 & e_2 \\ 1 & 1 & 1 & e_2 & e_2 & e_2 & e_1 & e_1 & e_1 \\ 1 & e_1 & e_2 & 1 & e_1 & e_2 & 1 & e_1 & e_2 \\ 1 & e_1 & e_2 & e_1 & e_2 & 1 & e_2 & 1 & e_1 \\ 1 & e_1 & e_2 & e_2 & 1 & e_1 & e_1 & e_2 & 1 \\ 1 & e_2 & e_1 & 1 & e_2 & e_1 & 1 & e_2 & e_1 \\ 1 & e_2 & e_1 & e_1 & 1 & e_2 & e_2 & e_1 & 1 \\ 1 & e_2 & e_1 & e_2 & e_1 & 1 & e_1 & 1 & e_2 \end{pmatrix}.$$

All the properties of the Chrestenson functions may be rephrased in terms of the matrices $\chi_m^{(p)}$, as was the case for the Walsh functions and Hadamard matrices (see §1.2.3).

Let us express the initial function and its Chrestenson spectrum as vectors, $\mathbf{\Phi} = (\Phi(0), \ldots, \Phi(p^m - 1))$, $\mathbf{S} = (S(0), \ldots, S(p^m - 1))$. Then, by (1.3.11),

$$\mathbf{S} = p^{-m}\overline{\chi_m^{(p)}}\mathbf{\Phi}$$

where $\overline{\chi_m^{(p)}}$ is the matrix complex-conjugate to $\chi_m^{(p)}$. This formula generalizes (1.2.14) and provides a simple construction of the spectrum if p and m are not too large. Calculation of the Chrestenson spectrum using formula (1.3.16) requires p^{2m} addition or subtraction operations (p^m operations per coefficient).

We now exhibit another construction of the Chrestenson spectrum, generalizing the method of Theorem 1.2.6 to the p-valued case and requiring the minimum number $m \cdot p^m(p - 1)$ of operations.

Given a natural number $\omega = \sum_{s=0}^{m-1} \omega^{(s)} p^{m-1-s}$, we set $\overleftarrow{\omega} = \sum_{s=0}^{m-1} \omega^{(s)} p^s$. If A is a symmetric $p^m \times p^m$ matrix, we let $\overleftarrow{A}$ denote the matrix whose ω-th row (column) is the $\overleftarrow{\omega}$-th row (column) of A. Consider the matrix

$$A_m^{(p)} = \left(\begin{array}{cccc} e_0^0 e_1^0 \dots e_{p-1}^0 & & & \mathbf{0} \\ \mathbf{0} & e_0^0 e_1^0 \dots e_{p-1}^0 & \dots & \\ & & & e_0^0 e_1^0 \dots e_{p-1}^0 \\ e_0^1 e_1^1 \dots e_{p-1}^1 & & & \mathbf{0} \\ \mathbf{0} & e_0^1 e_1^1 \dots e_{p-1}^1 & \dots & \\ & & & e_0^1 e_1^1 \dots e_{p-1}^1 \\ \cdot & \cdot & \cdot & \cdot \\ e_0^{p-1} e_1^{p-1} \dots e_{p-1}^{p-1} & & & \mathbf{0} \\ \mathbf{0} & e_0^{p-1} e_1^{p-1} \dots e_{p-1}^{p-1} & \dots & \\ & & & e_0^{p-1} e_1^{p-1} \dots e_{p-1}^{p-1} \end{array} \right) \begin{array}{l} \left.\vphantom{\begin{array}{c} a \\ a \\ a \end{array}}\right\} p^{m-1} \\ \\ \\ \\ \\ \\ \\ \end{array},$$

where $e_q = \exp(2q\pi i/p)$ $(q = 0, 1, \dots, p-1)$.

For $p = 2$, $A_m^{(p)}$ is simply the matrix A_m considered in §1.2.3.

THEOREM 1.3.6 [11].

$$\chi_m^{(p)} = \overleftarrow{(A_m^{(p)})^m}. \tag{1.3.17}$$

This theorem, generalizing formula (1.2.16), defines the factorization of the matrix $\overleftarrow{\chi}_m^{(p)}$ and, together with formula (1.3.16), defines the *fast Hadamard–Chrestenson transform*. Calculation of the Chrestenson spectrum using this transform requires $mp^m(p-1)$ addition or subtraction operations (multiplication of $\mathbf{\Phi}$ by $\overline{A_m^{(p)}}$ requires p^m operations). This is the minimum complexity of an algorithm for this purpose.

Another consequence of Theorem 1.3.6 is the following m-recursive procedure for calculating $\chi_m^{(p)}$, expressing it as a block matrix:

$$\overleftarrow{\chi}_m^{(p)} = \begin{pmatrix} e_0^0 \overleftarrow{\chi}_{m-1}^{(p)} & e_1^0 \overleftarrow{\chi}_{m-1}^{(p)} & \cdots & e_{p-1}^0 \overleftarrow{\chi}_{m-1}^{(p)} \\ e_0^1 \overleftarrow{\chi}_{m-1}^{(p)} & e_1^1 \overleftarrow{\chi}_{m-1}^{(p)} & \cdots & e_{p-1}^1 \overleftarrow{\chi}_{m-1}^{(p)} \\ \cdot & \cdot & \cdot & \cdot \\ e_0^{p-1} \overleftarrow{\chi}_{m-1}^{(p)} & e_p^{p-1} \overleftarrow{\chi}_{m-1}^{(p)} & \cdots & e_{p-1}^{p-1} \overleftarrow{\chi}_{m-1}^{(p)} \end{pmatrix}. \tag{1.3.18}$$

Formula (1.3.18) generalizes (1.2.17) to the p-valued case and may be used to generate the Hadamard–Chrestenson matrices $\chi_m^{(p)}$.

1.3.4. Each expansion coefficient of a system of p-valued logical functions relative to the Chrestenson basis utilizes the behavior of the system over its entire interval of definition. We now define a complete system of orthogonal $(p+1)$-valued functions generalizing the Haar functions, with the property that the corresponding expansion coefficients depend only on the local behavior of the original system. We shall refer to these functions as generalized Haar functions, denoting them by $\{M_{r,s}^{(p,q)}(z)\}$.

The *generalized Haar functions* $M_{r,s}^{(pq)}(z)$ are defined in terms of the generalized Rademacher functions $K_{r,s}^{(p)}(z) = \chi_{r,p^s}^{(p)}(z) = \exp(2\pi r i z^{(s)}/p)$ as follows:

$$M_{0,0}^{(p,1)}(z) \equiv 1,$$

$$M_{r,s}^{(p,q)}(z) = \begin{cases} K_{r,s}^{(p)}(z) & \text{if} \quad z \in [(q-1)\,p^{m-s}, qp^{m-s}), \\ 0 & \text{otherwise.} \end{cases} \tag{1.3.19}$$

It follows from (1.3.5), (1.3.19), (1.2.26) that the generalized Haar functions for $p = 2$ are simply the Haar functions of §1.2, $M_{1,s}^{(2,q)}(z) = H_s^{(q)}(z)$. The values of the generalized Haar functions for $p = 3$ and $m = 2$ are given in Table 1.3.4.

TABLE 1.3.4.

$M_{r,i}^{(3,q)}(z)$ \ z	0	1	2	3	4	5	6	7	8
$M_{0,0}^{(3,0)}(z)$	1	1	1	1	1	1	1	1	1
$M_{1,0}^{(3,1)}(z)$	1	1	1	e_1	e_1	e_1	e_2	e_2	e_2
$M_{2,0}^{(3,1)}(z)$	1	1	1	e_2	e_2	e_2	e_1	e_1	e_1
$M_{1,1}^{(3,1)}(z)$	1	e_1	e_2	0	0	0	0	0	0
$M_{1,1}^{(3,2)}(z)$	0	0	0	1	e_1	e_2	0	0	0
$M_{1,1}^{(3,3)}(z)$	0	0	0	0	0	0	1	e_1	e_2
$M_{2,1}^{(3,1)}(z)$	1	e_2	e_1	0	0	0	0	0	0
$M_{2,1}^{(3,2)}(z)$	0	0	0	1	e_2	e_1	0	0	0
$M_{2,1}^{(3,3)}(z)$	0	0	0	0	0	0	1	e_2	e_1

The generalized Haar functions form a complete orthogonal system. Any step function representing a p-valued logical function of m arguments admits an expansion in series of generalized Haar functions, containing at most p^m nonzero terms. In addition, the generalized Haar spectrum may be calculated by a procedure analogous to the fast Hadamard–Chrestenson transform. This procedure is described in the next theorem, which generalizes Theorem 1.2.9.

THEOREM 1.3.7. *Let $\Phi(z)$ be a step function representing a system of p-valued logical functions of m arguments, and $c_{r,m-s}^{(pq)}$ its expansion coefficients relative to the generalized Haar functions $M_{r,m-s}^{(p,q)}(z)$. Set*

$$a_0(t) = \Phi(t),$$

$$a_s(\tau p^{m-s} + t) = \sum_{\delta=0}^{p-1} e_\delta^{-\tau} a_{s-1}(pt + \delta) \tag{1.3.20}$$

$$(\tau = 0, 1, \ldots, p-1;\; s = 1, 2, \ldots, m;\; t = 0, 1, \ldots, p^{m-s} - 1),$$

where $e_\delta = \exp(2\pi\delta i/p)$, $i = \sqrt{-1}$, $\delta = 0, 1, \ldots, p-1$.

Then

$$c_{r,m-s}^{(p,q)} = p^{-s} a_s\big(r(p^{m-s} - 1) + q\big). \tag{1.3.21}$$

As stated above, this theorem generates an algorithm calculating the generalized Haar spectrum which requires $p \cdot p^m - p$ addition subtraction operations—this is the minimum possible number.

1.4. PROPERTIES OF DISCRETE WALSH AND CHRESTENSON TRANSFORMS

1.4.1. In § 1.1 we considered methods for representing LF by orthogonal series. We showed in § 1.2 and § 1.3 that if one uses the Walsh or Haar systems (for Boolean functions) and the Chrestenson or generalized Haar functions (for p-valued functions), the series thus obtained are finite. If the original system of LF is defined at p^m points, its spectrum will also have at most p^m nonzero values. Thus, the orthogonal-series representation of a system of LF defines a one-to-one mapping of the p^m-dimensional space of systems of LF into the p^m-dimensional space of spectral coefficients. We shall refer to the system in question as the "*original*," in keeping with the terminology used in Laplace transform theory [31], and the sequence of spectral coefficients $S(0), \ldots, S(p^{m-1})$ of the expansion of the system will be called the *image*, *transform*, or *spectrum* of the system relative to the appropriate basis. The mapping itself will be called the *Walsh or Chrestenson transformation.* The spectrum of the function $\Phi(z)$ relative to the basis in question will now be denoted by $S_{f(z)}(\omega)$, where $\Phi(z) = f(\delta)$ for $z \in [\delta, \delta + 1), \delta = 0, 1, \ldots, p^m - 1$.

1.4.2. In view of the fact that the Walsh and Chrestenson spectra will be used constantly in what follows as our main tool for the analysis and synthesis of DD, we list the most important properties of the Walsh and Chrestenson transforms below.

I. *Linearity.*

THEOREM 1.4.1. *Let*

$$\Phi(z) = \sum_{q=1}^{N} a_q \Phi_q(z) \tag{1.4.1}$$

where a_q are arbitrary numbers and $\Phi_q(z)$ $(q = 1, \ldots, N)$ step functions representing systems of LF $(s = 1, 2, \ldots, N)$.

Then

$$S_{f(z)}(\omega) = \sum_{q=1}^{N} a_q S_{f_q(z)}(\omega). \tag{1.4.2}$$

II. *Translation of argument in original and in transform.*

THEOREM 1.4.2. *For every* $\tau \in \{0, 1, \ldots, p^m - 1\}$,

$$S_{f(z \ominus \tau)}(\omega) = \overline{\chi_\tau^{(p)}(\omega)}\, S_{f(z)}(\omega) \qquad (\text{mod } p), \tag{1.4.3}$$

$$S_{f(z)\chi_\tau^{(p)}(z)}(\omega) = S_{f(z)}(\omega \ominus \tau) \qquad (\text{mod } p). \tag{1.4.4}$$

It is worth noting that Theorem 1.4.2 is analogous to the theorem of translation of arguments in the case of the Laplace transform [31], except that here the summation (translation) of arguments in both original and transform is carried out modulo p. (The translation of z or ω by an amount τ modulo p is defined by componentwise addition (mod p) of the p-ary expansions of z (or ω) and τ.)

III. *Involution property.*

THEOREM 1.4.3.

$$S_{\overline{S_{f(z)}}}(z) = p^{-m} f(z). \tag{1.4.5}$$

PROOF. Since

$$\overline{S_{f(z)}(\omega)} = p^{-m} \sum_{q=0}^{p^m - 1} f(q)\, \chi_\omega^{(p)}(q),$$

it follows in view of (1.4.1), (1.4.2), (1.3.8), (1.3.12) and (1.3.13) that

$$\begin{aligned}
S_{\overline{S_{f(z)}}}(z) &= p^{-m} \sum_{\omega=0}^{p^m - 1} \overline{S_{f(z)}(\omega)}\, \overline{\chi_z^{(p)}(\omega)} \\
&= p^{-m} \sum_{\omega=0}^{p^m - 1} \left(p^{-m} \sum_{q=0}^{p^m - 1} f(q)\, \chi_\omega^{(p)}(q) \right) \overline{\chi_z^{(p)}(\omega)} \\
&= p^{-2m} \sum_{\omega=0}^{p^m - 1} \sum_{q=0}^{p^m - 1} f(q)\, \chi_\omega^{(p)}(q \ominus z) = p^{-2m} \sum_{q=0}^{p^m - 1} f(q) \sum_{\omega=0}^{p^m - 1} \chi_\omega^{(p)}(q \ominus z) \\
&= p^{-m} f(z) \qquad (\text{mod } p). \quad \blacksquare
\end{aligned}$$

The involution property means that the Chrestenson transform of the Chrestenson transform of a step function $\Phi(z)$ is the original function, up to a normalizing factor p^{-m}.

IV. *Convolution theorem.*

THEOREM 1.4.4.

$$S_{f_1(z) \cdot f_2(z)}(\omega) = \sum_{\tau=0}^{p^m - 1} S_{f_1(z)}(\tau)\, S_{f_2(z)}(\omega \ominus \tau) \qquad (\text{mod } p). \tag{1.4.6}$$

Let $\Psi(z) = \sum_{\tau=0}^{p^m - 1} f_1(\tau)\, f_2(z \ominus \tau)\,(\text{mod } p)$. *Then*

$$S_{\Psi(z)}(\omega) = p^m S_{f_1}(\omega)\, S_{f_2}(\omega) \tag{1.4.7}$$

PROOF. By (1.4.1), (1.4.2), (1.3.8), (1.3.12) and (1.3.13),

$$\begin{aligned}
S_{f_1(z)f_2(z)}(\omega) &= p^{-m} \sum_{z=0}^{p^m-1} f_1(z)\, f_2(z)\, \overline{\chi_\omega^{(p)}(z)} \\
&= p^{-m} \sum_{z=0}^{p^m-1} \left(\sum_{\tau=0}^{p^m-1} S_{f_1(z)}(\tau)\, \chi_\tau^{(p)}(z) \right) \left(\sum_{\alpha=0}^{p^m-1} S_{f_2(z)}(\alpha)\, \chi_\alpha^{(p)}(z) \right) \overline{\chi_\omega^{(p)}(z)} \\
&= p^{-m} \sum_{\tau=0}^{p^m-1} \sum_{\alpha=0}^{p^m-1} S_{f_1(z)}(\tau)\, S_{f_2(z)}(\alpha) \sum_{z=0}^{p^m-1} \chi_{\tau\oplus\alpha\ominus\omega}^{(p)}(z) \\
&= p^{-m} \sum_{\tau=0}^{p^m-1} S_{f_1(z)}(\tau)\, S_{f_2(z)}(\omega \ominus \tau) \sum_{z=0}^{p^m-1} \chi_0^{(p)}(z) \\
&\quad + p^{-m} \sum_{\alpha \neq \omega\ominus\tau} S_{f_1(z)}(\tau) S_{f_2(z)}(\alpha) \sum_{z=0}^{p^m-1} \chi_{\tau\oplus\alpha\ominus\omega}^{(p)}(z) \\
&= \sum_{\tau=0}^{p^m-1} S_{f_1(z)}(\tau)\, S_{f_2(z)}(\omega \ominus \tau) \qquad (\mathrm{mod}\ p).
\end{aligned}$$

The proof of (1.4.7) is analogous to the proof of (1.4.6). ∎

COROLLARY. 1.4.1. *Let $f(z)$ be a Boolean function of m arguments and $S_{f(z)}(\omega)$ its Walsh spectrum. Then*

$$S_{f(z)}(\omega) = \sum_{\tau=0}^{2^m-1} S_{f(z)}(\tau)\, S_{f(z)}(\omega \oplus \tau) \qquad (\mathrm{mod}\ 2), \tag{1.4.8}$$

and conversely: if (1.4.8) is true, then $f(z)$ is a Boolean function.

Formula (1.4.8) is thus an alternative definition of Boolean functions. The corollary follows from Theorem 1.4.4, using the equality $f(z) = f(z)\, f(z)$, which is true if and only if $f(z)$ is a BF.

COROLLARY 1.4.2. *Let $f(z)$ be a Boolean function and $S_{f(z)}(\omega)$ its Walsh spectrum. Then*

$$\sum_{\omega=0}^{2^m-1} S_{f(z)}(\omega) \in \{0, 1\}$$

This is verified by summing both sides of (1.4.8) over ω.

Corrolary 1.4.2 is useful as a check on the correctness of calculations of Walsh spectra of BF by means of check sums $\sum_\omega S_{f(z)}(\omega)$.

V. *Linear* (mod *p*) *transformation of variables.*

Let p be a prime, and $\sigma = (\sigma_{is})$ $(i, s = 0, 1, \ldots, m-1)$ a matrix over the field $GF(p)$ of residues mod p, with nonvanishing determinant $|\sigma|_p$ over $GF(p)$ (i.e., $\sigma_{is} \in \{0, 1, \ldots, p-1\}$ and all multiplication and addition operations in the calculation of $|\sigma|_p$ are performed modulo p). Let $\sigma \otimes a$ $(a \otimes \sigma)$ be the number whose p-ary expansion vector is the product of σ and the p-ary expansion vector of the number a (in the appropriate order); all arithmetical operations are of course performed modulo p. For example,

$$\begin{pmatrix} 0 & 2 & 1 \\ 1 & 2 & 1 \\ 0 & 1 & 0 \end{pmatrix} \otimes 14 = \begin{pmatrix} 0 & 2 & 1 \\ 1 & 2 & 1 \\ 0 & 1 & 0 \end{pmatrix} \otimes \begin{pmatrix} 1 \\ 1 \\ 2 \end{pmatrix} = 16 \qquad (\text{mod } 3).$$

THEOREM 1.4.5. *Let* $\sigma_{is} \in \{0, 1, \ldots, p-1\}$ $(i, s = 0, 1, \ldots, m-1)$, $|\sigma|_p \neq 0$. *Then*

$$S_{f(\sigma \otimes z)}(\omega) = S_{f(z)}(\overleftarrow{\overleftarrow{\omega} \otimes \sigma^{-1}}). \tag{1.4.9}$$

The inversion of the matrix σ in (1.4.9) is again over the field of residues modulo p, and if $\omega = \sum_{s=0}^{m-1} \omega^{(s)} p^{m-1-s}$ $(\omega^{(s)} \in \{0, 1, \ldots, p-1\})$, then $\overleftarrow{\omega} = \sum_{s=0}^{m-1} \omega^{(s)} p^s$.

PROOF.

$$\sum_{s=0}^{m-1} \omega^{(m-1-s)} (\sigma \otimes z)^{(s)} = \sum_{s=0}^{m-1} \omega^{(m-1-s)} \sum_{q=0}^{m-1} \sigma_{sq} z^{(q)} = \sum_{q=0}^{m-1} z^{(q)} (\overleftarrow{\omega} \otimes \sigma)^{(q)}$$

$$= \sum_{q=0}^{m-1} z^{(q)} (\overleftarrow{\overleftarrow{\omega} \otimes \sigma})^{(m-1-q)} \qquad (\text{mod } p).$$

Hence

$$\chi_\omega^{(p)}(\sigma \otimes z) = \chi_{\overleftarrow{\overleftarrow{\omega} \otimes \sigma}}(z) \qquad (\text{mod } p). \tag{1.4.10}$$

It follows that

$$\begin{aligned} S_{f(\sigma \otimes z)}(\omega) &= \sum_{z=0}^{p^m-1} f(\sigma \otimes z) \overline{\chi_\omega^{(p)}(z)} \\ &= \sum_{z=0}^{p^m-1} f(\sigma \otimes z) \overline{\chi_\omega^{(p)}(\sigma^{-1} \otimes \sigma \otimes z)} \\ &= \sum_{z=0}^{p^m-1} f(\sigma \otimes z) \chi^{(p)}_{\overleftarrow{\overleftarrow{\omega} \otimes \sigma^{-1}}}(\sigma \otimes z) = S_{f(z)}(\overleftarrow{\overleftarrow{\omega} \otimes \sigma^{-1}}) \qquad (\text{mod } p). \quad ∎ \end{aligned}$$

COROLLARY 1.4.2. *Let* $\sigma_{is} \in \{0, 1, \ldots, p-1\}$ $(i, s = 0, 1, \ldots, m-1)$, $|\sigma|_p \neq 0$. *Let* σ *be an orthogonal matrix, i.e.,*

$$\bigoplus_{s=1}^{m} \sigma_{is}\sigma_{qs} = \begin{cases} 0 & \text{if} \quad i \neq q, \\ 1 & \text{if} \quad i = q. \end{cases} \qquad (\text{mod } p).$$

Then

$$S_{f(\sigma \otimes z)}(\omega) = S_{f(z)}(\overleftarrow{\sigma \otimes \overleftarrow{\omega}}) \qquad (\text{mod } p). \tag{1.4.11}$$

Formula (1.4.11) follows from (1.4.9), in view of the following identity for orthogonal matrices: for all α,

$$\alpha \otimes \sigma^{-1} = \sigma \otimes \alpha \qquad (\text{mod } p).$$

It follows from Theorem 1.4.5 that a nonsingular linear transformation of the argument of a function induces a linear transformation of the argument of its spectrum, and the transformation matrix for the spectrum is the inverse of that of the original transformation.

Since permutation of the arguments is a particular case of a linear transformation, Theorem 1.4.5 will be used constantly when we are looking for the "optimal" order of arguments, in linearization of systems of LF, analysis of various classes of LF, and so on.

EXAMPLE 1.4.1. Table 1.4.1 defines a function $f(z)$ and the function $f(\sigma \otimes z)$ (mod p) for $m = 4$, $p = 2$,

$$\sigma = \begin{pmatrix} 1 & 0 & 0 & 1 \\ 1 & 1 & 1 & 1 \\ 0 & 1 & 1 & 1 \\ 0 & 1 & 0 & 1 \end{pmatrix}, \qquad \sigma^{-1} = \begin{pmatrix} 0 & 1 & 1 & 0 \\ 1 & 1 & 1 & 1 \\ 0 & 0 & 1 & 1 \\ 1 & 1 & 1 & 0 \end{pmatrix}.$$

Also shown in the table are $S_{f(z)}(\omega)$ and $S_{f(\sigma\otimes z)}(\omega)$ (mod 2), illustrating Theorem 1.4.5.

REMARK. The properties of linearity, translation and convolution considered above for Walsh and Chrestenson transforms are analogs of corresponding properties for the ordinary Laplace transform. This observation makes it possible to utilize Walsh–Chrestenson functions in the analysis, synthesis and optimization of digital logical devices, in a role analogous to that of the exponential functions in the classical fields of radio-engineering, control theory (with regard to the Laplace transform) or the theory of sampled-data control systems (discrete Laplace transform).

1.5. AUTOCORRELATION AND CROSS-CORRELATION FUNCTIONS AND THEIR RELATION TO DISCRETE TRANSFORMS

1.5.1. In this section we are going to introduce the autocorrelation and cross-correlation functions, to be used subsequently for the analysis and synthesis of networks

TABLE 1.4.1.

z, ω	$f(z)$	$f(\sigma \otimes z)$	$16S_{f(z)}(\omega)$	$16S_{f(\sigma\otimes z)}(\omega)$
0	0	0	8	8
1	1	0	2	0
2	1	0	2	−4
3	1	0	0	−4
4	0	1	−2	0
5	1	1	0	0
6	0	1	0	0
7	1	1	−2	0
8	1	0	−2	−2
9	0	1	−4	2
10	0	0	0	−2
11	1	1	2	2
12	0	1	0	2
13	0	0	−2	−2
14	1	0	2	−2
15	0	1	−4	2

realizing LF; we shall also indicate the connection between these functions and the discrete transforms considered previously. These correlation functions are analogous to the classical correlation functions employed extensively in radio-engineering, theory of stochastic processes, etc.

1.5.2. Consider a system of p-valued logical functions

$$f^{(s)}(z^{(0)}, \ldots, z^{(m-1)}) \qquad (z^{(q)} \in \{0, 1, \ldots, p-1\}; q = 0, \ldots, m-1; s = 0, \ldots, k-1),$$

and the corresponding discrete function $f(z)$ (see § 1.1). We define the *autocorrelation function* (mod p) of the functions $f^{(s)}(z^{(0)}, \ldots, z^{(m-1)})$ or of the discrete function $f(z)$ by

$$B_{p,2}^{(f,f)}(\tau) = \sum_{z=0}^{p^m-1} f(z) f(z \ominus \tau) \qquad (\text{mod } p), \tag{1.5.1}$$

where $z \ominus \tau$ (mod p) denotes the number whose p-ary expansion is the component-wise difference of the numbers z and τ.

It should be clear from (1.5.1) that $B_{p,2}^{(f,f)}(\tau)$ is a convolution-type transform of the original function $f(z)$; note that the translation of the argument z by τ is understood modulo p.

The autocorrelation functions employed in radio-engineering systems for signal synchronization [98, 142] are special cases of (1.5.1) for $m = 1$, $p = N$, where N is the number of points at which the signal is defined, or $p = \infty$.

Now let $f_1(z)$ and $f_2(z)$ be two discrete functions, representing two systems of p-valued logical functions of m arguments. We define their *cross-correlation function* (mod p) as

$$B_{p,2}^{(f_1,f_2)}(\tau) = \sum_{z=0}^{p^m-1} f_1(z)\, f_2(z \ominus \tau) \qquad (\text{mod}\ p). \tag{1.5.2}$$

1.5.3. We now determine the relationship of the correlation functions to the Walsh and Chrestenson transforms of §§1.2, 1.3. In fact, we shall show that they may be expressed in terms of double transforms of the original functions, a representation similar to that of the usual correlation functions $B_{\infty,2}^{(f_1,f_2)}(\tau)$ as double Laplace transforms [98] or the representation furnished by the Wiener–Khinchin theorem in the theory of stochastic processes [98].

Let f be a discrete function. We let $\chi^{(p)}(f)$ denote its Chrestenson transform and $(\chi^{(p)})^{-1}(f)$ the inverse Chrestenson transform.

THEOREM 1.5.1. *Let f_1 and f_2 be discrete functions representing two systems of functions of p-valued logical functions of m arguments. Then*

$$B_{p,2}^{(f_1,f_2)} = p^m (\chi^{(p)})^{-1} \left(\chi^{(p)}(f_1)\, \overline{\chi^{(p)}(f_2)}\right), \tag{1.5.3}$$

where $\overline{\chi^{(p)}(f_2)}$ is the complex conjugate of $\chi^{(p)}(f_2)$.

PROOF. Denote the values of $\chi^{(p)}(f_1)$ and $\chi^{(p)}(f_2)$ at a point ω by $S^{(1)}(\omega)$ and $S^{(2)}(\omega)$, respectively. Then, in view of the fact that $f_1(z)$ and $f_2(z)$ are real functions,

$$S^{(1)}(\omega)\, \overline{S^{(2)}(\omega)} = p^{-2m} \sum_{z_1,z_2=0}^{p^m-1} f_1(z_1)\, f_2(z_2)\, \chi_\omega^{(p)}(z_1)\, \overline{\chi_\omega^{(p)}(z_2)}.$$

Hence, in view of Theorems 1.3.1, 1.3.3, 1.3.4, and 1.4.3,

$$\begin{aligned}
(\chi^{(p)})^{-1}\left(\chi^{(p)}(f_1)\, \overline{\chi^{(p)}(f_2)}\right)(\tau) &= \sum_{\omega=0}^{p^m-1} S^{(1)}(\omega)\, \overline{S^{(2)}(\omega)}\, \chi_\tau^{(p)}(\omega) \\
&= p^{-2m} \sum_{\omega=0}^{p^m-1} \sum_{z_1,z_2=0}^{p^m-1} f_1(z_1)\, f_2(z_2)\, \overline{\chi_\omega^{(p)}(z_1 \ominus z_2 \ominus \tau)} \\
&= p^{-2m} \sum_{z_1,z_2=0}^{p^m-1} f_1(z_1)\, f_2(z_2) \sum_{\omega=0}^{p^m-1} \overline{\chi_\omega^{(p)}(z_1 \ominus z_2 \ominus \tau)} \\
&= p^{-m} \sum_{z_1=0}^{p^m-1} f_1(z_1)\, f_2(z_1 \ominus \tau) \\
&\quad + p^{-2m} \sum_{z_2 \neq z_1 \ominus \tau} f_1(z_1)\, f_2(z_2) \sum_{\omega=0}^{p^m-1} \chi_\omega^{(p)}(z_1 \ominus z_2 \ominus \tau) \\
&= p^{-m} \sum_{z=0}^{p^m-1} f_1(z)\, f_2(z \ominus \tau) = p^{-m} B_{p,2}^{(f_1,f_2)}(\tau) \quad (\text{mod}\ p). \quad \blacksquare
\end{aligned}$$

In the binary case, the spectrum assumes only real values, so that

$$\overline{\chi^{(2)}(f)} = \chi^{(2)}(f) = 2^{-m}(\chi^{(2)})^{-1}(f),$$

and formula (1.5.3) then becomes*

$$B_{2,2}^{(f_1,f_2)} = 2^{2m}\chi^{(2)}(\chi^{(2)}(f_1)\cdot\chi^{(2)}(f_2)) = 2^{2m}W(W(f_1)\cdot W(f_2)). \qquad (1.5.4)$$

EXAMPLE 1.5.1. Table 1.5.1 defines a system of two BF for $m = 3$, the discrete function $f(z)$, the spectrum $S(\omega)$, $S^2(\omega)$, and the autocorrelation function $B_{2,2}^{(f,f)}(\tau)$.

TABLE 1.5.1.

$z^{(0)}$	$z^{(1)}$	$z^{(2)}$	$y^{(0)}$	$y^{(1)}$	z, ω, τ	$f(z)$	$8S(\omega)$	$64S^2(\omega)$	$B_{2,2}^{(f,f)}(\tau)$
0	0	0	0	1	0	1	13	169	29
0	0	1	1	0	1	2	5	25	24
0	1	0	1	1	2	3	−3	9	22
0	1	1	1	1	3	3	−3	9	22
1	0	0	1	0	4	2	1	1	16
1	0	1	0	0	5	0	−3	9	20
1	1	0	0	1	6	1	1	1	18
1	1	1	0	1	7	1	−3	9	18

It follows from Theorem 1.5.1 that the autocorrelation function is symmetric ("evenness" relation for autocorrelation functions):

$$B_{p,2}^{(f,f)}(\tau) = B_{p,2}^{(f,f)}(\bar{\tau}),$$

where $\tau \oplus \bar{\tau} = 0 \pmod{p}$. This property halves the computational work needed to determine autocorrelation functions in the case $p > 2$.

We now list the basic properties of the correlation functions, which determine convolution-type transforms of the original functions.

1.5.4. I. *Translation of argument of original function.*

THEOREM 1.5.2. *Let $f(z)$ and $\varphi(z)$ be discrete functions corresponding to two systems of p-valued logical functions of the same number of arguments, and*

$$f_\alpha(z) = f(z \ominus \alpha), \qquad \varphi_\alpha(z) = \varphi(z \ominus \alpha) \qquad (\text{mod } p). \qquad (1.5.5)$$

Then

$$B_{p,2}^{(f_\alpha,\varphi_\alpha)}(\tau) = B_{p,2}^{(f,\varphi)}(\tau). \qquad (1.5.6)$$

* Theorem 1.5.1 was published in [83]; for the case $p = 2$ (see (1.5.4)) it was in fact contained in [80], and also in [50, 150]. A generalization to arbitrary finite groups will be given in § 1.6 (see Theorem 1.6.14).

Theorem 1.5.2 implies that the correlation functions are invariant with respect to translation of argument in the original function. It can be shown that for autocorrelation functions the converse is also true: a nonnegative function is uniquely determined, up to translation of the argument, by its autocorrelation function.

In the case of Boolean functions, translation of the argument by α is equivalent to inversion of the arguments corresponding to nonzero components of the binary expansion of α. Thus the complexity of a network realizing a BF, relative to any basis, is completely determined by the autocorrelation function $B_{2,2}^{(f,f)}(\tau)$ (up to m inversion elements, where m is the number of arguments). One can thus solve any problem involved in the minimization of networks realizing systems of BF in terms of correlation characteristics.

II. *Linearity.*

THEOREM 1.5.3. *Let $f_s(s = 1, 2, \ldots, n)$ and $\varphi_q(q = 1, 2, \ldots, l)$ be discrete functions corresponding to systems of p-valued logical functions of m arguments; let $c_s(s = 1, 2, \ldots, n)$ and $d_q(q = 1, 2, \ldots, l)$ be arbitrary constants. Then*

$$B_{p,2}^{\left(\sum_{s=1}^{n} c_s f_s,\ \sum_{q=1}^{l} d_q \varphi_q\right)}(\tau) = \sum_{s=1}^{n} \sum_{q=1}^{l} c_s d_q B_{p,2}^{(f_s,\varphi_q)}(\tau). \tag{1.5.7}$$

COROLLARY 1.5.1. *Let f be a Boolean function of m arguments, $\sum_{z=0}^{2m-1} f(z) = N$, and $\bar{f}$ the inversion of f ($\bar{f} = 1 \oplus f \pmod 2$). Then*

$$B_{2,2}^{(f,f)}(\tau) = B_{2,2}^{(\bar{f},\bar{f})}(\tau) - 2^m + 2N. \tag{1.5.8}$$

This follows from (1.5.7) in view of the equality $\bar{f} = 1 - f$.

Corollary 1.5.1 simplifies the calculation of the autocorrelation function $B_{2,2}^{(f,f)}(\tau)$, replacing the direct calculation by evaluation of $B_{2,2}^{(\bar{f},\bar{f})}(\tau)$ when $N > 2^{m-1}$.

III. *Linear transformation of variables.*

THEOREM 1.5.4. *Let $f(z)$ and $\varphi(z)$ be discrete functions corresponding to two systems of p-valued logical functions of m arguments, $\sigma = (\sigma_{is})$ a p-ary matrix ($i, s = 0, 1, \ldots, m-1$), $|\sigma|_p \neq 0$ ($|\sigma|_p$ denotes the determinant of σ over $GF(p)$),*

$$f_\sigma(z) = f(\sigma \otimes z), \qquad \varphi_\sigma(z) = \varphi(\sigma \otimes z) \pmod p.$$

Then

$$B_{p,2}^{(f_\sigma,\varphi_\sigma)}(\tau) = B_{p,2}^{(f,\varphi)}(\sigma \otimes \tau) \pmod p. \tag{1.5.9}$$

PROOF.

$$B_{p,2}^{(f_\sigma,\varphi_\sigma)}(\tau) = \sum_{z=0}^{p^m-1} f_\sigma(z)\,\varphi_\sigma(z \ominus \tau) = \sum_{z=0}^{p^m-1} f(\sigma \otimes z)\,\varphi(\sigma \otimes (z \ominus \tau))$$

$$= \sum_{y=0}^{p^m-1} f(y)\,\varphi(y \ominus \sigma \otimes \tau) = B_{p,2}^{(f,\varphi)}(\sigma \otimes \tau) \qquad (\text{mod } p). \quad ▮$$

1.5.5. We now generalize the autocorrelation functions modulo p. Let $f(z)$ be the discrete function representing a system of p-valued logical functions of m arguments. Consider the class $\{B_{p,q}\}$ of autocorrelation functions defined as follows:

$$B_{p,q}(\tau) = \sum_{z=0}^{p^m-1} f(z)\,f(z \ominus \tau) \ldots f(z \ominus \underbrace{\tau \ominus \cdots \ominus \tau}_{q-1}). \tag{1.5.10}$$

Note that $B_{p,q}(\tau) = B_{p,2}^{(f,f)}(\tau)$.

We shall assume that $p \geqq q$ in (1.5.10), since when $p < q$

$$B_{p,q}(\tau) = B_{p,p}(\tau) \qquad (\tau = 0, 1, \ldots, p^m - 1). \tag{1.5.11}$$

The function $B_{p,q}(\tau)$ is the cross-correlation function of the q functions obtained by q successive translations (mod p) of the original function.

Of this class of correlation characteristics, the most frequently used in the sequel will be $\{B_{p,p}(\tau)\}$; these functions find application in the analysis and synthesis of networks realizing systems of p-valued logical functions.

EXAMPLE 1.5.2. Table 1.5.2 defines a function of two ternary arguments and its autocorrelation function $B_{3,3}(\tau)$.

To end this section, we remark that Theorems 1.5.2, 1.5.3 and 1.5.4 may be rephrased in terms of the characteristics $\{B_{p,q}\}$, on the assumption that $f_i = \varphi_i$.

TABLE 1.5.2.

$z^{(0)}$	$z^{(1)}$	$y^{(0)}$	z, τ	$f(z)$	$B_{3,3}(\tau)$
0	0	1	0	1	6
0	1	0	1	0	0
0	2	1	2	1	0
1	0	1	3	1	0
1	1	1	4	1	6
1	2	0	5	0	0
2	0	0	6	0	0
2	1	1	7	1	0
2	2	1	8	1	6

1.6. DISCRETE TRANSFORMS OVER AN ARBITRARY FINITE COMMUTATIVE GROUP

1.6.1. The purpose of this section is to describe a unified approach to the discrete transforms described in §§1.2, 1.3. This approach will yield more general versions of the properties of discrete transforms and correlation characteristics described in §§1.4, 1.5, and also enable us to construct new discrete transforms with given properties.

1.6.2. Let $G = (Z, \circ)$ be a commutative group, where Z is a set of N elements, $\circ$ an associative, commutative and invertible binary operation on Z (see §1.2.2).

An example of a commutative group is the set of all binary (p-ary) sequences of length m, with the operation of componentwise addition modulo 2 (modulo p). In this case $N = 2^m$ ($N = p^m$). The most familiar infinite commutative group is the set of all integers (including 0) with respect to addition.

We shall say that two groups $G_1(Z, \circ)$ and $G_2 = (Y, \Delta)$ are *isomorphic* if there exists a one-to-one mapping h of Z onto Y such that if $z_1 \circ z_2 = z_3 (z_1, z_2, z_3 \in Z)$ then

$$h(z_1) \Delta h(z_2) = h(z_3) \qquad (h(z_1), h(z_2), h(z_3) \in Y). \tag{1.6.1}$$

If the group operation is ordinary addition or multiplication, we shall speak of an additive or multiplicative group, respectively.

THEOREM 1.6.1. *Let $G = (Z, \circ)$, $Z = \{0, 1, \ldots, N-1\}$, be a finite commutative group. Then there exists a system of orthogonal functions*

$$\Psi_0^{(G)}(z), \Psi_1^{(G)}(z), \ldots, \Psi_{N-1}^{(G)}(z) \qquad (z \in Z).$$

which forms a multiplicative group isomorphic to G and is a complete orthogonal basis in the space F_G of all functions from Z to the complex numbers.

A proof of this theorem will be presented later in this section.

It follows from Theorem 1.6.1 that any function in F_G may be expressed as a linear combination of functions $\{\Psi_\omega^{(G)}(z)\}$, with the coefficients defined in the usual way as Fourier coefficients: if $f \in F_G$, then

$$f(z) = \sum_{\omega=0}^{N-1} S(\omega) \Psi_\omega^{(G)}(z) \tag{1.6.2}$$

and

$$S(\omega) = N^{-1} \sum_{z=0}^{N-1} f(z) \overline{\Psi_\omega^{(G)}(z)}. \tag{1.6.3}$$

The sequence $S(\omega)$ $(\omega = 0, 1, \ldots, N-1)$ is known as the *generalized Fourier transform* or *spectrum* of f.

Theorem 1.6.1 is a generalization of Theorems 1.2.3 and 1.3.5. Indeed, let $G = (\{0, 1\}^m, \oplus \pmod 2)$. Then $\{\Psi_\omega^{(G)}(z)\} = \{W_\omega(z)\}$. If $G = (\{0, 1, \ldots, p-1\}^m, \oplus \pmod p)$, we have $\{\Psi_\omega^{(G)}(z)\} = \{\chi_\omega^{(p)}(z)\}$.

Theorem 1.6.1 is also valid for some infinite groups. For example, if $G = (\{0, \pm 1, \pm 2, \ldots\}, +)$ is the additive group of integers, then

$$\{\Psi_\omega^{(G)}(z)\} = \{\exp(2\pi i\omega z)\} \qquad (i = \sqrt{-1})$$

is the system of exponential functions.

The space F_G of functions from G to the field of complex numbers contains many complete orthogonal bases. For example, if $G = (\{0, 1\}^m, \oplus \pmod 2)$ we may replace the Walsh basis by the Haar basis, and so on. It follows from Theorem 1.6.1, however, that among all these complete orthogonal bases there is exactly one, $\{\Psi_\omega^{(G)}(z)\}$, whose structure is identical to that of the group G. This is one of the main reasons for the extensive use of exponential functions in the analysis of discrete-time processes: as we have stated, this basis forms a multiplicative group isomorphic to the additive group of integers. The theory of the discrete Laplace transform, the basic tool of the theory of sampled-data control systems, is based on this approach.

In our context, Theorem 1.6.1 explains the expediency of the Walsh and Chrestenson bases as a tool for the analysis and synthesis of networks realizing Boolean and p-valued logical functions.

Our immediate problem is this: given a group G, how can we construct the isomorphic complete orthogonal multiplicative system $\{\Psi_\omega^{(G)}(z)\}$? The method proposed below is essentially a constructive proof of Theorem 1.6.1. Before proceeding to the construction, however, we recall a few important propositions from the theory of group characters.

A mapping χ (not necessarily one-to-one) of a group $G = (Z, \circ)$ into the multiplicative group of (nonzero) complex numbers is called a *homomorphism* if, whenever $z_1 \circ z_2 = z_3$,

$$\chi(z_1)\,\chi(z_2) = \chi(z_3) \qquad (z_1, z_2, z_3 \in Z). \tag{1.6.4}$$

Any homomorphism of a group G into the multiplicative group of complex numbers is known as a *character* of G. Let $\chi_\omega(z)$ denote the ω-th character of G; $\chi_0(z)$ will denote the character such that $\chi_0(z) = 1$ for any $z \in Z$, the *identity character*.

EXAMPLE 1.6.1. Let $G = (\{0, 1, 2\}^2, \oplus \pmod 3)$. Then the function $\chi_\omega(z)$ defined by Table 1.6.1 is a character of G (here $e_1 = \exp(2\pi i/3)$, $e_2 = \exp(4\pi i/3)$, $i = \sqrt{-1}$). Indeed, taking, say $z_1 = (0, 2)$, $z_2 = (1, 1)$, we have $z_1 \oplus z_2 = (0, 2) \oplus (1, 1) = (1, 0) \pmod 3$, and $\chi_\omega(z_1)\,\chi_\omega(z_2) = e_2 \cdot e_2 = e_1 = \chi_\omega(z_1 \oplus z_2) \pmod 3$.

The reader should note that this character is the Chrestenson function $\chi_4^{(3)}(z)$ (see § 1.3.2). It will be clear from what follows that every character of this group is a Chrestenson function, and every Chrestenson function $\chi_\omega^{(p)}(z)$ is a character of $G = (\{0, 1, \ldots, p-1\}^m, \oplus \pmod p)$.

TABLE 1.6.1.

z	$z^{(0)}$	$z^{(1)}$	$\chi_\omega(z)$
0	0	0	1
1	0	1	e_1
2	0	2	e_2
3	1	0	e_1
4	1	1	e_2
5	1	2	1
6	2	0	e_2
7	2	1	1
8	2	2	e_1

THEOREM 1.6.2. *Let G be a finite commutative group. Then the set of all characters of G is a multiplicative group isomorphic to G.*

Theorem 1.6.2 is a corollary of Theorem 1.6.3, to be proved later in this section.

Using Theorem 1.6.2, one can show that the set of characters $\{\chi_\omega(z)\}$ ($\omega = 0, 1, \ldots, N-1$) of a group $G = (Z, \circ)$ ($Z = \{0, 1, \ldots, N-1\}$) is the complete orthogonal system whose existence is asserted in Theorem 1.6.1. In other words:

$$1)\ \sum_{z=0}^{N-1} \chi_\omega(z)\overline{\chi_s(z)} = \begin{cases} 0 & \text{if} \quad \omega \neq s, \\ N & \text{if} \quad \omega = s \end{cases} \tag{1.6.5}$$

(*orthogonality of characters*), and

2) if

$$\sum_{z=0}^{N-1} f(z)\overline{\chi_\omega(z)} = 0 \qquad (\omega = 0, \ldots, N-1),$$

then

$$f(z) \equiv 0 \qquad (z = 0, \ldots, N-1) \tag{1.6.6}$$

(*completeness of characters*).

Thus, whenever we are analyzing the properties of functions defined on finite groups (in particular, logical functions) or synthesizing functions of this type with prescribed properties, we may view the functions as being defined not on the original group but on its character group; this is extremely convenient, in view of the orthogonality of the characters and the fact that the group operation is now ordinary multiplication. By the value of a function $f(z)$ on a character $\chi_\omega(z)$ we mean the quantity $S(\omega) = N^{-1}\sum_{z=0}^{N-1} f(z)\overline{\chi_\omega(z)}$. The numbers $S(\omega)$ ($\omega = 0, 1, \ldots, N-1$) form the spectrum (Fourier transform) of the original function. Thus, when concerned with problems of analysis, synthesis and optimization of digital networks, we may go over from originals to transforms, as done, for example, in automatic control theory.

1.6.3. We can now proceed to methods for constructing the character group of a given group G and describe some of the basic properties of characters. As mentioned previously, the method proposed below proves Theorem 1.6.2 and hence, in view of the orthogonality and completeness relations (1.6.5), (1.6.6), also Theorem 1.6.1.

Let $G = (Z, \circ)$ be a finite commutative group. For any $z \in Z$, the set of elements $e = z^0, z^1, z^2, \ldots$, where $z^n = \underbrace{z \circ z \circ \cdots \circ z}_{n}$, is a group; this group is known as a *cyclic subgroup* of G and z is called its *generator*.

A standard theorem of group theory states that any finite commutative group is a direct product of cyclic subgroups. This means that there exist elements $\gamma_0, \gamma_1, \ldots, \gamma_{m-1} \in Z$ such that, for any $z \in Z$, there exist numbers $z^{(0)}, z^{(1)}, \ldots, z^{(m-1)}$, where $0 \leqq z^{(s)} \leqq N - 1$ $(s = 0, 1, \ldots, m)$ (N is the number of elements in Z), such that

$$z = \gamma_0^{z^{(0)}} \circ \gamma_1^{z^{(1)}} \circ \cdots \circ \gamma_{m-1}^{z^{(m-1)}}. \tag{1.6.7}$$

Note that the elements γ_s $(s = 0, 1, \ldots, m - 1)$ are independent of z; they are known as *generators*, forming a *basis* of G, and m is called the *dimension* of G.

Throughout the sequel, we shall assume that for any z the numbers $z^{(s)}$ $(s = 0, 1, \ldots, m - 1)$ in (1.6.7) are minimal. In that case the numbers $n_s = \max_{z \in Z} z^{(s)} + 1$ will be called the *orders* of the appropriate subgroups.

For example, if $G = (\{0, 1, \ldots, p - 1\}^m, \oplus \text{ (mod p)})$, then the vectors

$$(\underbrace{0, \ldots, 0}_{s}, z^{(s)}, 0, \ldots, 0) \qquad (z^{(s)} = 0, 1, \ldots, p - 1)$$

form a cyclic subgroup of G, of order p, with generator $\gamma_s = (\underbrace{0, \ldots, 0}_{s}, 1, 0, \ldots, 0)$.

The group G is the direct product of all such subgroups. In this case, then,

$$\gamma_s^{z^{(s)}} = z^{(s)}\gamma_s \qquad (0 \leqq z^{(s)} \leqq p - 1), \tag{1.6.8}$$

and it follows from (1.6.7) that for any $z \in \{0, 1, \ldots, p - 1\}^m$ there exist numbers $z^{(s)}$ (here $z^{(s)}$ is simply the s-th component of z) such that

$$z = \bigoplus_{s=0}^{m-1} z^{(s)}\gamma_s \qquad (\text{mod } p). \tag{1.6.9}$$

THEOREM 1.6.3. *Let $G = (\{0, 1, \ldots, N - 1\}, \circ)$ be a finite commutative group which is a direct product of m cyclic subgroups of order n_s $(s = 0, 1, \ldots, m - 1)$ with generators γ_s; let $\chi_\omega(z)$ be the ω-th character of G. Then*

$$\chi_\omega(z) = \exp\left(2\pi i \sum_{s=0}^{m-1} \frac{\omega^{(s)} z^{(s)}}{n_s} \right) \qquad (i = \sqrt{-1}), \tag{1.6.10}$$

where

$$\left.\begin{aligned} z &= \gamma_0^{z^{(0)}} \circ \cdots \circ \gamma_{m-1}^{z^{(m-1)}} \\ \omega &= \gamma_0^{\omega^{(0)}} \circ \cdots \circ \gamma_{m-1}^{\omega^{(m-1)}} \end{aligned}\right\} \qquad (0 \leqq \omega^{(s)}, z^{(s)} \leqq n_s - 1). \tag{1.6.11}$$

The mapping $h: \chi_\omega \to \omega$ *is an isomorphism of the character group of* G *onto* G.

PROOF. We first show that the function $\chi_\omega(z)$ defined by (1.6.10), (1.6.11) is indeed a character of G. Let $z_1 \circ z_2 = z_3$, where

$$z_q = \gamma_0^{z_q^{(0)}} \circ \cdots \circ \gamma_{m-1}^{z_q^{(m-1)}} \qquad (q = 1, 2, 3).$$

Then it follows from (1.6.11), in view of the definition of n_s, that

$$z_3^{(s)} = z_1^{(s)} \oplus z_2^{(s)} \quad (\text{mod } n_s) \qquad (s = 0, \dots, m-1), \tag{1.6.12}$$

$$\begin{aligned} \chi_\omega(z_1 \circ z_2) &= \exp\left(2\pi i \sum_{s=0}^{m-1} \frac{\omega^{(s)}(z_1^{(s)} \oplus z_2^{(s)})}{n_s}\right) \\ &= \exp\left(2\pi i \sum_{s=0}^{m-1} \frac{\omega^{(s)} z_1^{(s)}}{n_s}\right) \cdot \exp\left(2\pi i \sum_{s=0}^{m-1} \frac{\omega^{(s)} z_2^{(s)}}{n_s}\right) \\ &= \chi_\omega(z_1) \cdot \chi_\omega(z_2) \qquad (\text{mod } n_s). \end{aligned}$$

Thus the function in question is indeed a character of G.

Now the set $\{\chi_\omega(z)\}$ is a multiplicative group, with

$$\chi_\omega^{-1}(z) = \exp\left(2\pi i \sum_{s=0}^{m-1} \frac{(n_s - \omega^{(s)})\, z^{(s)}}{n_s}\right) = \chi_{\bar\omega}(z),$$

where

$$\bar\omega = \gamma_0^{(n_0 - \omega^{(0)})} \circ \cdots \circ \gamma_{m-1}^{(n_{m-1} - \omega^{(m-1)})}$$

and

$$\chi_\omega(z)\, \chi_{\bar\omega}(z) = \chi_0(z) \equiv 1.$$

We now show that the character group is isomorphic to G, under the isomorphism $h: \chi_\omega \to \omega$. Let $z, \omega_1, \omega_2, \omega_3 \in G$ and

$$\omega_q = \gamma_0^{\omega_q^{(0)}} \circ \cdots \circ \gamma_{m-1}^{\omega_q^{(m-1)}} \qquad (q = 1, 2, 3; \omega_q^{(s)} \in \{0, 1, \dots, n_s - 1\}.$$

Then, if $\omega_3 = \omega_1 \circ \omega_2$, we have $\omega_3^{(s)} = \omega_1^{(s)} \oplus \omega_2^{(s)}$ (mod n_s) ($s = 0, 1, \dots, n_s - 1$) and

$$\begin{aligned} \chi_{\omega_3}(z) &= \exp\left(2\pi i \sum_{s=0}^{m-1} \frac{\omega_3^{(s)} z^{(s)}}{n_s}\right) = \exp\left(2\pi i \sum_{s=0}^{m-1} \frac{(\omega_1^{(s)} \oplus \omega_2^{(s)})\, z^{(s)}}{n_s}\right) \\ &= \exp\left(2\pi i \sum_{s=0}^{m-1} \frac{\omega_1^{(s)} z^{(s)}}{n_s}\right) \cdot \exp\left(2\pi i \sum_{s=0}^{m-1} \frac{\omega_2^{(s)} z^{(s)}}{n_s}\right) = \chi_{\omega_1}(z) \cdot \chi_{\omega_2}(z) \qquad (\text{mod } n_s). \end{aligned}$$

Conversely, if $\chi_{\omega_3}(z) = X_{\omega_1}(z) \cdot \chi_{\omega_2}(z)$, we set $z = \gamma_s$ $(s = 0, 1, \dots, m-1)$, to obtain

$$\exp\left(2\pi i \frac{\omega_3^{(s)}}{n_s}\right) = \exp\left(\frac{2\pi}{n_s} i\omega_1^{(s)}\right) \cdot \exp\left(\frac{2\pi}{n_s} i\omega_2^{(s)}\right)$$

whence it follows that $\omega_3^{(s)} = \omega_1^{(s)} \oplus \omega_2^{(s)} \pmod{n_s}$ and $\omega_3 = \omega_1 \circ \omega_2$. ∎

Formulas (1.6.10) and (1.6.11) generate a simple procedure for constructing the characters, thus furnishing a complete multiplicative system of functions isomorphic to the original group.

In addition, these formulas yield a direct proof of Theorem 1.6.2.

If $G = (\{0, 1, \dots, p-1\}^m, \oplus \pmod p)$, then $n_s = p$ $(s = 0, 1, \dots, m-1)$; $\omega^{(s)}$, $z^{(s)}$ are the components of the p-ary expansions of ω, z, and the set of characters $\{\chi_\omega(z)\}$ is precisely the Chrestenson system $\{\omega_\omega^{(p)}(z)\}$ (see § 1.3).

The reader should note that for the sake of convenience we have indexed the characters in a manner slightly different from that adopted previously for the Chrestenson functions; this of course does not affect the validity of the exposition.

Using Theorems 1.6.1–1.6.3, we can generalize the properties of Walsh functions (§ 1.2) and Chrestenson functions (§ 1.3) to arbitrary finite commutative groups (completeness and orthogonality, finiteness of representing series, symmetry of index and argument, translation theorem, etc.). The same applies to the properties of the corresponding discrete transforms (§ 1.4: linearity, translation of original and transform, involution property, convolutions, and so on).

EXAMPLE 1.6.2. Consider the group $G = (\{0, 1, 2, 3, 4, 5\}, \circ)$, where $\circ$ is defined by Table 1.6.2 ($N = 6, e = 0$). We construct the characters $\chi_\omega^{(G)}(z)$ of G $(z \in \{0, 1, \dots, 5\})$. G contains two cyclic subgroups $G_0 = (\{0, 3\}, \circ)$ and $G_1 = (\{0, 1, 1^2 = 2\}, \circ)$, with generators $\gamma_0 = 3$, $\gamma_1 = 1$, respectively, and it is the direct product of these subgroups ($n_0 = 2$, $n_1 = 3$). The representation of the elements $z \in G$ in terms of powers of the generators of the cyclic groups (see (1.6.11)) is shown in Table 1.6.3.

TABLE 1.6.2.

$\circ$	0	1	2	3	4	5
0	0	1	2	3	4	5
1	1	2	0	4	5	3
2	2	0	1	5	3	4
3	3	4	5	0	1	2
4	4	5	3	1	2	0
5	5	3	4	2	0	1

TABLE 1.6.3.

z, ω	$z^{(0)}, \omega^{(0)}$	$z^{(1)}, \omega^{(1)}$
0	0	0
1	0	1
2	0	2
3	1	0
4	1	1
5	1	2

TABLE 1.6.4.

$z \backslash \omega$	0	1	2	3	4	5
0	1	1	1	1	1	1
1	1	e_1	e_2	1	e_1	e_2
2	1	e_2	e_1	1	e_2	e_1
3	1	1	1	-1	-1	-1
4	1	e_1	e_2	-1	$-e_1$	$-e_2$
5	1	e_2	e_1	-1	$-e_2$	e_1

The characters $\chi_\omega^{(G)}(z)$ of G are, by (1.6.10),

$$\chi_\omega^{(G)}(z) = \exp\left(2\pi i\left(\frac{\omega^{(0)}z^{(0)}}{2} + \frac{\omega^{(1)}z^{(1)}}{3}\right)\right).$$

Table 1.6.4 lists the values of these characters, $e_1 = \exp(2\pi i/3), e_2 = e_1^2 = \exp(4\pi i/3)$. A direct examination of Table 1.6.4 readily shows that $\chi_\omega^{(G)}(z) = \chi_z^{(G)}(\omega), \sum_{z=0}^{s} \chi_\omega^{(G)}(z) = 0$ if $\omega \neq 0$, $\sum_{z=0}^{s} \chi_{\omega_1}^{(G)}(z)\,\overline{\chi_{\omega_2}^{(G)}(z)} = 0$ if $\omega_1 \neq \omega_2$, and $\chi_\omega^{(G)}(z_1)\cdot\chi_\omega^{(G)}(z_2) = \chi_\omega^{(G)}(z_1 \circ z_2)$.

We can also generalize the auto- and cross-correlation functions to any group $G = (\{0, 1, \ldots, N-1\}, \circ)$. Thus, if $f_1(z), f_2(z)$ are two functions defined on the group, their cross-correlation function is defined as

$$B_{\circ,2}^{(f_1,f_2)}(\tau) = \sum_{z=0}^{N-1} f_1(z)\, f_2(z \circ \tau^{-1}), \tag{1.6.13}$$

where $\tau \circ \tau^{-1} = e$ is the identity of the group.

Let $\chi(f)$ denote the generalized Fourier transform of a function f defined on $\{0, 1, \ldots, N-1\}$, i.e., the sequence of expansion coefficients of f in characters of G. We may assume that $\chi(f)$, like f, is defined on $\{0, 1, \ldots, N-1\}$. We then have the following generalization of Theorem 1.5.1 to an arbitrary finite commutative group (analog of the Wiener–Khinchin theorem [98]).

THEOREM 1.6.4. *Let $G = (\{0, 1, 2, \ldots, N-1\}, \circ)$ be a finite commutative group and f_1, f_2 functions defined on $\{0, 1, \ldots, N-1\}$. Then*

$$B_{\circ,2}^{(f_1,f_2)} = N \cdot \chi^{-1}(\chi(f_1)\,\overline{\chi(f_2)}), \tag{1.6.14}$$

where $\chi^{-1}()$ is the inverse transform.

Theorem 1.6.4 enables us to generalize to arbitrary groups the properties of the correlation functions considered in §1.5.

1.6.4. We now generalize the discrete transforms, originally defined by

$$f(z) = \sum_{\omega=0}^{N-1} S(\omega)\, \Psi_\omega(z), \tag{1.6.15}$$

$$S(\omega) = N^{-1} \cdot \sum_{z=0}^{N-1} f(z)\, \overline{\Psi_\omega(z)}, \tag{1.6.16}$$

where $\{\Psi_\omega(z)\}$ is a basis and z ranges over the elements of a finite group G, known as the *argument group*. The values of $f(z)$, $S(\omega)$, $\Psi_\omega(z)$ lie in the field of complex numbers. (Recall that a *field* is a triple $R = (Z, \oplus, \odot)$, where $(Z, \oplus)$ is a commutative group with identity denoted by 0, $(Z \setminus \{0\}, \odot)$* is a commutative group, and for any $x, y, z \in Z$,

$$x \odot (y \oplus z) = (x \odot y) \oplus (x \odot z) \quad \text{and} \quad x \odot 0 = 0.)$$

We shall now introduce transforms for which all the operations in (1.6.15), (1.6.16) are interpreted not over the field of complex numbers but over a finite field (Galois field). We shall denote the field with N elements by $GF(N)$. The transforms thus defined will be referred to as *Walsh–Galois*, *Haar–Galois* and *Chrestenson–Galois transforms*, in accordance with the choice of the basis $\{\Psi_\omega(z)\}$. (In the case of Chrestenson–Galois transforms, as applied to p-valued logic, we have the additional condition that p be a divisor of $N - 1$.)

Let us consider the important special case in which the field is the field R of residues modulo some prime number Q, $R = GF(Q)$. (Recall that the field of residues modulo Q is the field whose elements are the numbers $0, 1, \ldots, Q - 1$, the operations of addition and multiplication being performed modulo Q.) Let

$$Q > \max\left(\max_z f(z), p\right)$$

(where we are dealing with p-valued logic, and p is a divisor of $Q - 1$); assume all operations in (1.6.15), (1.6.16) to be modulo Q. We then define a discrete transform over $GF(Q)$:

$$f(z) = \bigoplus_{\omega=0}^{N-1} S(\omega) \otimes \Psi_\omega(z) \qquad (\text{mod } Q), \tag{1.6.17}$$

$$S(\omega) = N^{-1} \otimes \bigoplus_{z=0}^{N-1} f(z) \otimes \Psi_{\omega^{-1}}(z) \qquad (\text{mod } Q), \tag{1.6.18}$$

where $\omega \circ \omega^{-1} = e$, $\circ$ denoting the operation in G.

The use of transforms over finite fields $GF(Q)$ turns out be convenient for the design of networks realizing LF (see §6.3), since when this approach is adopted the number

* $A \setminus B$ denotes the set of all elements of A not in B.

of p-ary digits of the code of $S(\omega)$ is at most $]\log_p Q[$, and this number may be considerably smaller than that needed for expansions over the complex field. (Here and below $]a[$ denotes the smallest integer $\geqq a$.)

As an example, let us construct the Walsh spectrum of the BF defined by Table 1.6.5 over the real field, on the one hand, and over $GF(3)$, on the other (Table 1.6.5). It is clear from the table that the length of $S(\omega)$ is decreased from 5 to $]\log_2 3[= 2$.

All the properties of the Walsh, Haar, etc. transforms remain valid for the Walsh–Galois, Haar–Galois, etc. versions.

Of the many discrete transforms over finite fields, one of the most important for analysis and synthesis is the special case of the Chrestenson–Galois transform with basis $\{\Psi_\omega(z)\} = \{\chi_\omega^{(p)}(z)\}$ (see §1.3), over the field $GF(Q^n)$ [37, 38]. This special case arises when $p = Q^n - 1$ and $m = 1$, and is known as the *Laplace–Galois transform*, used in analysis and synthesis of linear switching networks with memory described by discrete periodic time functions (with period at most $Q^n - 1$). (A *linear network* consists exclusively of mod p adders and delay lines; n is the number of delay lines.) The use of Laplace–Galois transforms in analysis and synthesis of linear networks with memory is discussed in detail in [37–39, 44].

In the general case, then, any discrete transform may be characterized by the algebraic structure (group, field, etc.) of the argument set G of the original function and the algebraic structure of the value set R of the basis functions and the corresponding expansion coefficients. In the cases considered hitherto G is a finite commutative group, while R is the field of real numbers for the Haar and Walsh transforms, the field of complex numbers for the generalized Haar transform and the Chrestenson transform, and a finite (Galois) field for the Haar–Galois, Walsh–Galois, etc. transforms.

It is sometimes convenient to work with discrete functions defined on an infinite set G. The set G is usually the set of discrete times (sampling times) at which the behavior of the object is observed. This set is treated as the additive group of integers.

If the set of values of z is the additive group G of integers and R (the field of values

TABLE 1.6.5.

$z^{(0)}$	$z^{(1)}$	$z^{(2)}$	$z, \omega)$	$f(z)$	$8S(\omega)$	$S(\omega)$ (mod 3)
0	0	0	0	1	5	1
0	0	1	1	0	−1	1
0	1	0	2	1	1	2
0	1	1	3	0	−1	1
1	0	0	4	1	3	0
1	0	1	5	1	1	2
1	1	0	6	1	−1	1
1	1	1	7	0	1	2

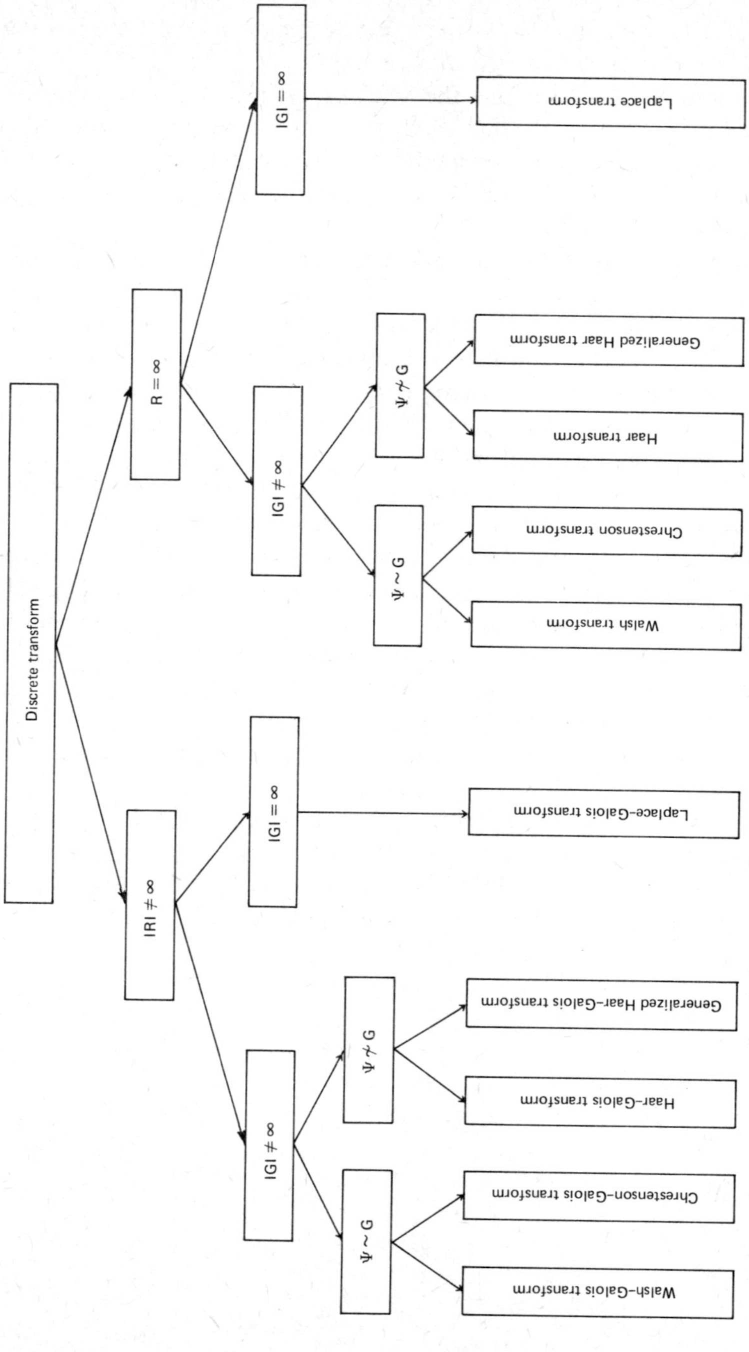
Discrete transform
R = ∞
|G| = ∞
Laplace transform
|G| ≠ ∞
Ψ ≁ G
Generalized Haar transform
Haar transform
Ψ ~ G
Chrestenson transform
Walsh transform
|R| ≠ ∞
|G| = ∞
Laplace–Galois transform
|G| ≠ ∞
Ψ ≁ G
Generalized Haar–Galois transform
Haar–Galois transform
Ψ ~ G
Chrestenson–Galois transform
Walsh–Galois transform

FIGURE 1.6.1.

of the basis functions and the coefficients) is the complex field, the transform defined by the multiplicative basis isomorphic to G is the familiar discrete Laplace transform [31].

Since digital devices with finite memory capacity are described by periodic time functions, the argument group G may always be taken finite, its order coinciding with the maximum period of the functions describing the operation of the device. This explains why this book is concerned only with discrete transforms, over finite argument groups. A general classification of all discrete transforms described in this chapter is given in Figure 1.6.1. In the figure $|G|$, $|R|$ denote the number of elements in G and R, respectively, $\Psi \sim G$ ($\Psi \nsim G$) denotes isomorphism (nonisomorphism) of the expansion basis $\{\Psi\}$ and the argument group G.

As a final remark, we note that the properties of the Walsh, Chrestenson, etc. transforms and the corresponding correlation characteristics are readily generalized to all the discrete transforms listed in Figure 1.6.1.

Chapter Two

SPECTRAL METHODS IN ANALYSIS AND OPTIMIZATION PROBLEMS OF ALGEBRA OF LOGIC

In this chapter we shall study spectral methods for solving the decision problems of some important classes of logical functions,* and also develop spectral methods for solving optimization problems in the algebra of logic (including the linearization and approximation of systems of logical functions).

A special section will be devoted to analysis of the spectral complexity of Boolean functions, deriving bounds for the number of nonzero coefficients in their orthogonal expansions. Our main tool throughout the chapter will be the Walsh transform.

2.1. SPECTRAL ANALYSIS OF BOOLEAN FUNCTIONS

2.1.1. The problem of analysis of Boolean functions (BF) is to decide whether a given function belongs to some standard class (linear, self-dual, threshold, etc., functions). The reason for the importance of this problem is that standard methods of network synthesis are available for the standard classes of BF, and these "special-purpose" methods are usually incomparably more efficient than universal methods. Thus, the synthesis stage of network design is frequently preceded by analysis of an appropriate system of BF.

Below we shall evolve decision procedures for some important classes of BF using their Walsh spectra and autocorrelation functions. (Tables of spectra and autocorrelation functions of classes of BF widely used in engineering practice are given in the Appendix.)

The analysis methods described in this section may be generalized to many-valued logical functions by using Chrestenson transforms and the appropriate autocorrelation functions.

All the BF considered in this section will be assumed to be completely specified; if $f(z) = f(z^{(0)}, \dots, z^{(m-1)})$, we shall denote its Walsh spectrum by $S(\omega)$ or $S(\omega^{(0)}, \dots, \omega^{(m-1)})$, where $\omega = \sum_{s=0}^{m-1} \omega^{(s)} 2^{m-1-s}$ ($\omega^{(s)} \in \{0, 1\}$), and its autocorrelation function $\sum_{z=0}^{2m-1} f(z) f(z \oplus \tau)$ (mod 2) by $B(\tau)$ or $B(\tau^{(0)}, \dots, \tau^{(m-1)})$, where $\tau = \sum_{s=0}^{m-1} \tau^{(s)} 2^{m-1-s} (\tau^{(s)} \in \{0, 1\})$.

* The *decision problem* of a class of functions is: to devise a procedure (or algorithm) whereby, given any function, one can determine ("recognize") whether it is or is not a member of the class (the rigorous definition is formulated in the framework of mathematical logic).

2.1.2. *Linear functions.* A BF $f(z^{(0)}, \ldots, z^{(m-1)})$ is said to be *linear* if and only if there exist numbers $c^{(s)} \in \{0, 1\}$ $(s = 0, 1, \ldots, m - 1)$ such that

$$f(z^{(0)}, \ldots, z^{(m-1)}) = \bigoplus_{s=0}^{m-1} c^{(s)} z^{(s)} \qquad (\text{mod } 2). \tag{2.1.1}$$

The class of linear functions is closed under superposition: the superposition of any two linear BF is again a linear BF. Any linear BF of m arguments can be realized by a network of at most $m - 1$ two-input mod 2 adders. Any k linear BF of m arguments may be realized by a network using a number of adders of the order of $mk/\log_2 m$ (see §3.2).

THEOREM 2.1.1. *Let* $f(z) = \bigoplus_{s=0}^{m-1} c^{(s)} z^{(s)}$ (mod 2). *Then*

$$S(\omega) = \begin{cases} \frac{1}{2} & \textit{if} \quad \omega = 0, \\ -\frac{1}{2} & \textit{if} \quad \omega = \overleftarrow{c}, \\ 0 & \textit{otherwise}, \end{cases} \tag{2.1.2}$$

where $\overleftarrow{c} = \sum_{s=0}^{m-1} c^{(s)} 2^s$, *and*

$$B(\tau) = 2^{m-2}(W_{\overleftarrow{c}}(\tau) + 1) \tag{2.1.3}$$

PROOF. By the definition of Walsh functions,

$$\bigoplus_{s=0}^{m-1} c^{(s)} z^{(s)} = \tfrac{1}{2}(1 - W_{\overleftarrow{c}}(z)) \qquad (\text{mod } 2).$$

Hence, using (1.2.3) and (1.2.6), (1.2.7), we obtain (2.1.2). Formula (2.1.3) is derived from (2.1.2) using Theorem 1.5.1. ▮

The converse is also valid, in the sense that if either of conditions (2.1.2) or (2.1.3) holds the function $f(z)$ may be expressed in the form (2.1.1).

Thus, in order to decide whether a function is linear, we need only calculate its spectrum; if the spectrum contains only one nonzero point $\omega = \overleftarrow{c}$ (apart from the point $\omega = 0$), the BF is linear and has the form $\bigoplus_{i=0}^{m-1} c^{(s)} z^{(s)}$ (mod 2).

EXAMPLE 2.1.1. Consider the BF defined in Table 2.1.1. Since

$$S(\omega) = \begin{cases} -\frac{1}{2} & \text{if} \quad \omega = 0, \\ -\frac{1}{2} & \text{if} \quad \omega = 7, \\ 0 & \text{otherwise}, \end{cases}$$

it follows that $f(z)$ is linear and $\overleftarrow{c} = 7$.

TABLE 2.1.1.

$z^{(0)}z^{(1)}z^{(2)}$	$f(z)$	ω, τ	$S(\omega)$	$B(\tau)$
0 0 0	0	0	0.5	4
0 0 1	1	1	0	0
0 1 0	1	2	0	0
0 1 1	0	3	0	4
1 0 0	1	4	0	0
1 0 1	0	5	0	4
1 1 0	0	6	0	4
1 1 1	1	7	−0.5	0

2.1.3. *Self-dual and anti-self-dual functions.* A BF is said to be *self-dual* (*anti-self-dual*) if and only if

$$f(z^{(0)}, \ldots, z^{(m-1)}) = \overline{f(\overline{z^{(0)}}, \ldots, \overline{z^{(m-1)}})} \tag{2.1.4}$$

$$\left(f(z^{(0)}, \ldots, z^{(m-1)}) = f(\overline{z^{(0)}}, \ldots, \overline{z^{(m-1)}})\right) \tag{2.1.5}$$

The bar over a function or an argument denotes inversion: $\bar{z} = 1 - z = 1 \oplus z$ (mod 2).

The class of self-dual BF is closed under superposition.

Since self-duality and anti-self-duality are invariant with respect to translation of arguments, these classes may be characterized in terms of autocorrelation functions.

THEOREM 2.1.2. *A BF $f(z) = f(z^{(0)}, \ldots, z^{(m-1)})$ is self-dual (anti-self-dual) if and only if*

$$B(2^m - 1) = \sum_{z=0}^{2^m - 1} f(z) - 2^{m-1} \tag{2.1.6}$$

$$\left(B(2^m - 1) = \sum_{z=0}^{2^m - 1} f(z)\right). \tag{2.1.7}$$

PROOF. To prove (2.1.6), one observes that if condition (2.1.4) holds then $f(z) = \overline{f(\bar{z})} = 1 - f(\bar{z})$ (where $\bar{z} = (\overline{z^{(0)}}, \ldots, \overline{z^{(m-1)}})$), and $f(z)\, f(\bar{z}) = 0$, $\overline{f(z)}\, \overline{f(\bar{z})} = 0$ for all $z \in \{0, 1, \ldots, 2^m - 1\}$. Then

$$\sum_{z=0}^{2^m - 1} f(z)\, f(\bar{z}) + \sum_{z=0}^{2^m - 1} \overline{f(z)}\, \overline{f(\bar{z})} = 2 \sum_{z=0}^{2^m - 1} f(z)\, f(\bar{z}) + 2^m - \sum_{z=0}^{2^m - 1} f(z) = 0,$$

and

$$B(2^m - 1) = \sum_{z=0}^{2^m - 1} f(z)\, f(\bar{z}) = \sum_{z=0}^{2^m - 1} f(z) - 2^{m-1}.$$

Conversely, by (2.1.6),

$$2B(2^m - 1) + 2^m - 2 \sum_{z=0}^{2^m-1} f(z) = \sum_{z=0}^{2^m-1} f(z)\, f(\bar{z}) + \sum_{z=0}^{2^m-1} \overline{f(z)}\, \overline{f(\bar{z})} = 0,$$

and $f(z)\,f(\bar{z}) = 0$, $\overline{f(z)}\,\overline{f(\bar{z})} = 0$. Hence $f(z) = \overline{f(\bar{z})}$. The proof of (2.1.7) is similar. ▮

We have thus reduced the verification of (anti-)self-duality to calculation of the autocorrelation function at the point $2^m - 1$.

For example, for the BF of Example 2.1.1 ($m = 3$) we have $B(7) = 0$, so that it is self-dual.

2.1.4. *Partially self-dual and partially anti-self-dual functions.*

A BF $f(z^{(0)}, \ldots, z^{(m-1)})$ is said to be *partially (anti-)self-dual* if and only if there exists $\alpha = \sum_{s=0}^{m-1} \alpha^{(s)} 2^{m-1-s}$ such that

$$f(z^{(0)}, \ldots, z^{(m-1)}) = \overline{f\left((z^{(0)})^{\alpha^{(0)}}, \ldots, (z^{(m-1)})^{\alpha^{(m-1)}}\right)} \tag{2.1.8}$$

$$\left(f(z^{(0)}, \ldots, z^{(m-1)}) = f((z^{(0)})^{\alpha^{(0)}}, \ldots, (z^{(m-1)})^{\alpha^{(m-1)}})\right), \tag{2.1.9}$$

where

$$(z^{(s)})^{\alpha^{(s)}} = \begin{cases} z^{(s)} & \text{if} \quad \alpha^{(s)} = 1, \\ \overline{z^{(s)}} & \text{if} \quad \alpha^{(s)} = 0. \end{cases}$$

A BF satisfying (2.1.8) ((2.1.9)) will also be called $\bar{\alpha}$-*(anti-)self-dual*, where $\bar{\alpha} = \sum_{s=0}^{m-1} \overline{\alpha^{(s)}} 2^{m-1-s}$ and $\overline{\alpha^{(s)}} = 1 - \alpha^{(s)}$. (According to this definition, a BF is (anti-) self-dual if it is $(2^m - 1)$-(anti-)self-dual.)

For the same BF $f(z)$, there may be several numbers $\alpha_1, \alpha_2, \ldots, \alpha_g$ such that $f(z)$ is α_s-self-dual ($s = 1, 2, \ldots, g$) and $\beta_1, \beta_2, \ldots, \beta_h$ such that $f(z)$ is β_s-anti-self-dual ($s = 1, 2, \ldots, h$).

Note that any BF is 0-anti-self-dual, and if $f(z)$ is both β_s-anti-self-dual and β_q-anti-self-dual then it is also $(\beta_s \oplus \beta_q)$-anti-self-dual. Indeed, in that case it follows from (2.1.9) that $f(z) = f(z \oplus \beta_s) = f(z \oplus \beta_q)$ and $f(z) = (z \oplus \beta_s \oplus \beta_q) \pmod 2$ for every $z \in \{0, 1, \ldots, 2^m - 1\}$.

We shall call β an *(anti-)self-duality point* for $f(z)$ if and only if $f(z)$ is β-(anti-)self-dual. The set of anti-self-duality points for any BF is a group, and the number of such points (the order of the group) is a power of 2. The group of anti-self-duality points will be called the *anti-self-duality group* or *inertia group*. If $f(z^{(0)}, \ldots, z^{(m-1)})$ is 2^s-anti-self-dual, then it is independent of $z^{(s)}$.

THEOREM 2.1.3. *A function* $f(z) = f(z^{(0)}, \ldots, z^{(m-1)})$ *has self-duality points* $\alpha_1, \ldots, \alpha_g$ *and anti-self-duality points* $\beta_1, \ldots, \beta_h$ *if and only if*

$$B(\alpha_s) = \sum_{z=0}^{2^m-1} f(z) - 2^{m-1} \qquad (s = 1, \ldots, g),$$

$$B(\beta_q) = \sum_{z=0}^{2^m-1} f(z) \qquad (q = 1, \ldots, h) \qquad (1 \leqq g + h \leqq 2^m). \tag{2.1.10}$$

Theorem 2.1.3 is a generalization of Theorem 2.1.2 (in the latter $\alpha = 2^m - 1$ or $\beta = 2^m - 1$), and the proof is analogous.

COROLLARY 2.1.2. *For any BF* $f(z)$, *the number of points at which the autocorrelation function takes the value* $\sum_{z=0}^{2^m-1} f(z)$ *is a power of* 2, *and the corresponding autocorrelation readings form a group.*

Corollary 2.1.2 generates a simple procedure for checking the validity of a calculation of the autocorrelation function of a BF.

Thus, if $\sum_z f(z) = 2^{m-1}$, the zeros of the autocorrelation function define the self-duality points. In any case, the maximum readings $\sum_{z=0}^{2^m-1} f(z)$ define the anti-self-duality points.

As an example of analysis for partial anti-self-duality, let us consider a BF constituting the characteristic function of a linear code.

A *linear* (m, k)*-code* V is defined by a generating matrix

$$\begin{pmatrix} V_1 \\ V_2 \\ \vdots \\ V_k \end{pmatrix}$$

where $V_s = (V_s^{(0)}, \ldots, V_s^{(m-1)})$ is a binary vector of length m. The code V is the set of all linear combinations $\bigoplus_{s=1}^{k} c_s V_s \pmod 2$, where $c_s \in \{0, 1\}$. The *characteristic function* $f_V(z)$ *of* V is defined by

$$f_V(z) = \begin{cases} 1 & \text{if } z \in V, \\ 0 & \text{if } z \notin V. \end{cases} \tag{2.1.11}$$

THEOREM 2.1.4. $f(z)$ *is the characteristic function of a linear* (m, k)*-code* V *if and only if*

$$S(\omega) = 2^{-m} \prod_{r=1}^{k} (W_{\overleftarrow{v}_r} + 1), \tag{2.1.12}$$

where

$$\overleftarrow{v}_r = \sum_{q=0}^{m-1} V_r^{(q)} 2^q,$$

or if and only if

$$B(\tau) = \begin{cases} 2^k & \text{if} \quad \tau \in V, \\ 0 & \text{if} \quad \tau \notin V. \end{cases} \tag{2.1.13}$$

PROOF. We prove (2.1.12) by induction on k. If $k = 1$, we have

$$f_V(z) = \begin{cases} 1 & \text{if} \quad z = 0, v_1, \\ 0 & \text{otherwise,} \end{cases}$$

and $S(\omega) = 2^{-m}(W_{\tilde{v}_1}(\omega) + 1)$.

Now let $f_{V_1,\ldots,V_{k-1}}(z)$ be the characteristic function of the code with generating matrix

$$\begin{pmatrix} V_1 \\ V_2 \\ \vdots \\ V_{k-1} \end{pmatrix},$$

$S_{V_1,\ldots,V_{k-1}}(\omega)$ its spectrum and f_{V_k} the characteristic function of the code whose generating matrix is V_k. Let

$$S_{V_1,\ldots,V_{k-1}}(\omega) = 2^{-m} \prod_{r=0}^{k-1} (W_{\tilde{v}_r}(\omega) + 1).$$

Since $f_V(z)$ is the characteristic function of the code with generating matrix

$$\begin{pmatrix} V_1 \\ V_2 \\ \vdots \\ V_k \end{pmatrix}$$

if and only if

$$f_V(z) = f_{V_1,\ldots,V_k}(z) = \sum_{\tau=0}^{2^m-1} f_{V_k}(\tau)\, f_{V_1,\ldots,V_{k-1}}(z \oplus \tau) \qquad (\text{mod } 2),$$

it follows by the convolution theorem for the original function (1.4.7) that (2.1.12) is true.

To prove (2.1.13), we note that a linear (m, k)-code is a group with respect to $\oplus$ (mod 2), and $f_V(z)$ is the characteristic function of an (m, k)-code V if and only if $\sum_{z=0}^{2^m-1} f_V(z) = 2^k$ and, if $f_V(z) = 1$, then

$$f_V(z) = \begin{cases} f_V(z \oplus \tau) & \text{if} \quad \tau \in V, \\ \overline{f_V(z \oplus \tau)} & \text{otherwise.} \end{cases} \qquad (\text{mod } 2)$$

Thus the vectors τ of the code are anti-self-duality points for $f_V(z)$.

2.1.5. *Quadratic forms.* A BF $f(z^{(0)}, \ldots, z^{(m-1)})$ is called a *quadratic form* if and only if there exists $c_{qs}, \alpha^{(q)} \in \{0, 1\}$, such that

$$f(z^{(0)}, \ldots, z^{(m-1)}) = \bigoplus_{\substack{q,s=0 \\ q<s}}^{m-1} c_{qs}(z^{(q)})^{\alpha(q)} (z^{(s)})^{\alpha(s)} \pmod 2. \tag{2.1.14}$$

The design of a network realizing a quadratic form (c_{qs}) requires at most $L_{\&} = \sum_{s=0}^{m-1} \sum_{q=0}^{s-1} c_{qs}$ two-input AND gates (logical multiplication) and $L_{\oplus} = L_{\&} - 1$ two-input mod 2 adders.

A quadratic form in $m = 2s$ variables is said to be *nonrepetitive* if each of its variables $z^{(q)}$ appears in (2.1.14) exactly once.

An example of a nonrepetitive quadratic form is

$$f_s(z) = \bigoplus_{q=0}^{s-1} z^{(q)} z^{(q+s)} \pmod 2 \qquad (m = 2s).$$

This function may also be expressed as

$$\bigoplus_{q=0}^{s-1} z^{(q)} z^{(q+s)} = \frac{1}{2}\left(1 - (-1)^{\bigoplus_{q=1}^{s-1} z^{(q)} z^{(q+s)}}\right) = \frac{1}{2}(1 - W_{a(z)}(b(z))), \tag{2.1.15}$$

where

$$a(z) = \sum_{q=0}^{s-1} z^{(q)} 2^q, \qquad b(z) = \sum_{q=0}^{s-1} z^{(q+s)} 2^{s-1-q}$$

(see §1.2).

Any nonrepetitive quadratic form in $m = 2s$ variables may be realized by a network of s two-input AND gates and $s - 1$ two-input mod 2 adders.

THEOREM 2.1.5. *A BF is a nonrepetitive quadratic form in $m = 2s$ variables if and only if its autocorrelation function $B^{(s)}(\tau)$ satisfies the condition*

$$B^{(s)}(\tau) = \begin{cases} 2^{2s-1} - 2^{s-1} & \text{for } \tau = 0, \\ 2^{2s-2} - 2^{s-1} & \text{for } \tau \neq 0. \end{cases} \tag{2.1.16}$$

We first prove a lemma:

LEMMA 2.1.1. *Let* $\varphi_1(z) = \bigoplus_{q=0}^{s-1} a_q z^{(q)}$, $\varphi_2(z) = \bigoplus_{q=0}^{s-1} b_q z^{(q)}$, $a \neq b$, $a, b \neq 0$. *Then for any* $\tau \in \{0, 1, \ldots, 2^s - 1\}$

$$B^{(\varphi_1,\varphi_2)}(\tau) = \sum_{z=0}^{2^s-1} \varphi_1(z)\varphi_2(z \oplus \tau) = B^{(\varphi_1,\bar{\varphi}_2)}(\tau) = 2^{s-2}. \tag{2.1.17}$$

The equality $B^{(\varphi_1,\varphi_2)}(\tau) = 2^{s-2}$ follows from Theorem 2.1.1 and formula (1.5.4). For the other equality, we have

$$B^{(\varphi_1,\bar{\varphi}_2)}(\tau) = \sum_{z=0}^{2^s-1} \varphi_1(z)(1 - \varphi_2(z \oplus \tau)) = \sum_{z=0}^{2^s-1} \varphi_1(z) - B^{(\varphi_1,\varphi_2)}(\tau)$$

$$= 2^{s-1} - 2^{s-2} = 2^{s-2}.$$

PROOF OF THEOREM 2.1.5. We first show that if $f_s(z) = \bigoplus_{q=0}^{s-1} z^{(q)}z^{(q+s)}$ is a nonrepetitive quadratic form, its autocorrelation function $B^{(s)}(\tau)$ satisfies (2.1.16). Consider the function

$$Q(z^{(0)}, \dots, z^{(s-1)}, \tau)$$

$$= \sum_{z^{(s)},\dots,z^{(2s-1)}\in\{0,1\}} \left(\bigoplus_{q=0}^{s-1} z^{(q)}z^{(q+s)} \right)\left(\bigoplus_{q=0}^{s-1} (z^{(q)} \oplus \tau^{(q)})(z^{(q+s)} \oplus \tau^{(q+s)}) \right).$$

Let $a(z) = \sum_{q=0}^{s-1} z^{(q)}2^{(q)} \neq 0$, $\tau = 0$. Then for any $z^{(0)}, \dots, z^{(s-1)}$

$$Q(z^{(0)}, \dots, z^{(s-1)}, \tau) = \sum_{z^{(s)},\dots,z^{(2s-1)}\in\{0,1\}} \left(\bigoplus_{q=0}^{s-1} z^{(q)}z^{(q+s)} \right)^2$$

$$= \sum_{z^{(s)},\dots,z^{(2s-1)}\in\{0,1\}} \bigoplus_{q=0}^{s-1} z^{(q)}z^{(q+s)} = 2^{s-1}.$$

Let $a(z) \neq 0$, $\tau \neq 0$. Then for any $z^{(0)}, \dots, z^{(s-1)}$, by Lemma 2.1.1,

$$Q(z^{(0)}, \dots, z^{(s-1)}, \tau) = 2^{s-2}.$$

Thus,

$$Q(z^{(0)}, \dots, z^{(s-1)}, \tau) = \begin{cases} 2^{s-1} & \text{if } a(z) \neq 0, \tau = 0, \\ 2^{s-2} & \text{if } a(z) \neq 0, \tau \neq 0, \\ 0 & \text{if } a(z) = 0. \end{cases} \tag{2.1.18}$$

Next, we have

$$B^{(s)}(\tau) = \sum_{z=0}^{2s-1} f_s(z)\, f_s(z \oplus \tau) = \sum_{z^{(0)},\dots,z^{(s-1)}\in\{0,1\}} Q(z^{(0)}, \dots, z^{(s-1)}, \tau). \tag{2.1.19}$$

Formula (2.1.16) now follows from (2.1.18), (2.1.19).

Thus formula (2.1.16) holds for $f_s(z)$. Now any form of $2s$ variables may be obtained from $f_s(z)$ by a suitable permutation and translation of the arguments. The translation leaves the autocorrelation function invariant. A permutation is equivalent to a linear transformation of the arguments (with matrix containing exactly one unit in each row and column). A linear transformation of the arguments in $f_s(z)$ induces the same linear transformation of the arguments in the autocorrelation function (see Theorem 1.5.4), but since $B^{(s)}(\tau) = \text{const}\ (\tau \neq 0)$ (see 2.1.19)) it follows that $B^{(s)}(\tau)$ is invariant under linear transformations. Thus any nonrepetitive quadratic form has the autocorrelation function (2.1.16).

Conversely, if the autocorrelation function satisfies (2.1.16), the corresponding BF will be a nonrepetitive quadratic form, since the autocorrelation determines the initial BF uniquely up to a translation of the argument, and the class of nonrepetitive quadratic forms is closed under translations. ∎

Summarizing Theorem 2.1.15, we see that in order to determine whether a given BF is a nonrepetitive quadratic form of $m = 2s$ variables we need only check that $B(\tau) = 2^{2s-2} - 2^{s-1}$ for $\tau \neq 0$ and $B(0) = 2^{2s-1} - 2^{s-1}$.

EXAMPLE 2.1.2. Table 2.1.2 defined the BF $f(z) = z^{(0)}z^{(3)} \oplus z^{(1)}z^{(2)}$; the autocorrelation is $B(\tau) = 2$ for $\tau \neq 0$.

TABLE 2.1.2.

z, τ	$f(z)$	$B(\tau)$
0	0	6
1	0	2
2	0	2
3	0	2
4	0	2
5	0	2
6	1	2
7	1	2
8	0	2
9	1	2
10	0	2
11	1	2
12	0	2
13	1	2
14	1	2
15	0	2

It is a useful observation that nonrepetitive quadratic forms are the BF with the "best" correlation characteristics, in the following sense. The quality of the correlation functions involved in transmission of synchronizing signals,

$$B_{\infty,2}(\tau) = \sum_{z=0}^{N-1} f(z) f(z - \tau) \quad \text{and} \quad B_{N,2}(\tau) = \sum_{z=0}^{N-1} f(z) f(z \ominus \tau) \qquad (\text{mod } N)$$

(where N is the number of points at which the function is specified), is defined for given N and signal power (number of units) $\sum_{z=0}^{N-1} f(z)$ by $\Delta_\infty = B_{\infty,2}(0) -$

$\max_{\tau \neq 0} B_{\infty,2}(\tau)$ or $\Delta_N = B_{N,2}(0) - \max_{\tau \neq 0} B_{N,2}(\tau)$, respectively. These quantities determine the error-correcting capability in transmission of synchronizing signals using $f(z)$ [98, 142]. Determination of the best synchronizing codes $f(z)$, maximizing Δ_∞ or Δ_N for given "length" N and "power" $\sum_{z=0}^{N-1} f(z)$, is a highly important and difficult problem. From this standpoint, the best quality is that of a BF with $B_{\infty,2}(\tau) = \text{const}\ (\tau \neq 0)$ or $B_{N,2}(\tau) = \text{const}\ (\tau \neq 0)$ (if such a BF exists). A function whose autocorrelation function is "most nearly" a constant for $\tau \neq 0$ is closest to a pseudo-random sequence of the given length.

With these remarks in mind, let us consider the autocorrelation functions $B^{(s)}(\tau)$ $(s = 1, 2, \ldots)$ of nonrepetitive quadratic forms in $m = 2s$ variables.

By Theorem 2.1.5, we have

$$B^{(s)}(\tau) = 2^{2s-2} - 2^{s-1} = \text{const}\ (\tau \neq 0),$$

and

$$\Delta_2^{(s)} = B^{(s)}(0) - \max_{\tau \neq 0} B^{(s)}(\tau) = 2^{2s-2}.$$

Now the quadratic form $f_s(z) = \bigoplus_{i=0}^{s-1} z^{(i)} z^{(i+s)}$ satisfies the condition: $f_s(z) = 0$ for $0 \leqq z < 2^s$ and, if a zero value of $f_s(z)$ corresponds to the absence of a signal at the transmitter output (passive zero codes [142]), we may assume that the transmitted code is of length $2^{2s} - 2^s$.

Thus, if we define an (N, b, Δ_2)-code to be a code of length N such that $B(0) = b$ and $\Delta_2 = B(0) - \max_{\tau \neq 0} B(\tau)$, then the nonrepetitive quadratic forms in $2s$ arguments constitute the class of optimal $(2^s(2^s - 1), 2^{s-1}(2^s - 1), 2^{2s-2})$-codes.

Some of these optimal codes are shown in Table 2.1.3. Note that codes defined by nonrepetitive quadratic forms have the same number of zeros and units.

TABLE 2.1.3.

s	Code length N	No. of units $B^{(s)}(0)$	$\Delta_2^{(s)}$
1	2	1	1
2	12	6	4
3	56	28	16
4	240	120	64
5	992	496	256

2.2. ANALYSIS AND SYNTHESIS OF THRESHOLD ELEMENT NETWORKS

2.2.1. In recent years, threshold elements have been used quite widely in digital data-processing systems. By a *threshold element* (TE) with *weights* $d_0, d_1, \ldots, d_{m-1}$

and *threshold* T we mean an element realizing the function

$$f(z^{(0)}, \ldots, z^{(m-1)}) = \operatorname{sign}\left(\sum_{s=0}^{m-1} d_s z^{(s)} - T \right) \tag{2.2.1}$$

where*

$$\operatorname{sign} z = \begin{cases} 0 & \text{if} \quad z \leqq 0, \\ 1 & \text{if} \quad z > 0. \end{cases} \tag{2.2.2}$$

(The numbers $d_0, d_1, \ldots, d_{m-1}$, T in (2.2.1) are assumed to be integers.)

The class $\Pi(m)$ of Boolean functions of m arguments for which there exist d_s and T satisfying (2.2.1) is called the class of single-threshold BF. It is quite difficult to determine whether a given function $f(z^{(0)}, \ldots, z^{(m-1)})$ belongs to class $\Pi(m)$; at present no analytical methods are available for this purpose. Tables have been constructed [29, 161] which enable one to determine whether a BF is in $\Pi(m)$ according to $m + 1$ certain characteristics of the function, and to solve the so-called *threshold synthesis problem*—to determine weights d_s $(s = 0, 1, \ldots, m - 1)$ and threshold T for a given single-threshold BF which minimize the quantity $T + \sum_{s=0}^{m-1} d_s$.

There are fairly sophisticated methods for synthesis of networks involving one TE. The situation is considerably inferior for TE networks realizing functions not in $\Pi(m)$; and in fact methods are completely lacking for synthesis of TE networks in which the number of TE inputs exceeds the number of arguments of the function realized by the entire network.

In view of this situation, we propose in this section a spectral characterization of $\Pi(m)$; we shall pay special attention to spectral methods for synthesis of TE networks realizing both completely and partially specified BF.

Thanks to the isomorphism between the arguments $z^{(s)}$ and the Rademacher functions $R_{s+1}(z)$ (see §1.2.2), we may treat the representation of a BF in $\Pi(m)$ in the form (2.2.1) as its expansion in an orthogonal series of Walsh functions. In this interpretation, we consider series containing $m + 1$ terms; in order to calculate the value of the BF, we need only the sign of the sum of the series, i.e., it suffices to consider the sign digit of the adder computing the sum. Thus the spectral analysis and synthesis methods developed in the preceding sections may be utilized for this narrow class $\Pi(m)$ of Boolean functions.

2.2.2. We now define the concept of k-monotonicity, which proves very useful for the identification of single-threshold functions and the synthesis of TE networks.

* The reader is warned that we are using two different "sign" functions (see the definition of sgn a in the footnote on p. 6).

Two BF f_1 and f_2 of m arguments are said to be *comparable* if

$$f_1(z^{(0)}, z^{(1)}, \ldots, z^{(m-1)}) \leqq f_2(z^{(0)}, z^{(1)}, \ldots, z^{(m-1)}) \tag{2.2.3}$$

or

$$f_1(z^{(0)}, z^{(1)}, \ldots, z^{(m-1)}) \geqq f_2(z^{(0)}, z^{(1)}, \ldots, z^{(m-1)}) \tag{2.2.4}$$

for all $(z^{(0)}, \ldots, z^{(m-1)})$.

Fix k $(1 \leqq k \leqq m)$ arguments of a function $f(z^{(0)}, \ldots, z^{(m-1)})$. The function f is said to be *k-comparable* if any two BF obtained by substituting the constants 0, 1 for the fixed k arguments are comparable and this property is independent of which k arguments are fixed and how. The function f is said to be *n-monotone* $(1 \leqq n \leqq m)$ if it is k-comparable for all $k = 1, 2, \ldots, n$; an m-monotone function will be called *completely monotone.*

THEOREM 2.2.1 [161]. *If $f(z^{(0)}, \ldots, z^{(m-1)})$ is $[m/2]$-monotone, it is completely monotone. (Recall that $[a]$ denotes the largest integer $\leqq a$.)*

The next theorem establishes the connection between monotonicity and the concept of a single-threshold function.

THEOREM 2.2.2 [161]. *If $f \in \Pi(m)$, then f is completely monotone; if $m \leqq 6$ and f is completely monotone, then $f \in \Pi(m)$.*

A completely monotone BF which is not single-threshold was constructed by Moore [161] for $m = 12$. To date there is no similar counterexample for $6 < m < 12$.

We now discuss the use of spectral methods for identification of 1,2-monotone functions, leading to corresponding necessary conditions for a BF to be single-threshold.

Denote

$$\Pi_f(q) = \sum_{\|z\| = q} f(z), \tag{2.2.5}$$

$$\Pi_s(q) = 2^m \sum_{\|\omega\| = q} |S(\omega)|, \tag{2.2.6}$$

$$\Pi_B(q) = \sum_{\|\tau\| = q} B(\tau) \qquad (q = 0, 1, \ldots, m), \tag{2.2.7}$$

where $\|z\|$, $\|\omega\|$, $\|\tau\|$ denote the number of units in the binary expansions of the numbers z, ω, τ; $S(\omega)$ is the spectrum and $B(\tau)$ the autocorrelation function of $f(x)$.

THEOREM 2.2.3. *A function $f(z^{(0)}, \ldots, z^{(m-1)})$ is 1-monotone if and only if*

$$\sum_{q=0}^{m-1} \Pi_f(q)(m - q) = \tfrac{1}{2}\Pi_B(1) \tag{2.2.8}$$

or

$$\Pi_S(1) + \Pi_B(1) = m\Pi_B(0). \tag{2.2.9}$$

Thus, either of conditions (2.2.8) or (2.2.9) may be used to check whether a given BF is 1-monotone.

EXAMPLE 2.2.1. Table 2.2.1 defines a BF $f(z^{(0)}, \ldots, z^{(3)})$ for which $\Pi_f(0) = \Pi_f(1) = 0$, $\Pi_f(2) = 3$, $\Pi_f(3) = 4$, $\Pi_f(4) = 1$, $\Pi_B(0) = 8$, $\Pi_B(1) = 20$, $\Pi_S(1) = 12$. Since both conditions (2.2.8) and (2.2.9) are satisfied, f is 1-monotone.

TABLE 2.2.1.

z, ω, τ	$f(z)$	$16S(\omega)$	$B(\tau)$	$\varphi(z)$	$16S(\omega)$	$B(\tau)$
0	0	8	8	0	5	5
1	0	−4	6	0	−3	2
2	0	−4	6	0	−3	2
3	0	0	4	0	1	2
4	0	−2	4	0	−3	2
5	0	−2	4	0	1	2
6	1	2	4	0	1	2
7	1	2	4	1	1	0
8	0	−2	4	0	−3	2
9	1	2	4	0	1	2
10	0	−2	4	0	1	2
11	1	2	4	1	1	0
12	1	0	2	0	1	2
13	1	0	2	1	1	0
14	1	0	2	1	1	0
15	1	0	2	1	−3	0

THEOREM 2.2.4. *A 1-monotone function $f(z^{(0)}, \ldots, z^{(m-1)})$ is 2-monotone if and only if*

$$2^m \sum_{p=1}^{m-1} (m-p)\,|S(2^{i_p})| + \Pi_B(2) = \binom{m}{2}\Pi_B(0), \tag{2.2.10}$$

where $i_1, i_2, \ldots, i_{m-1} \in \{0, 1, \ldots, m-1\}$ are defined by the condition

$$|S(2^{i_1})| \geqq |S(2^{i_2})| \geqq \cdots \geqq |S(2^{i_{m-1}})|. \tag{2.2.11}$$

We thus have a condition (2.2.10) providing a check on the 2-monotonicity of a BF.

EXAMPLE 2.2.2. Turning again to the function $f(z^{(0)}, \ldots, z^{(3)})$ of Table 2.2.1, we have: $2^4S(2^0) = -4$, $2^4S(2^1) = -4$, $2^4S(2^2) = -2$, $2^4S(2^3) = -2$; $\Pi_B(0) = 8$,

$\Pi_B(2) = 22$. Thus $i_1 = 0, i_2 = 1, i_3 = 2, i_4 = 3$ and

$$2^4 \sum_{p=1}^{3} (4-p)|S(2^{i_p})| + \Pi_B(2) = 12 + 8 + 2 + 22 = 44; \qquad \binom{4}{2}\Pi_B(0) = 48.$$

Thus condition (2.2.10) is not satisfied, and f is not 2-monotone; hence it is not a single-threshold function.

EXAMPLE 2.2.3. The function $\varphi(z^{(0)}, \ldots, z^{(3)})$ defined in Table 2.2.1 has $\Pi_S(1) = 12$, $\Pi_B(1) = 8$, $\Pi_B(0) = 5$; thus (2.2.9) holds and φ is 1-monotone. Furthermore, $2^4 S(2^0) = 2^4 S(2^1) = 2^4 S(2^2) = 2^4 S(2^3) = -3$; $\Pi_B(2) = 12$; $i_1 = 0, i_2 = 1, i_3 = 2, i_4 = 3$ and

$$2^4 \sum_{p=1}^{3} (4-p)|S(2^{i_p})| + \Pi_B(2) = 9 + 6 + 3 + 12 = 30, \qquad \binom{4}{2}\Pi_B(0) = 30.$$

Thus condition (2.2.10) is satisfied and so φ is 2-monotone. But in this case $[m/2] = 2$, and so, by Theorems 2.2.1 and 2.2.2, φ is a single-threshold function.

It is an extremely laborious matter to check k-monotonicity when $k \geqq 3$; in practice, therefore, one normally makes use of tables [29, 161], according to which a BF may be identified as being of class $\Pi(m)$ on the basis of the $m+1$ characteristics $S(0), S(2^0), \ldots, S(2^{m-1})$ (these characteristics are often called *Chow parameters* [23]). The tables also provide values of integers d_s and T that minimize $\sum_{s=0}^{m-1} d_s + T$ (see (2.2.1)). If $(S(0), a_0, a_1, \ldots, a_{m-1})$, where $a_i = S(2^i)\,(i = 0, \ldots, m-1)$, conform to these tables, so that the corresponding BF is single-threshold, we shall write $(S(0), a_0, a_1, \ldots, a_{m-1}) \in \Pi$.

Since the relative abundance* of single-threshold BF tends to zero quite rapidly as $m \to \infty$ [161], it is naturally important to consider the analysis and synthesis of networks containing one TE and additional equipment, and also networks of several TE.

2.2.3. Let us consider the class of linear-threshold networks—networks consisting of a block of mod 2 adders and one threshold element (see Figure 2.2.1).

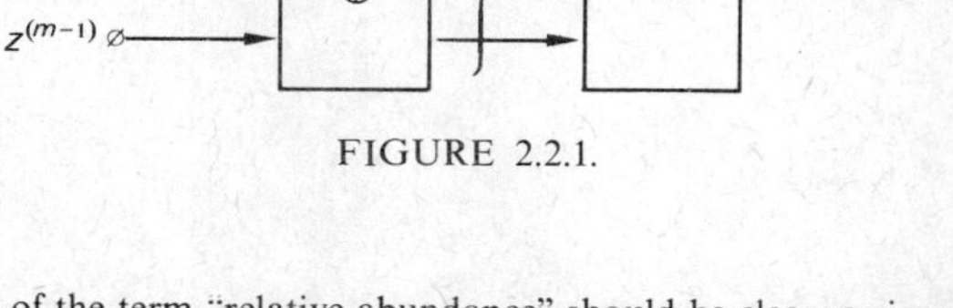

FIGURE 2.2.1.

* The intuitive meaning of the term "relative abundance" should be clear; a rigorous interpretation may be given only in the context of probability or measure theory.

THEOREM 2.2.5. *Any BF may be realized by a linear-threshold network.*

PROOF. The assertion follows from the fact that any BF may be represented by a finite series of Walsh functions and from the isomorphism between the multiplicative group of Walsh functions and the group of linear BF. ∎

The number n of inputs the TE may exceed the number m of arguments of the BF.

We now consider the important special case $n = m$.

THEOREM 2.2.6. *A BF may be realized by a linear-threshold network in which the number of TE inputs is equal to the number m of arguments of the BF if and only if there exist* $\omega_0, \omega_1, \ldots, \omega_{m-1}$ *such that* $(S(0), S(\omega_0), S(\omega_1), \ldots, S(\omega_{m-1})) \in \Pi$ *and the matrix* (ω) *whose rows are the binary expansions of the numbers* ω_i $(i = 0, 1, \ldots, m-1)$ *is nonsingular over* $GF(2)$: $|\omega|_2 \neq 0$.

Note that in this case the linear block "$\oplus$" in Figure 2.2.1 computes the product of $(z^{(0)}, \ldots, z^{(m-1)})$ by the matrix $(\overleftarrow{\omega})$ (where $\overleftarrow{\omega}_i = \sum_{s=0}^{m-1} \omega_i^{(s)} 2^s$ if $\omega_i = \sum_{s=0}^{m-1} \omega_i^{(s)} 2^{m-1-s}$). The proof of Theorem 2.2.6 is based on Theorem 1.4.5 for the case $p = 2$.

EXAMPLE 2.2.4. Consider the function $f(z^{(0)}, z^{(1)}, z^{(2)})$ defined by Table 2.2.2, and its spectrum $S(\omega)$. Setting $\omega_0 = 1, \omega_1 = 2, \omega_2 = 7$, we have $(S(0), S(1), S(2), S(7)) \in \Pi$ and

$$|\omega|_2 = \begin{vmatrix} 0 & 0 & 1 \\ 0 & 1 & 0 \\ 1 & 1 & 1 \end{vmatrix}_2 = 1.$$

Thus $f(z^{(0)}, z^{(1)}, z^{(2)})$ can be realized by a linear-threshold network with $n = m = 3$ (see Figure 2.2.2).

TABLE 2.2.2.

z, ω	$f(z)$	$8S(\omega)$
0	0	5
1	1	1
2	1	-1
3	1	-1
4	1	-1
5	0	-1
6	0	1
7	1	-3

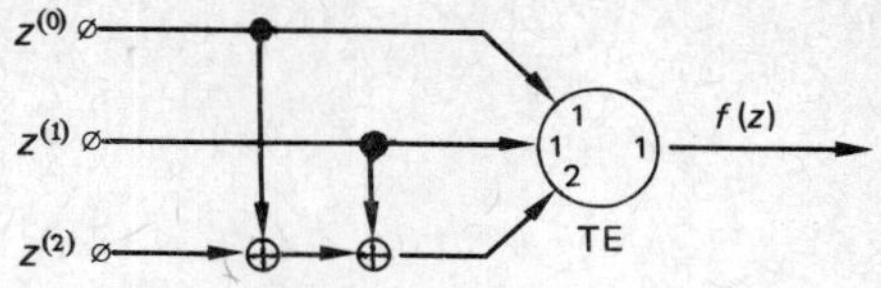

FIGURE 2.2.2.

Since not every BF is realizable by a linear-threshold network with $n \leqq m$, we are naturally interested in upper bounds for the number n of TE inputs for an arbitrary BF. The results presented below are relevant to both completely and incompletely specified BF.

Let $B^{(f_0,f_0)}$, $B^{(f_1,f_1)}$ denote the autocorrelation functions of the characteristic functions of occurrence of 0 and 1, respectively, in the original function f, and $B^{(f_0,f_1)}$ their cross-correlation function. By the *characteristic function of occurrence of* i ($i \in \{0, 1\}$) *in a BF* $f(z)$ we mean the BF $f_i(z)$ such that $f_i(z) = 1$ if and only if $f(z) = i$. It is clear that for incompletely specified functions $f \neq f_1$, $\bar{f} \neq f_0$ (where $\overline{f(z)}$ *is* undefined if $f(z)$ is undefined), and for any z we have $f_0(z) + f_1(z) \leqq 1$. If $f(z)$ is completely specified, however, then $f = f_1$ and $B^{(f_1,f_1)}(\tau) = B(\tau)$.

THEOREM 2.2.7. *Any Boolean function f defined at $N(f)$ points is realizable by a linear-threshold network with n TE inputs, where*

$$n \leqq N(f) - \max\left\{\tfrac{1}{2}\max_{\tau \neq 0}\left(B^{(f_0,f_0)}(\tau) + B^{(f_1,f_1)}(\tau)\right), \max_{\tau \neq 0} B^{(f_0,f_1)}(\tau)\right\}. \qquad (2.2.12)$$

For the proof, we need two lemmas, which are also of independent interest.

LEMMA 2.2.1. *Let f be a Boolean function, undefined at $d(f)$ points. Then there exists a completion $\tilde{f}$ of f such that*

$$\overline{L^W}(\tilde{f}) \geqq d(f), \qquad (2.2.13)$$

where $\overline{L^W}(\tilde{f})$ is the number of zero coefficients in the Walsh expansion of $\tilde{f}$.

The proof follows easily from the linear independence of the rows of the Hadamard matrix and from (1.2.14).

LEMMA 2.2.2. *Let f be a Boolean function, undefined at $d(f)$ points. Suppose that there are $d_1(f) + d_0(f)$ points $p_1, \ldots, p_{d_1(f)}, q_1, \ldots, q_{d_0(f)}$, at which f is defined, such that*

$$\begin{aligned} f(2p_j) &\neq f(2p_j + 1) \qquad (j = 1, \ldots, d_1(f)),\\ f(2q_s) &= f(2q_s + 1) \qquad (s = 1, \ldots, d_0(f)), \end{aligned} \qquad (2.2.14)$$

Then there exists a function $\varphi(z)$ such that, if $f(z^*)$ is defined,

$$\operatorname{sign}\varphi(z^*) = f(z^*), \tag{2.2.15}$$

$$\overline{L^W}(\varphi) \geqq d(f) + \max\{d_1(f), d_0(f)\}. \tag{2.2.16}$$

PROOF. We shall prove that there exists a function $\varphi = \varphi_1$ satisfying (2.2.15), such that

$$\overline{L^W}(\varphi_1) \geqq d(f) + d_1(f). \tag{2.2.17}$$

(The proof that there exists a function φ_0 satisfying (2.2.15) such that $\overline{L^W}(\varphi_0) \geqq d(f) + d_0(f)$ is similar.)

We note first that for any $\varphi(z)$

$$\overline{L^W}(\varphi(z)) = \overline{L^W}(\varphi(2z) + \varphi(2z+1)) + \overline{L^W}(\varphi(2z) - \varphi(2z+1)). \tag{2.2.18}$$

This follows from the factorization (1.2.16) of the Hadamard matrix or from Theorem 1.2.6.

We now see that for any $y_1, \ldots, y_{d_1(f)}$ there exists $\varphi_1(z)$ such that, if $f(z^*)$ is defined,

$$\begin{cases} \operatorname{sign}\varphi_1(z^*) = f(z^*), \\ \varphi_1(2p_j) + \varphi_1(2p_j+1) = y_j \qquad (j = 1, \ldots, d_1(f)), \end{cases} \tag{2.2.19}$$

and $\varphi_1(z^*) = f(z^*)$ if $z^* \neq 2p_j, 2p_j + 1$ $(j = 1, \ldots, d_1(f))$. In view of (2.2.19), we may assume that the function $\varphi_1(2z) + \varphi_1(2z+1)$ is undefined at the points p_j, $j = 1, \ldots, d_1(f)$, and hence, by Lemma 2.2.1,

$$\overline{L^W}(\varphi_1(2z) + \varphi_1(2z+1)) = d_1(f). \tag{2.2.20}$$

Now the functions $\varphi_1(2z) + \varphi_1(2z+1)$ and $\varphi_1(2z) - \varphi_1(2z+1)$ are also undefined at no less than $d(f)$ points where $f(2z)$ or $f(2z+1)$ is undefined. Hence, again using Lemma 2.2.1 and (2.2.18), we obtain (2.2.17). ▮

PROOF OF THEOREM 2.2.7. We first note that

$$n \leqq 2^m - \overline{L^W}(f). \tag{2.2.21}$$

This follows from the isomorphism of the group of linear BF and the group of Walsh functions.

In addition, for any BF $f(z)$, there always exist linear transformations σ_1 and σ_0 of z such that, if we define BF $f_{\sigma_1}(z)$ and $f_{\sigma_0}(z)$ by

$$f_{\sigma_1}(\sigma_1 \otimes z) = f_{\sigma_0}(\sigma_0 \otimes z) = f(z) \qquad (\text{mod } 2),$$

then the number of points z such that $f_{\sigma_1}(2z) \neq f_{\sigma_1}(2z+1)$ and $f_{\sigma_0}(2z) = f_{\sigma_0}(2z+1)$ is equal to $\max_{\tau \neq 0} B^{(f_0, f_1)}(\tau)$ and $\frac{1}{2}\max_{\tau \neq 0}\left(B^{(f_0, f_0)}(\tau) + B^{(f_1, f_1)}(\tau)\right)$, respectively.

(In view of Theorem 1.5.4, the matrices σ_1 and σ_0 may be found from the conditions

$$\sigma_1 \otimes \tau_1 = 1, \qquad \sigma_0 \otimes \tau_0 = 1 \qquad (\text{mod } 2),$$

where $\max_{\tau \neq 0} B^{(f_0,f_1)}(\tau) = B^{(f_0,f_1)}(\tau_1)$ and

$$\max_{\tau \neq 0} \left(B^{(f_0,f_0)}(\tau) + B^{(f_1,f_1)}(\tau)\right) = B^{(f_0,f_0)}(\tau_0) + B^{(f_1,f_1)}(\tau_0).)$$

Since $\overline{L^W}(f(\sigma \otimes z)) = \overline{L^W}(f(z))$ for any σ (see Theorem 1.4.5) and $N(f) = 2^m - d(f)$, formula (2.2.12) now follows from (2.2.21) and Lemma 2.2.2. ∎

COROLLARY 2.2.1. *For any completely specified Boolean function of m arguments,*

$$n \leqq 2^m - \max\left\{2^{m-1} - B(0) + \max_{\tau \neq 0} B(\tau), B(0) - \min_{\tau \neq 0} B(\tau)\right\}. \tag{2.2.22}$$

This follows from (2.1.2) with $N(f) = 2^m$, using the fact that

$$B^{(f_0,f_0)}(\tau) = 2^m - 2B(0) + B(\tau), \tag{2.2.23}$$

$$B^{(f_0,f_1)}(\tau) = B(0) - B(\tau). \tag{2.2.24}$$

EXAMPLE 2.2.5. Table 2.2.3 defines a completely specified BF $f(z^{(0)}, z^{(1)}, z^{(2)})$ and the corresponding $B(\tau)$. It is clear from the table and from (2.2.22) that

$$n \leqq 2^3 - \max\left\{4 - 5 + \max_{\tau \neq 0} B(\tau), 5 - \min_{\tau \neq 0} B(\tau)\right\} = 5.$$

TABLE 2.2.3.

z, τ	$f(z)$	$B(\tau)$
0	0	5
1	0	4
2	1	2
3	1	2
4	1	2
5	1	2
6	0	4
7	1	4

REMARK. Thanks to the isomorphism between the linear BF and the Walsh functions (Theorem 1.2.5), formulas (2.2.12) and (2.2.22) yield upper bounds for the number of nonzero coefficients in the Walsh expansions of the completions of incompletely specified BF.

To conclude this section, we point out that the spectral methods considered above may also be used in synthesis and analysis of networks of many-valued threshold elements, as described, e.g., in [7].

2.3. COMPLEXITY OF LOGICAL FUNCTIONS

2.3.1. In the following sections of this chapter we shall be studying some optimization problems for LF and spectral methods for their solution. Before solving an optimization problem, we must have criteria for the complexity of systems of LF. This section is therefore devoted to some basic results in regard to complexity of LF. Most of our discussion is devoted to binary logic.

The reader should note that although this section is essentially auxiliary in nature, it is necessary for a good understanding of the subsequent sections.

2.3.2. The *complexity* of a system of LF is defined as the sum of complexities of the individual functions. The latter concept will now be defined.

By the *complexity* $L(f)$ of a BF f we mean the minimum number of two- and one-input elements necessary for a network realizing f (i.e., $L(f)$ is the complexity of a minimal realization of f). Yablonskii [75] has conjectured that the labor involved in determining the exact value of $L(f)$ is roughly of the same order of magnitude as that required to construct a minimal network for f by the brute-force method. Since this conjecture seems highly reasonable, the theory of network complexity [104, 141, 143] deals mainly with asymptotic ($m \to \infty$) estimates for the complexity of functions in various classes. Shannon was the first to point out that although almost all functions admit only the most complex network realization, the functions actually used in practice are quite simple [141]. Since the exact complexity of a BF is generally impossible to determine, one introduces special functionals on the set of BF, known as *complexity criteria.* These criteria in effect provide an upper bound for the complexity, and the level curves of each criterion induce a classification of functions according to their complexity.

To be precise: a *complexity criterion* on the set of BF is a functional $\pi(f)$ with the following property. There exists a monotone real-valued function $\theta_\pi(t)$ such that, for any sequence of functions $f_m(z^{(0)}, \ldots, z^{(m-1)})$, $m = 1, 2, \ldots,$ for which $\lim_{m\to\infty} \pi(f_m)/m \log_2 m = \infty$, the following two conditions are satisfied:

a) Any function f of m arguments such that $\pi(f) \leqq \pi(f_m)$ has complexity

$$L(f) \lesssim Q_\pi(\pi(f_m)) \tag{2.3.1}$$

(we write $\alpha_m \lesssim \beta_m$ if $\overline{\lim}_{m\to\infty} \alpha_m/\beta_m \leqq 1$, and $\alpha_m \sim \beta_m$ if $\alpha_m \lesssim \beta_m$ and $\beta_m \lesssim \alpha_m$).

b) For almost all such functions,

$$L(f) \sim Q_\pi(\pi(f_m)), \tag{2.3.2}$$

i.e., the relative abundance of f's not satisfying this condition tends to zero as $m \to \infty$. (Henceforth we shall use this terminology freely: we shall say that a proposition A holds for *almost all BF of m arguments* if and only if the relative abundance of BF of m arguments for which A holds tends to 1 as $m \to \infty$.)

The function $\theta_\pi(t)$ of the definition enables us to estimate the actual complexity of a BF in terms of the criterion, and $\theta_\pi(\pi(f_m))$ will be called the *complexity of* f_m *relative to* π, or the π*-complexity* of f_m.

To facilitate the comparison of criteria, it is convenient to normalize them in such a way that $\theta_\pi(t)$ is the same for all criteria π: we set

$$\theta_\pi(t) = \frac{t}{\log_2 t} \tag{2.3.3}$$

since, under quite broad assumptions, the asymptotic estimates for complexity will have the form $\pi(f_m)/\log_2\pi(f_m)$ [143]. Note that the values of complexity criteria are independent of the choice of the basis system of elements participating in the design of the network.

An example of a very simple complexity criterion is 2^m, where m is the number of arguments on which the BF depends essentially. Denote this criterion by $\xi(f)$. The choice of this functional as a criterion is justified by the following proposition (first proved by Shannon and Lupanov [104, 141]). For almost all BF of m arguments, the complexity $L(m)$ of a minimal network realizing the BF satisfies the asymptotic estimate

$$L(m) \sim \frac{2^m}{m}. \tag{2.3.4}$$

We now present another important sequence of complexity criteria.

If $\sum_{i=0}^{k-1} t_i = 1$ $(0 \leqq t_i \leqq 1)$, we let $H(t_0, t_1, \ldots, t_{k-1})$ denote the *entropy function* $-\sum_{i=0}^{k-1} t_i \log_2 t_i$ (where $0 \log_2 0 = 0$ by convention). Then

$$0 \leqq H(t_0, t_1, \ldots, t_{k-1}) \leqq \log_2 k. \tag{2.3.5}$$

We can represent the function $f_m(z^{(0)}, \ldots, z^{(m-1)})$ by a vector of length 2^m: $(f_m(0, 0, \ldots, 0), f_m(0, 0, \ldots, 1), \ldots, f_m(1, 1, \ldots, 1))$. Divide this vector into 2^{m-r} non-overlapping "subvectors" of length 2^r. Set $p_i(f_m) = l_i(f_m)/2^{m-r}$, where $l_i(f_m)$ is the number of distinct subvectors of the i-th type in f_m. Now define a functional

$$\mu_r(f_m(z^{(0)}, \ldots, z^{(m-1)})) = \begin{cases} 2^{m-r} H(p_0, p_1, \ldots, p_{R-1}) & \text{if } r \leqq m, \\ 0 \quad \text{if } r > m, \end{cases} \tag{2.3.6}$$

where $R = 2^{2^r}$.

The values of the functionals μ_r $(r = 0, 1, \ldots)$ are obtained by substituting constants for $m - r$ arguments of the original function in all possible ways. Thus the values of μ_1 are defined by substituting constants in all possible ways for $m - 1$ arguments;

each such substitution determines a pair of readings of the original function whose arguments differ in only one position, and μ_1 is defined by the entropy function of the frequencies with which the various pairs appear.

It is evident that the values of the criteria μ_r depend on the order of the arguments. If the order of the arguments of f_m is T, let us denote $\mu_r(f_m)$ by $\mu_r^T(f_m)$, and

$$\hat{\mu}_r(f_m) = \min_{\{T\}} \mu_r^T(f_m). \tag{2.3.7}$$

THEOREM 2.3.1. [143]. *The functionals μ_r and $\hat{\mu}_r$ are complexity criteria. Moreover, for every r, $\hat{\mu}_r$ is stronger than $\hat{\mu}_{r-1}$, in the sense that $\hat{\mu}_r$ gives a sharper upper bound than $\hat{\mu}_{r-1}$.*

Theorem 2.3.1 shows that the "power" of the criterion $\hat{\mu}_r$ increases with increasing r, but on the other hand it becomes more difficult to calculate: to calculate $\hat{\mu}_r$ one needs $\binom{m}{m-r} \cdot 2^m$ operations (one has to examine the $\binom{m}{m-r} = \binom{m}{r}$ different arrangements of the $m - r$ arguments, and the complexity of each alternative is 2^m, in accordance with the number of readings of f_m).

EXAMPLE 2.3.1. Let $f_m(z^{(0)}, \ldots, z^{(m-1)})$ be a sequence of *functionally-separable* BF, i.e., functions of the form

$$f_m(z^{(0)}, \ldots, z^{(m-1)}) = F_m(z^{(0)}, \ldots, z^{(m-r-1)}, g_m(z^{(m-r)}, \ldots, z^{(m-1)})) \tag{2.3.8}$$

where F_m and g_m are BF.

Let us calculate $\mu_r(f_m)$. Substitution of constants for $z^{(0)}, \ldots, z^{(m-r-1)}$ yields a total of four functions: 1, 0, g_m, $\bar{g}_m$; thus the 2^r-component subvectors in the representation of f_m may have only four forms, whose frequencies we denote by p_0, p_1, p_2, p_3. Then, by (2.3.6),

$$\mu_r(f_m) = 2^{m-r} H(p_0, p_1, p_2, p_3) \leqq 2^{m-r+1},$$

whence, in view of (2.3.3),

$$L(f_m) \lesssim \frac{2^{m-r+1}}{m-r+1}, \tag{2.3.9}$$

and this bound is asymptotically the best possible [143].

The properties of the criteria $\hat{\mu}_r$ ($r = 0, 1, \ldots$) yield information on the structure of the most complex functions. In particular, the most complex BF have approximately the same number of zeros and ones, and substitution of constants for $m - r$ arbitrary arguments in the most complex BF of m arguments yields all BF of r arguments; moreover, the number of such BF of each type are approximately the same.

2.3.3. The next complexity criterion we consider is quite widely used in practice. We first recall a few simple definitions from Boolean algebra.

We call z^* a *constituent of unity* of a Boolean function $f(z)$ if $f(z^*) = 1$, and a *constituent of zero* if $f(z^*) = 0$. Constituents z_1^* and z_2^* are said to *match* if $z_1^* \oplus z_2^* = 2^i \pmod 2$.

A BF $\varphi(z) = \prod_{s=1}^{q} (z^{(i_s)})^{\alpha_{i_s}}$ ($q \leqq m$, $z^\alpha = z$ if $\alpha = 1$, $z^\alpha = \bar{z} = 1 - z$ if $\alpha = 0$) is called a *prime implicant* of a BF $f(z^{(0)}, \ldots, z^{(m-1)})$ if $\varphi(z) = 1$ always implies $f(z) = 1$; the integer q will be called the *length* of $\varphi(z)$. We shall also say that the prime implicant $\varphi(z)$ *covers* $\sum_z \varphi(z)$ constituents of unity of $f(z)$ (a constituent of unity z^* is *covered* by $\varphi(z)$ if $\varphi(z^*) = 1$).

A set of prime implicants $\varphi_1(z), \ldots, \varphi_l(z)$ of a BF $f(z)$ will be called a *minimal complete set* of implicants if any constituent of unity of $f(z)$ is covered by at least one implicant $\varphi_i(z)$ and the sum of lengths of $\varphi_1(z), \ldots, \varphi_l(z)$ is minimal. The disjunction (logical sum) of all prime implicants in a minimal complete set is the *minimal disjunctive form* of $f(z)$, which in turn determines an economical realization of $f(z)$ [60, 70].

The construction of minimal complete sets of implicants is one of the most important and difficult problems in the theory of network synthesis [60, 65, 70].

We now introduce another complexity criterion for $f(z) = f(z^{(0)}, \ldots, z^{(m-1)})$. Let $\tilde{\eta}(f)$ denote the number of unordered pairs of matching constituents of unity or zero in $f(z)$, i.e.,

$$\tilde{\eta}(f) = \sum_{\|\tau\|=1} \sum_{z=0}^{2^m-1} \left(f(z) f(z \oplus \tau) + \overline{f(z)}\, \overline{f(z \oplus \tau)} \right) \pmod 2, \quad (2.3.10)$$

where $\|\tau\|$ is the number of units in the binary expansion of τ. From now on we shall only compare the complexity of pairs of BF (f_1, f_2) such that

$$\sum_{z=0}^{2^m-1} f_1(z) = \sum_{z=0}^{2^m-1} f_2(z).$$

Therefore, if $\|f\| = \sum_{z=0}^{2m-1} f(z)$ is known, the number

$$\sum_{\|\tau\|=1} \sum_{z=0}^{2^m-1} \overline{f(z)}\, \overline{f(z \oplus \tau)}$$

of matching constituents of zero may easily be calculated in terms of the number

$$\sum_{\|\tau\|=1} \sum_{z=0}^{2^m-1} f(z) f(z \oplus \tau)$$

of matching constituents of unity. Indeed, we have

$$\sum_{\|\tau\|=1} \sum_{z=0}^{2^m-1} \overline{f(z)}\,\overline{f(z \oplus \tau)} = \sum_{\|\tau\|=1} \sum_{z=0}^{2^m-1} (1 - f(z))(1 - f(z \oplus \tau))$$

$$= \sum_{\|\tau\|=1} \sum_{z=0}^{2^m-1} (1 - f(z) - f(z \oplus \tau) + f(z)\, f(z \oplus \tau))$$

$$= 2^m - 2\|f\| + \sum_{\|\tau\|=1} \sum_{z=0}^{2^m-1} f(z)\, f(z \oplus \tau).$$

We may thus characterize the complexity of a BF by the quantity $\eta(f)$, defined as the number of pairs of matching constituents of unity:

$$\eta(f) = \sum_{\|\tau\|=1} \sum_{z=0}^{2^m-1} f(z)\, f(z \oplus \tau). \tag{2.3.11}$$

An experimental check of this criterion shows that in most cases it provides a quite satisfactory estimate of the complexity of network realizations of $f(z)$. In particular, it can be shown [83] that the number of nonoverlapping pairs of matching constituents of unity in $f(z)$ (for which $\|f\| = 2^{m-1}$) is equal to $\frac{1}{4}(2^{m+1}\eta(f)/m)^{1/2}$ for almost all $f(z)$ that depend on m arguments. Consequently, at least $\frac{1}{2}(2^{m+1}\eta(f)/m)^{1/2}$ constituents of unity are covered by $\frac{1}{4}(2^{m+1}\eta(f)/m)^{1/2}$ prime implicants of length $m-1$.

REMARK. It would be more appropriate to call the criterion η (and its generalizations to many-valued logic; see § 2.3.4) not the complexity but the *simplicity* of the BF, since for almost all BF a decrease in η implies an increase in the complexity of the minimal realization. We shall nevertheless continue to use the term "complexity," since the only important factor for our purposes is the monotonicity of the complexity of the minimal realization of f as a function of $\eta(f)$.

2.3.4. The complexity criteria $\xi(f)$, $\hat{\mu}_r(f)$ $(r = 0, 1, \ldots)$ and $\eta(f)$ considered above are readily generalized to many-valued logical functions. We shall do this only for $\eta(f)$.

If $f(z) = f(z^{(0)}, \ldots, z^{(m-1)})$ is a p-valued logical function, we define its complexity $\eta(f)$ to be the number of sets $\{z_1, \ldots, z_p\}$ such that

$$\begin{cases} f(z_1) = f(z_2) = \cdots = f(z_p), \\ z_2 \ominus z_1 = z_3 \ominus z_2 = \cdots = z_p - z_{p-1} = p^s \quad (\text{mod } p). \end{cases} \tag{2.3.12}$$

i.e., the p-ary expansions of the numbers $z_i \ominus z_{i+1}$ (mod p) $(i = 1, 2, \ldots, p-1)$ have the form $(\underbrace{0, \ldots, 0}_{s}, 1, 0, \ldots, 0)$.

A final remark. The specific complexity criteria ξ, $\hat{\mu}_r$ $(r = 0, 1, \ldots)$ and η were selected in view of the fact that, on the one hand, they are used quite frequently, and on the other they provide a good illustration of the effective use of spectral methods in optimization problems.

2.4. LINEARIZATION OF SYSTEMS OF BOOLEAN FUNCTIONS

2.4.1. The remainder of this chapter discusses spectral methods for solving certain optimization problems. We shall use the languages of spectral and correlation characteristics. In contrast to the methods of classical algebra of logic, the solutions will be analytical, in the sense that calculation of the final result does not require an exhaustive search of all possible alternatives (in other words, the number of elementary operations will not be an exponential function but only a linear function of the number of points at which the system of functions is defined).

The topic studied in this section and the next is linearization of LF: the problem of representing a given system of LF as the superposition of a system of linear LF and a residual nonlinear part of minimal complexity. Any such representation determines the design of a network realizing the original system as a series connection of two blocks: linear and nonlinear. For a system of p-valued logical functions, the linear block consists of mod p adders and networks implementing multiplication by a constant. Estimates for the complexity of the linear block will be given in §§ 3.2, 3.4: we shall show that the complexity of the linear block increases asymptotically no faster than $m^2/\log_p m$ as $m \to \infty$, where m is the number of arguments, whereas the complexity of the nonlinear block is almost always an exponentially increasing function of m (see (2.3.4)). For this reason, the complexity of the linear block may be ignored in linearization problems.

Throughout this section and in § 2.5 we shall assume that all systems of LF are completely specified.

This section is devoted to the case of greatest practical importance—Boolean functions; we shall employ the complexity criteria ξ, $\hat{\mu}$, η introduced in §2.3. Recall that the complexity of a system of functions is the sum of complexities of the BF occurring therein.

A rigorous formulation of the linearization problem is as follows. Consider a system of BF $f(z) = \{f^{(i)}(z^{(0)}, \ldots, z^{(m-1)})\}$ $(i = 0, 1, \ldots, k-1)$. Let Ξ denote the class of all nonsingular (mod 2) $m \times m$ matrices and α any one of our complexity criteria. Problem: Given α, devise an algorithm which, for any $f(z)$, computes a matrix $\sigma \in \Xi$ which minimizes $\alpha(f_\sigma)$, where f_σ is defined by

$$f_\sigma(\sigma \otimes z) = f(z) \qquad (\text{mod } 2). \tag{2.4.1}$$

We shall refer to this as the linearization problem relative to α.

2.4.2. We first consider the linearization problem for the most easily calculated criterion, ξ. Recall that $\xi(\varphi) = 2^n$, where φ is a BF depending essentially on exactly

n of its arguments (φ is said to *depend inessentially on* $z^{(i)}$ if

$$\varphi(z^{(0)}, \ldots, z^{(i-1)}, 0, z^{(i+1)}, \ldots, z^{(m-1)}) = \varphi(z^{(0)}, \ldots, z^{(i-1)}, 1, z^{(i+1)}, \ldots, z^{(m-1)})$$

for all $z^{(0)}, \ldots, z^{(i-1)}, z^{(i+1)}, \ldots, z^{(m-1)}$).

Let $f(z) = \{f^{(i)}(z^{(0)}, \ldots, z^{(m-1)})\}$ $(i = 0, 1, \ldots, k-1)$ be a system of BF depending on all their arguments. Set

$$B(\tau) = \sum_{i=0}^{k-1} B_i(\tau) = \sum_{i=0}^{k-1} \sum_{z=0}^{2^m-1} f^{(i)}(z)\, f^{(i)}(z \oplus \tau) \qquad (\text{mod } 2).$$

In order to determine a matrix σ_ξ minimizing $\xi(f_\sigma)$ (see (2.4.1)) we shall use the *total autocorrelation functions* $B(\tau)$ thus defined.

Note that for any τ we have $B(\tau) \leqq B(0)$. Let $G_I(f)$ denote the set of all τ such that $B(\tau) = B(0) = \sum_{i=0}^{k-1} \sum_{z=0}^{2^m-1} f^{(i)}(z)$. It is clear that $G_I(f)$ is a group (with respect to $\oplus$ (mod 2)) which, in accordance with §2.1.4, we call the *inertia group* of the system f.

Let $\tau_0, \tau_1, \ldots, \tau_{a(f)-1}$ be an arbitrary *basis* of $G_I(f)$ (i.e., a maximal set of elements of $G_I(f)$ linearly independent modulo 2).

THEOREM 2.4.1. *Let T be an arbitrary nonsingular matrix ($T \in \Xi$) whose columns are some basis of the inertia group $G_I(f)$ of the system $\{f^{(i)}(z^{(0)}, \ldots, z^{(m-1)})\}$ $(i = 0, \ldots, k-1)$ of BF which depend essentially on all their arguments. Then*

$$\min_{\sigma \in \Xi} \xi(f_\sigma) = \xi(f_{\sigma_\xi}) = k2^{m-a(f)} \qquad (2.4.2)$$

and

$$\sigma_\xi \otimes T = E_m = \begin{pmatrix} 1 & & & \\ & 1 & & 0 \\ 0 & & \ddots & \\ & & & 1 \end{pmatrix} \quad (\text{mod } 2) \qquad (2.4.3)$$

(*i.e.*, $\sigma_\xi = T^{-1}$ (mod 2)).

PROOF. The proof falls into two parts.

1. We first prove that $\xi(f_{T^{-1}}) = k \cdot 2^{m-a(f)}$. Let $B_\sigma(\tau)$ be the total autocorrelation function of the system f_σ (see (2.4.1)). Then, by Theorem 1.5.4, if τ_s is a basis element of $G_I(f)$,

$$B_{T^{-1}}(T^{-1} \otimes \tau_s) = B(\tau_s) = \sum_{i=0}^{k-1} \sum_{z=0}^{2^m-1} f^{(i)}(z) \qquad (s = 0, 1, \ldots, a(f)-1) \qquad (\text{mod } 2).$$

But by the definition of T the binary expansion of $T^{-1} \otimes \tau_s$ (mod 2) contains only one unit; hence, it follows that $f^{(i)}(z)$ $(i = 0, 1, \ldots, k-1)$ is independent of $a(f)$

arguments, i.e.,

$$\xi(f_{T^{-1}}^{(i)}) = 2^{m-a(f)}, \qquad \xi(f_{T^{-1}}) = \sum_{i=0}^{k-1} \xi(f_{T^{-1}}^{(i)}) = k2^{m-a(f)}.$$

2. We now prove that there is no $\sigma \in \Xi$ such that

$$\xi(f_\sigma) < k2^{m-a(f)}.$$

Indeed, were σ a matrix satisfying this condition, there would exist $\beta_0, \beta_1, \ldots, \beta_{a(f)-1+\Delta}$ ($\Delta \geq 1$) such that $B_\sigma(\beta_s) = \sum_{i=0}^{k-1} \sum_{z=0}^{2m-1} f^{(i)}(z)$ and the binary expansion of β_s contains only one unit ($s = 0, 1, \ldots, a(f) - 1 + \Delta$). Next, since $\sigma^{-1} \otimes \beta_s \neq \sigma^{-1} \otimes \beta_q$ (mod 2) ($s \neq q$) and

$$B(\sigma^{-1} \otimes \beta_s) = B_\sigma(\beta_s) = \sum_{i=0}^{k-1} \sum_{z=0}^{2m-1} f^{(i)}(z) \pmod 2 \quad (s = 0, 1, \ldots, a(f) - 1 + \Delta),$$

it follows that the system $\{\sigma^{-1} \otimes \beta_0, \ldots, \sigma^{-1} \otimes \beta_{a(f)-1+\Delta}\}$ is linearly dependent (since $\sigma^{-1} \otimes \beta_s \in G_I(f)$ and the basis of $G_I(f)$ contains only $a(f)$ elements). But since $\sigma^{-1} \in \Xi$ this contradicts the assumption that $\{\beta_0, \beta_1, \ldots, \beta_{a(f)-1+\Delta}\}$ is linearly independent. ▮

Theorem 2.4.1 provides a procedure for determining a ξ-optimal linear transformation σ_ξ of the arguments of a system of BF and also an estimate for the ξ-complexity of the nonlinear part corresponding to σ_ξ (see (2.4.2)). To settle these questions, we need only construct an arbitrary basis of the inertia group $G_I(f)$ of the system.

EXAMPLE 2.4.1. Table 2.4.1 defines a system $f^{(0)}, f^{(1)}$ ($m = 4$, $k = 2$) and also its total autocorrelation function $B(\tau)$. It follows from the table that

$$G_I(f) = \{(0, 0, 0, 0), (0, 1, 0, 1), (1, 0, 1, 0), (1, 1, 1, 1)\}.$$

As a basis we take $\tau_0 = (0, 1, 0, 1)$, $\tau_1 = (1, 0, 1, 0)$ ($a(f) = 2$). Now set

$$T = \begin{pmatrix} 1 & 0 & 0 & 1 \\ 0 & 1 & 1 & 0 \\ 0 & 0 & 0 & 1 \\ 0 & 0 & 1 & 0 \end{pmatrix}.$$

Then

$$\sigma_\xi = T^{-1} = \begin{pmatrix} 1 & 0 & 1 & 0 \\ 0 & 1 & 0 & 1 \\ 0 & 0 & 0 & 1 \\ 0 & 0 & 1 & 0 \end{pmatrix}.$$

The functions $f_{T^{-1}}^{(0)}$ and $f_{T^{-1}}^{(1)}$ are shown in Table 2.4.1. It is evident that $f_{T^{-1}}^{(0)}$ and $f_{T^{-1}}^{(1)}$ do not depend essentially on $z^{(2)}$, $z^{(3)}$. Consequently, by (2.4.2),

$$\xi(f_{\sigma_\xi}) = \xi(f_{T^{-1}}^{(0)}) + \xi(f_{T^{-1}}^{(1)}) = 4 + 4 = 8.$$

TABLE 2.4.1.

z, τ	$z^{(0)}$	$z^{(1)}$	$z^{(2)}$	$z^{(3)}$	$f^{(0)}$	$f^{(1)}$	B	$f_{T^{-1}}^{(0)}$	$f_{T^{-1}}^{(1)}$
0	0	0	0	0	0	0	16	0	0
1	0	0	0	1	1	0	8	0	0
2	0	0	1	0	1	0	8	0	0
3	0	0	1	1	1	1	8	0	0
4	0	1	0	0	1	0	8	1	0
5	0	1	0	1	0	0	16	1	0
6	0	1	1	0	1	1	8	1	0
7	0	1	1	1	1	0	8	1	0
8	1	0	0	0	1	0	8	1	0
9	1	0	0	1	1	1	8	1	0
10	1	0	1	0	0	0	16	1	0
11	1	0	1	1	1	0	8	1	0
12	1	1	0	0	1	1	8	1	1
13	1	1	0	1	1	0	8	1	1
14	1	1	1	0	1	0	8	1	1
15	1	1	1	1	0	0	16	1	1

2.4.3. We now consider linearization with respect to the criteria $\hat{\mu}$ (see §2.3). Recall that for a BF $f(z^{(0)}, \ldots, z^{(m-1)})$

$$\mu_r(f) = \begin{cases} 2^{m-r}H(p_0, \ldots, p_{R-1}) & \text{if } r \leqq m, \\ 0 & \text{if } r > m, \end{cases} \tag{2.4.4}$$

where $R = 2^{2^r}$, $H(p_0, \ldots, p_{R-1}) = -\sum_{i=0}^{R-1} p_i \log_2 p_i$ and $p_i = l_i 2^{-m+r}$ $(0 \log_2 0 = 0)$, with l_i the number of subvectors of length 2^r of the i-th type in the vector $f = (f(0), \ldots, f(2^m - 1))$. The criterion $\hat{\mu}_r(f)$ is then defined as the minimum of $\mu_r(f)$ over all possible orders of the binary arguments $z^{(i)}$ of f.

Since $\hat{\mu}_0$ depends only on the number of unit values of each function, which remains invariant under any nonsingular linear transformation, if follows via (2.4.1) that for any σ

$$\hat{\mu}_0(f_\sigma) = \hat{\mu}_0(f),$$

and so the linearization problem relative to $\hat{\mu}_0$ is trivial.

We therefore proceed to the linearization problem for $\hat{\mu}_1$ the most simply evaluated criterion for which the problem is not trivial. Denote the corresponding optimal linear transformation by σ_{μ_1}:

$$\hat{\mu}_1(f_{\sigma_{\mu_1}}) = \min_{\sigma\in\Xi} \hat{\mu}_1(f_\sigma).$$

Let $f^{(i)}(z^{(0)}, \ldots, z^{(m-1)})$ $(i = 0, 1, \ldots, k-1)$ be a given system and $S_i(\tau)$, $B_i(\tau)$ the Walsh spectrum and autocorrelation function of $f^{(i)}(z^{(0)}, \ldots, z^{(m-1)})$; we set $S(\omega) = \sum_{i=0}^{k-1} S_i(\omega)$, $B(\tau) = \sum_{i=0}^{k-1} B_i(\tau)$.

THEOREM 2.4.2. *Let*

$$\max_{\omega\neq 0} |S(\omega)| = |S(\omega')|, \qquad \min_{\tau\neq 0} B(\tau) = B(\tau'), \tag{2.4.5}$$

and

$$\left.\begin{aligned} (00\ldots01) \otimes \tilde{\sigma} &= \overleftarrow{\omega'} \\ \tilde{\sigma} \otimes \tau' &= \begin{pmatrix} 0 \\ 0 \\ \vdots \\ 0 \\ 1 \end{pmatrix} \end{aligned}\right\} \pmod 2, \tag{2.4.6}$$

where $\overleftarrow{\alpha} = (\alpha_{m-1}, \ldots, \alpha_0)$ *if* $\alpha = (\alpha_0, \ldots, \alpha_{m-1})$.

Then $\sigma_{\mu_1} = \tilde{\sigma}$.

PROOF. We first observe that by the definitions of $\hat{\mu}_1$, $S(\omega)$ and $B(\tau)$,

$$\hat{\mu}_1(f) = \sum_{i=0}^{k-1} \hat{\mu}_1(f^{(i)}) = 2^{m-1}H(p_0, p_1, p_2, p_3), \tag{2.4.7}$$

where

$$\begin{aligned} p_0 &= 1 - 2^{-m+1}B(0) + 2^{-m}B(1), \\ p_1 &= 2^{-m}(B(0) - B(1) - 2^m S(2^{m-1})), \\ p_2 &= 2^{-m}(B(0) - B(1) + 2^m S(2^{m-1})), \\ p_3 &= 2^{-m}B(1) \end{aligned} \tag{2.4.8}$$

In view of (2.4.7), (2.4.8), we may express $\hat{\mu}_1(f)$ as

$$\hat{\mu}_1(f) = F(B(1), S(2^{m-1}), B(0), m)$$

and by Theorems 1.4.5, 1.5.4,

$$\hat{\mu}_1(f_\sigma) = \sum_{i=0}^{k-1} \hat{\mu}_1(f_\sigma^{(i)}) = F(B(\sigma \otimes 1, S(\overleftarrow{2^{m-1}} \otimes \sigma^{-1}), B(0), m).$$

The functional F is minimized by the matrix σ defined in (2.4.6). The value $\hat{\mu}_1(f_{\sigma_{\mu_1}})$ is determined by substituting $B(\tau')$ for $B(1)$ and $S(\omega')$ for $S(2^{m-1})$ in (2.4.7), (2.4.8), where τ' and ω' are defined in (2.4.5). ∎

EXAMPLE 2.4.2. Table 2.4.2 defines a BF $f(z^{(0)}, \ldots, z^{(3)})$ and the corresponding $S(\omega)$, $B(\tau)$; we have $\mu_1(f) = \hat{\mu}_1(f) = 8H(3/8, 2/8, 3/8) \simeq 12.4$.

It follows from the table that $\omega' = 7$, $\tau' = 14$; then, by (2.4.6),

$$\sigma_{\mu_1} = \begin{pmatrix} 1 & 1 & 0 & 0 \\ 0 & 1 & 1 & 0 \\ 0 & 0 & 0 & 1 \\ 1 & 1 & 1 & 0 \end{pmatrix}.$$

The function $f_{\sigma_{\mu_1}}(z)$, where $f_{\sigma_{\mu_1}}(\sigma_{\mu_1} \otimes z) = f(z) \pmod 2$, is also given in Table 2.4.2; we have $\mu_1(f_{\sigma_{\mu_1}}) = \hat{\mu}_1(f_{\sigma_{\mu_1}}) = 8H(7/8, 1/8) \simeq 4.4$. Thus linearization brings about a decrease in μ_1 by a factor of 3.3, and in $\hat{\mu}_1$ by a factor of 2.8.

TABLE 2.4.2.

z, ω, τ	$f(z)$	$16S(\omega)$	$B(\tau)$	$f_{\sigma_{\mu_1}}(z)$
0	1	8	8	1
1	1	2	6	0
2	0	−2	2	1
3	0	0	2	0
4	1	2	2	1
5	0	0	2	0
6	1	0	6	1
7	1	6	6	0
8	0	0	2	1
9	0	2	2	0
10	0	−2	6	1
11	1	0	6	0
12	1	2	6	0
13	1	0	6	1
14	0	0	0	1
15	0	−2	2	0

2.4.4. We now proceed to linearization relative to η (see §2.3.3). Recall that the criterion $\eta(f)$ for a BF $f^{(i)}(z^{(0)}, \ldots, z^{(m-1)})$ is defined by

$$\eta(f^{(i)}) = \sum_{\|\tau\|=1} \sum_{z=0}^{2^m-1} f^{(i)}(z)\, f^{(i)}(z \oplus \tau) \qquad \pmod 2, \tag{2.4.9}$$

where $\|\tau\|$ is the number of units in the binary expansion of τ.

Given a system of BF $f(z) = \{f^{(0)}, \ldots, f^{(k-1)}(z)\}$, we wish to find a linear transformation σ_η such that

$$\max_{\sigma \in \Xi} \eta(f_\sigma) = \eta(f_{\sigma_\eta}), \tag{2.4.10}$$

where f_σ is defined as before by

$$f_\sigma(\sigma \otimes z) = f(z) \qquad (\text{mod } 2),$$

and Ξ is the class of all nonsingular matrices ($\sigma \in \Xi$ if and only if $|\sigma|_2 = 1$).

Let $T = (\tau_{qs})$, where $\tau_{qs} \in \{0, 1\}$ $(q, s = 0, 1, \ldots, m-1)$; let $B_i(\tau)$ be the autocorrelation function of $f^{(i)}(z)$ and $B(\tau) = \sum_{i=0}^{k-1} B_i(\tau)$. Set

$$B(T) = \sum_{s=0}^{m-1} B\left(\sum_{q=0}^{m-1} \tau_{q,s} 2^{m-1-q}\right). \tag{2.4.11}$$

THEOREM 2.4.3. *Let*

$$\max_{T \in \Xi} B(T) = B(T_\eta). \tag{2.4.12}$$

Then

$$\sigma_\eta \otimes T_\eta = E_m = \begin{pmatrix} 1 & & & \\ & 1 & 0 & \\ 0 & & \ddots & \\ & & & 1 \end{pmatrix} \qquad (\text{mod } 2). \tag{2.4.13}$$

PROOF. For any function $f_\sigma^{(i)}(z)$ $(\sigma \in \Xi)$, where $f_\sigma^{(i)}(\sigma \otimes z) = f^{(i)}(z)$, let $B_{\sigma,i}(\tau)$ denote the autocorrelation function and $B_\sigma(\tau) = \sum_{i=0}^{k-1} B_{\sigma,i}(\tau)$. Then, by Theorem 1.5.4,

$$B_\sigma(\sigma \otimes \tau) = B(\tau). \tag{2.4.14}$$

In addition, by the definition of η (see (2.4.9)), we see that for any f_σ

$$\eta(f_\sigma) = B_\sigma(E_m). \tag{2.4.15}$$

In view of (2.4.10)–(2.4.13), formulas (2.4.14) and (2.4.15) imply that

$$\begin{aligned} \max_{\sigma \in \Xi} \eta(f_\sigma) &= \max_{\sigma \in \Xi} B_\sigma(E_m) = \max_{\sigma \in \Xi} B_\sigma(\sigma \otimes \sigma^{-1}) = \max_{\sigma \in \Xi} B(\sigma^{-1}) = \max_{T \in \Xi} B(T) \\ &= B(T_\eta) = B_{\sigma_\eta}(\sigma_\eta \otimes T_\eta) = B_{\sigma_\eta}(E_m) = \eta(f_{\sigma_\eta}) \qquad (\text{mod } 2). \quad \blacksquare \end{aligned}$$

Theorem 2.4.3 generates a simple procedure for determining the η-optimal linear transformation σ_η. One first sets up the autocorrelation function $B(\tau) = \sum_{i=0}^{k-1} B_i(\tau)$ and then finds m linearly independent readings over $GF(2)$ such that the sum of $B(\tau)$

over them is maximal. These m readings may be found recursively as follows. Supposing that we already have s readings $(1 \leqq s \leqq m)$ $\tau_0, \tau_1, \ldots, \tau_{s-1}$, we find τ_s from the condition

$$B(\tau_s) = \max_{\tau \notin Q_s} B(\tau), \tag{2.4.16}$$

where Q_s is the set of all linear combinations (mod 2) of the vectors $\tau_0, \tau_1, \ldots, \tau_{s-1}$ and the vector $(0, 0, \ldots, 0)$. It can be shown [83] that the vectors $\tau_0, \tau_1, \ldots, \tau_{m-1}$ thus determined are the columns of the matrix T_η defined by (2.4.12). The η-optimal matrix σ_η is now determined from (2.4.13).

EXAMPLE 2.4.3. In Table 2.4.3 we have a system $f^{(0)}(z)$, $f^{(1)}(z)$ $(m = 4)$ and the functions $B_0(\tau)$, $B_1(\tau)$, $B(\tau) = B_0(\tau) + B_1(\tau)$. We have $\eta(f) = \eta(f^{(0)}) + \eta(f^{(1)}) = 3 + 1 = 4$. Using the procedure described above, we deduce from (2.4.16) that

$$T_\eta = \begin{pmatrix} 1 & 1 & 0 & 0 \\ 0 & 0 & 0 & 1 \\ 0 & 0 & 1 & 0 \\ 0 & 1 & 1 & 1 \end{pmatrix}, \qquad \sigma_\eta = \begin{pmatrix} 1 & 1 & 1 & 1 \\ 0 & 1 & 1 & 1 \\ 0 & 0 & 1 & 0 \\ 0 & 1 & 0 & 0 \end{pmatrix}.$$

The functions $f^{(0)}_{\sigma_\eta}(z)$ and $f^{(1)}_{\sigma_\eta}(z)$ are shown in Table 2.4.3. For this example, we have $\eta(f_{\sigma_\eta}) = \eta(f^{(0)}_{\sigma_\eta}) + \eta(f^{(1)}_{\sigma_\eta}) = 4 + 4 = 8$, and the η-optimal linear transformation doubles the number of matching pairs of constituents of unity.

Using the autocorrelation function $B(\tau)$, we can estimate the criteria $\eta(f)$ and

TABLE 2.4.3.

z, τ	$f^{(0)}(z)$	$f^{(1)}(z)$	$B_0(\tau)$	$B_1(\tau)$	$B(\tau)$	$f^{(0)}_{\sigma_\eta}(z)$	$f^{(1)}_{\sigma_\eta}(z)$
0	0	0	5	5	10	0	0
1	0	0	2	0	2	0	1
2	1	1	0	0	0	1	0
3	1	0	2	2	4	1	0
4	0	0	2	0	2	0	0
5	0	1	2	2	4	0	0
6	1	0	0	2	2	0	0
7	0	0	0	2	2	0	0
8	1	0	2	2	4	1	0
9	0	0	2	2	4	0	1
10	0	0	2	0	2	1	1
11	1	1	2	2	4	0	1
12	0	0	0	2	2	0	0
13	0	1	2	0	2	0	0
14	0	1	2	2	4	1	1
15	0	0	0	2	2	0	0

$\eta(f_{\sigma_\eta})$; the quantity $\frac{1}{2}B(\tau_s)$, where τ_s is determined by the above procedure, is equal to the number of pairs of constituents of unity for f_{σ_η} which are matched in the s-th argument. For example, if

$$B(\tau_0) = B(\tau_1) = \cdots = B(\tau_s) = \sum_{i=0}^{k-1} \sum_{z=0}^{2^m-1} f^{(i)}(z), \tag{2.4.17}$$

it can be shown [83] that as $m \to \infty$ the complexity $L(f_{\sigma_\eta})$ of a minimal network realizing f_{σ_η} satisfies the asymptotic estimate

$$L(f_{\sigma_\eta}) \sim \frac{k2^{m-s-1}}{m-s-1}. \tag{2.4.18}$$

The linearization procedure we have described is analytical and readily programmed for a computer. The program contains about 250 instructions, linearization of five BF of 10 variables requiring less than four minutes [83].

2.4.5. The linearization procedure of this section may be utilized in spectral analysis of BF to identify certain important classes of functions.

As an example, let us consider the class of partially linear BF. A BF $f(z) = f(z^{(0)}, \ldots, z^{(m-1)})$ is said to be *partially linear* if and only if

$$f(z^{(0)}, \ldots, z^{(m-1)}) = \varphi(z^{(0)}, \ldots, z^{m-b-1}, l(z^{(m-b)}, \ldots, z^{(m-1)})) \tag{2.4.19}$$

where $l(z^{(m-b)}, \ldots, z^{(m-1)}) = \bigoplus_{i=m-b}^{m-1} c_i z^{(i)}$ $(c_i \in \{0, 1\})$ (mod 2). It is apparent that once a function has been identified as partially linear the design of a network realization becomes much simpler.

Given a BF f, we let $G_I(f)$ denote the group of inertia points, i.e., $\beta \in G_I(f)$ if and only if $f(z) = f(z \oplus \beta)$ (mod 2) (see §2.1.4). Set $a(f) = \log_2|G_I(f)|$, where $|G_I(f)|$ is the order of the group $G_I(f)$; we shall assume throughout that f depends essentially on all $z^{(i)}$ $(i = 0, 1, \ldots, m-1)$.

THEOREM 2.4.4. *$f(z^{(0)}, \ldots, z^{(m-1)})$ satisfies* (2.4.19) *if and only if*

$$a(f) = b - 1, \tag{2.4.20}$$

and if $\beta = (\beta^{(0)}, \ldots, \beta^{(m-1)}) \in G_I(f)$, then $\beta^{(i)} = 0$ $(i = 0, \ldots, m-b-1)$.

PROOF. If f satisfies (2.4.19) (or, conversely, (2.4.20)), the matrices σ_η and T_η $(\sigma_\eta \otimes T_\eta = E_m \text{ (mod 2)})$ have the form shown in Figure 2.4.1 (the nonzero portion of the matrix is hatched in the figure). The $a(f)$ last columns of T_η are $a(f)$ linearly independent vectors in $G_I(f)$ and the function φ of (2.4.19) is defined by

$$\varphi(z) = f_{\sigma_\eta}(z). \quad \blacksquare \tag{2.4.21}$$

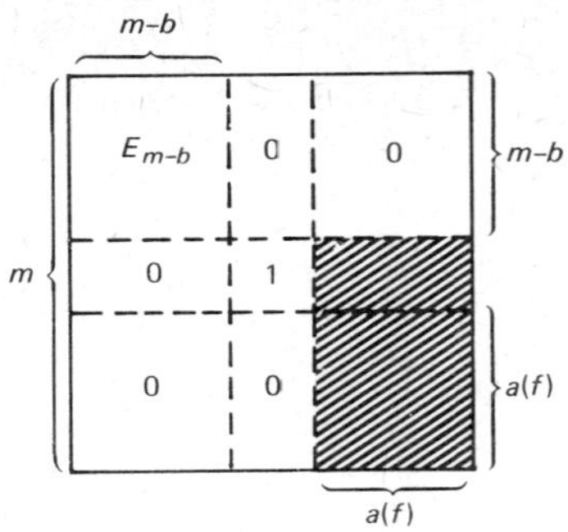

FIGURE 2.4.1.

From Theorem 2.4.4 se see, setting $b = m$, that f is a linear function of all its m arguments if and only if $a(f) = m - 1$.

EXAMPLE 2.4.4. Table 2.4.4 defines a function $f(z)$ and its autocorrelation function $B(\tau)$ ($m = 4$). It follows from the table that $B(0) = B(3) = \sum_{z=0}^{15} f(z) = 8$, and so, by Theorem 2.1.3, $G_I(f) = \{(0,0,0,0),(0,0,1,1)\}$, $f(z) = f(z \oplus (0,0,1,1))$ (mod 2), $a(f) = 1$. The assumptions of Theorem 2.4.4 hold with $b = 2$, and we have

$$T_\eta = \left(\begin{array}{cc|cc} \multicolumn{2}{c|}{E_2} & \multicolumn{2}{c}{0} \\ \hline & & 1 & 1 \\ \multicolumn{2}{c|}{0} & 0 & 1 \end{array}\right),$$

TABLE 2.4.4.

z, τ	$f(z)$	$B(\tau)$	$f_{\sigma_\eta}(z)$
0	0	8	0
1	1	0	0
2	1	0	1
3	0	8	1
4	0	4	0
5	1	4	0
6	1	4	1
7	0	4	1
8	0	4	0
9	1	4	0
10	1	4	1
11	0	4	1
12	1	4	1
13	0	4	1
14	0	4	0
15	1	4	0

$\sigma_\eta = T_\eta$. The BF $f_{\sigma_\eta}(z)$ (where $f_{\sigma_\eta}(\sigma_\eta \otimes z) = f(z) \pmod 2$) is also shown in Table 2.4.4. Again using the table, we see that

$$f(z) = f_{\sigma_\eta}(\sigma_\eta \otimes z) = f_{\sigma_\eta}(z^{(0)}, z^{(1)}, z^{(2)} \oplus z^{(3)}) = z^{(0)}z^{(1)} \oplus z^{(2)} \oplus z^{(3)} \qquad \pmod 2.$$

Thus a decision procedure for partially linear functions may be based on the group $G_I(f)$, or, in other words, on the autocorrelation function $B(\tau)$ (since $\beta \in G_I(f)$ if and only if $B(\beta) = \sum_z f(z)$; see Theorem 2.1.3).

To end this section, we remark that the linearization methods considered above are also useful in connection with other problems, related to optimal ordering of the arguments in BF or linearization of systems of BF (see, e.g., [55]).

2.5. LINEARIZATION OF SYSTEMS OF MANY-VALUED LOGICAL FUNCTIONS

2.5.1. We first generalize the linearization method of §2.4.2 (criterion ξ) to systems of p-valued logical functions, $p \geqq 2$, where p is a prime.

Let $f(z^{(0)}, \dots, z^{(m-1)})$ be a p-valued logical function. We shall say that f *depends inessentially on* $z^{(i)}$ if, for any fixed $z^{(0)}, \dots, z^{(i-1)}, z^{(i+1)}, \dots, z^{(m-1)}$,

$$f(z^{(0)}, \dots, z^{(i-1)}, r, z^{(i+1)}, \dots, z^{(m-1)}) = f(z^{(0)}, \dots, z^{(i-1)}, s, z^{(i+1)}, \dots, z^{(m-1)})$$

for any $r, s \in \{0, 1, \dots, p-1\}$.

If n is the number of p-ary arguments on which f depends essentially, we set $\xi(f) = p^n$. For a system of k functions $f = \{f^{(0)}, \dots, f^{(k-1)}\}$, we set

$$\xi(f) = \sum_{i=0}^{k-1} \xi(f^{(i)}).$$

Consider a system of logical functions (LF)

$$f(z) = \{f^{(i)}(z^{(0)}, \dots, z^{(m-1)})\} \qquad (i = 0, 1, \dots, k-1; z^{(s)} \in \{0, 1, \dots, p-1\}),$$

which depend essentially on all their arguments. Let Ξ_p be the set of all $m \times m$ matrices which are nonsingular modulo p. In other words, $\sigma = (\sigma_{ij}) \in \Xi_p$ if and only if $|\sigma|_p \neq 0$, where $|\sigma|_p$ is the determinant of σ over $GF(p)$, $\sigma_{ij} \in \{0, 1, \dots, p-1\}$.

Our problem is to devise an algorithm computing a matrix $\sigma = \sigma_\xi \in \Xi_p$ which, for each $f(z)$, minimizes $\xi(f_\sigma)$, where

$$f_\sigma(\sigma \otimes z) = f(z) \qquad \pmod p. \tag{2.5.1}$$

For each of the original LF $f^{(i)}(z)$, we define a system of characteristic functions:

$$f_t^{(i)}(z) = \begin{cases} 1 & \text{if } f^{(i)}(z) = t, \\ 0 & \text{if } f^{(i)}(z) \neq t, \end{cases} \tag{2.5.2}$$

autocorrelation functions modulo p

$$B_t^{(i)}(\tau) = \sum_{z=0}^{p^m-1} f_t^{(i)}(z)\, f(z \ominus \tau) \qquad (\text{mod } p), \tag{2.5.3}$$

and total autocorrelation function

$$B_{p,2}(\tau) = \sum_{t=0}^{p-1} \sum_{i=0}^{k-1} B_t^{(i)}(\tau). \tag{2.5.4}$$

We now define: $\tau_s \in G_I(f)$ if and only if

$$B_{p,2}(\tau_s) = B_{p,2}(0) = k \cdot p^m. \tag{2.5.5}$$

$G_I(f)$ is a group with respect to $\oplus$ (mod p), which we call the *inertia group* of the original system of LF. The motivation for this term is that, as follows from (2.5.2)–(2.5.5), if $\tau \in G_I(f)$, then $f(z) = f(z \oplus \tau)$ (see also §2.1). Let $a(f)$ be the number of elements in any basis of $G_I(f)$. The generalized version of Theorem 2.4.1 is as follows.

THEOREM 2.5.1. *Let T be an arbitrary nonsingular matrix ($T \in \Xi_p$) whose columns constitute a basis of the inertia group of a system of k p-valued logical functions ($p \geqq 2$) which depend essentially on all their m arguments. Then*

$$\min_{\sigma \in \Xi_p} \xi(f_\sigma) = \xi(f_{\sigma_\xi}) = k \cdot p^{m-a(f)}, \tag{2.5.6}$$

$$\sigma_\xi \otimes T = E_m = \begin{pmatrix} 1 & & & 0 \\ & 1 & & \\ & & \ddots & \\ 0 & & & 1 \end{pmatrix} \qquad (\text{mod } p). \tag{2.5.7}$$

The proof is analogous to that of Theorem 2.4.1. The only difference is that for p-valued logical functions we must replace each function by its system of characteristic functions (2.5.2), and the arguments are now translated not modulo 2 but modulo p in the formula for the autocorrelation functions $B_{p,2}(\tau)$.

EXAMPLE 2.5.1. Table 2.5.1 defines a system $f^{(0)}$, $f^{(1)}$ with $p = 3$, $k = 2$, $m = 2$. The table also shows the nonvanishing characteristic functions $f_0^{(0)}, f_2^{(0)}, f_1^{(1)}, f_2^{(1)}$, the autocorrelations $B_0^{(0)}, B_2^{(0)}, B_1^{(1)}, B_2^{(1)}$ and the total autocorrelation function $B_{3,2}$.

An examination of the table shows that

$$G_I(f) = \{(0,0), (1,1), (2,2)\}.$$

The basis will be $\tau_0 = (1, 1)$ $\big(a(f) = 1\big)$. Now set $T = \begin{pmatrix} 1 & 1 \\ 0 & 1 \end{pmatrix}$; then $\sigma_\xi = T^{-1} =$

TABLE 2.5.1

z, τ	$z^{(0)}$	$z^{(1)}$	$f^{(0)}$	$f^{(1)}$	$f_0^{(0)}$	$f_2^{(0)}$	$f_1^{(1)}$	$f_2^{(1)}$	$B_0^{(0)}$	$B_2^{(0)}$	$B_1^{(1)}$	$B_1^{(2)}$	$B_{3,2}$	$f_{T^{-1}}^{(0)}$	$f_{T^{-1}}^{(1)}$
0	0	0	0	1	1	0	1	0	3	6	6	3	18	0	1
1	0	1	2	2	0	1	0	1	0	3	3	0	6	0	1
2	0	2	2	1	0	1	1	0	0	3	3	0	6	0	1
3	1	0	2	1	0	1	1	0	0	3	3	0	6	2	1
4	1	1	0	1	1	0	1	0	3	6	6	3	18	2	1
5	1	2	2	2	0	1	0	1	0	3	3	0	6	2	1
6	2	0	2	2	0	1	0	1	0	3	3	0	6	2	2
7	2	1	2	1	0	1	1	0	0	3	3	0	6	2	2
8	2	2	0	1	1	0	1	0	3	6	6	3	18	2	2

$\begin{pmatrix} 1 & 2 \\ 0 & 1 \end{pmatrix}$ (mod 3). The functions $f_{T^{-1}}^{(0)}, f_{T^{-1}}^{(1)}$ are also shown in Table 2.5.1, from which it follows that they depend inessentially on $z^{(1)}$. Therefore, by (2.5.2), $\xi(f_{\sigma_\xi}) = \xi(f_{T^{-1}}^{(0)}) + \xi(f_{T^{-1}}^{(1)}) = 3 + 3 = 6$.

REMARK. Using the inertia group concept, one can also generalize Theorem 2.4.4 (decision procedure for partially linear LF) to p-valued logic.

2.5.2. Our next project is to generalize the linearization methods of §2.4 (criterion η) to p-valued logical functions. Recall (see §2.3.4) that in this case the criterion η for a LF $f^{(i)}(z)$ is defined as the number of sets of p values $z_1, z_2, \ldots, z_p$ of the argument vector z such that

$$\begin{cases} f^{(i)}(z_1) = f^{(i)}(z_2) = \cdots = f^{(i)}(z_p) \\ z_2 \ominus z_1 = z_3 \ominus z_2 = \cdots = z_p \ominus z_{p-1} = p^s \qquad (\text{mod } p). \end{cases} \tag{2.5.8}$$

As usual, the complexity of a system of functions is the sum of complexities of the individual functions.

The linearization problem relative to η is to find a matrix $\sigma = \sigma_\eta$ such that

$$\eta(f_{\sigma_\eta}) = \max_{\sigma \in \Xi_p} \eta(f_\sigma) \tag{2.5.9}$$

where $\sigma = (\sigma_{q,s})$, $\sigma_{q,s} \in \{0, 1, \ldots, p-1\}$.

As before, we construct the system of characteristic functions $\{f_t^{(i)}(z)\}$, but now, instead of the functions $B_{2,2}(\tau)$, we construct a class of autocorrelation functions $B_{p,p}(\tau)$:

$$B_{p,p}(\tau) = \sum_{t=0}^{p-1} B_t(\tau) = \sum_{t=0}^{p-1} \sum_{i,z} f_t^{(i)}(z)\, f_t^{(i)}(z \ominus \tau) \cdots f_t^{(i)}(z \ominus \overbrace{\tau \ominus \cdots \ominus \tau}^{p-1}) \quad (\text{mod } p). \tag{2.5.10}$$

In analogy to the case of Boolean functions, the algorithm computing σ_η makes use of the functions $B_{p,p}(\tau)$. Let $T = (\tau_{q,s})$ $(\tau_{p,s} \in \{0, 1, \ldots, p-1\}; q, s = 0, 1, \ldots, m-1)$. We set

$$B_{p,p}(T) = \sum_{s=0}^{m-1} B_{p,p}\left(\sum_{q=0}^{m-1} \tau_{q,s} p^{m-1-q}\right). \tag{2.5.11}$$

THEOREM 2.5.2. *Let*

$$\max_{T \in \Xi_p} B_{p,p}(T) = B_{p,p}(T_\eta). \tag{2.5.12}$$

Then

$$\sigma_\eta \otimes T_\eta = E_m = \begin{pmatrix} 1 & & & \\ & 1 & 0 & \\ 0 & & \ddots & \\ & & & 1 \end{pmatrix} \pmod{p}. \tag{2.5.13}$$

PROOF. First, by the definitions of $\eta(f)$, $B_{p,p}$, σ_η and T_η, we have

$$\eta(f) = B_{p,p}(E_m) \quad \text{and} \quad \eta(f_{\sigma_\eta}) = B_{p,p}(T_\eta).$$

Second, the analog of Theorem 1.5.4 for the function $B_{p,p}(\tau)$ is true: if $f(z)$ corresponds to $B_{p,p}(\tau)$, then $f(\sigma \otimes z)$ $(\sigma = (\sigma_{ij}), \sigma_{ij} \in \{0, 1, \ldots, p-1\}, \sigma \in \Xi_p)$ corresponds to $B_{p,p}(\sigma \otimes \tau) \pmod{p}$. These two remarks directly yield Theorem 2.5.2. ▮

Theorem 2.5.2 generalizes Theorem 2.4.3 to p-valued logic. The matrix T_η may be determined from $B_{p,p}(\tau)$ by a recursive procedure analogous to that described in §2.4.4.

It is readily seen from (2.5.10) that

$$B_{p,p}(\tau) = B_{p,p}(\bar{\tau}), \tag{2.5.14}$$

where $\tau \oplus \bar{\tau} = (0, \ldots, 0) \pmod{p}$. This remark yields a substantial simplification of the construction of $B_{p,p}(\tau)$; in fact, it halves the number of operations necessary to calculate $B_{p,p}(\tau)$ in the case $p > 2$.

EXAMPLE 2.5.2. Table 2.5.2 defines a ternary logical function of two variables ($p = 3$, $m = 2$, $k = 1$) and also gives the autocorrelation functions $B_0(\tau)$, $B_1(\tau)$, the characteristic functions f_0, f_1 $(f_2(z) = 0)$, and the function $B_{3,3}(\tau)$; we have $\eta(f) = 0$. Using the table and formulas (2.5.12)–(2.5.13), we see that

$$T_\eta = \begin{pmatrix} 1 & 1 \\ 0 & 1 \end{pmatrix}, \qquad \sigma_\eta = \begin{pmatrix} 1 & 2 \\ 0 & 1 \end{pmatrix}$$

and $\eta(f_{\sigma_\eta}) = 3$.

TABLE 2.5.2

z, τ	$f(z)$	$B_0(\tau)$	$B_1(\tau)$	$B_{3,3}(\tau)$	$f_{\sigma_n}(z)$
0	1	3	6	9	1
1	0	0	0	0	1
2	1	0	0	0	1
3	1	0	0	0	1
4	1	0	6	6	1
5	0	0	0	0	1
6	0	0	0	0	0
7	1	0	0	0	0
8	1	0	6	6	0

The linearization procedures considered here may be extended to systems of incompletely specified BF or many-valued LF. Linearization methods employing other complexity criteria have been considered in [90], which also describes a generalization of the procedures to nonprime p.

2.6. POLYNOMIAL APPROXIMATION OF SYSTEMS OF LOGICAL FUNCTIONS

2.6.1. The linearization methods of §§2.4, 2.5 were concerned with structures involving linear and nonlinear blocks connected in series, minimizing the complexity of the nonlinear part. In this section we shall discuss structures in which the linear and nonlinear blocks are connected in parallel; the minimized functional will again be the complexity of the nonlinear part.

Another difference between this and the preceding sections is that, whereas previously our main tools were the correlation functions, the methods of polynomial approximation to be considered here illustrate the effective use of spectral characteristics in optimization problems.

We shall measure the complexity of a given completely specified BF $f(z^{(0)}, \ldots, z^{(m-1)})$ (depending essentially on all its arguments) in terms of the simplest criterion $\xi(f)$ (see §2.3): $\xi(f) = 2^m$.

In this context, we shall use the term "linear function" not only for a linear function proper but also for its inversion; in other words, $l_i(z)$ is a linear function if and only if

$$l_i(z) = \bigoplus_{s=0}^{m-1} l_i^{(s)} z^{(s)} \oplus l_i^{(m)} \qquad (\text{mod } 2), \tag{2.6.1}$$

where $l_i^{(s)} \in \{0, 1\}$, $s = 0, 1, \ldots, m$.

2.6.2. We can now state the problem of polynomial approximation for a completely specified BF. We shall say that a BF f is *approximated* by a product of q linear functions $l_0, \ldots, l_{q-1}$ if

$$f(z) = f_P(z) \cdot \prod_{i=0}^{q-1} l_i(z). \tag{2.6.2}$$

The function $P(z) = \prod_{i=0}^{q-1} l_i(z)$ is a polynomial of degree q, which we shall call a *polynomial approximation* of f.

If f has a polynomial approximation $P(z) = \prod_{i=0}^{q-1} l_i(z)$, it may be realized by a network of the type shown in Figure 2.6.1 (the symbol $\dashrightarrow \otimes \rightarrow$ signifies a (logical) multiplication element).

The reader should note that for given f and a given polynomial approximation $P(z)$ there may exist several functions f_P satisfying (2.6.2). Let $\mathscr{F}_P$ denote the set of all such functions, and set

$$\min_{f_P \in \mathscr{F}_P} \xi(f_P) = \xi(\tilde{f}_P). \tag{2.6.3}$$

Denote the set of all approximations $P(z)$ of f by π_f.

We shall say that an approximation $P(z) = P_\xi(z)$ is *optimal* for f if it minimizes the complexity of the residual part f_P:

$$\min_{P(z) \in \pi_f} \xi(\tilde{f}_P) = \xi(\tilde{f}_{P_\xi}). \tag{2.6.4}$$

Our problem is thus to determine an optimal approximation $P_\xi(z)$ for a given function f from condition (2.6.4).

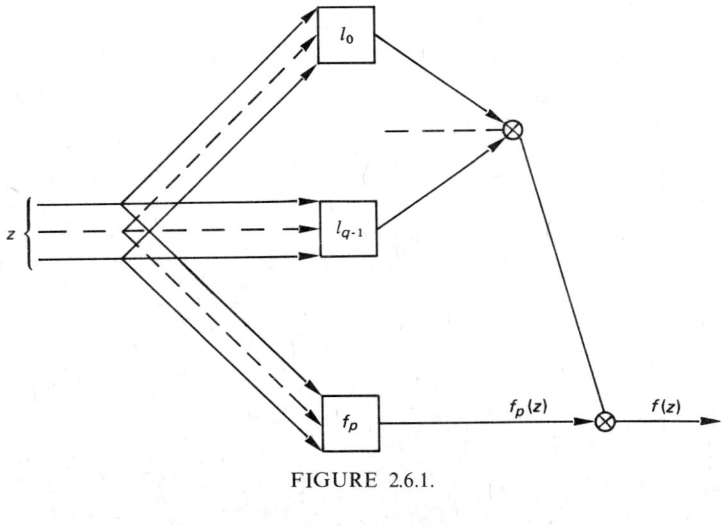

FIGURE 2.6.1.

If $f(z) = f(z^{(0)}, \ldots, z^{(m-1)})$, we shall call $l_i = (l_i^{(0)}, \ldots, l_i^{(m-1)})$ $(l_i^{(s)} \in \{0, 1\}, s = 0, \ldots, m-1)$ a *linearity point* if

$$l_i(z) = \bigoplus_{s=0}^{m-1} l_i^{(s)} z^{(s)} \oplus l_i^{(m)} \qquad (\text{mod } 2)$$

is an approximation of f.

THEOREM 2.6.1. *A BF $f(z^{(0)}, \ldots, z^{(m-1)})$ has linearity points $l_0, l_1, \ldots, l_{Q(f)-1}$ if and only if*

$$|S_f(\overleftarrow{l_i})| = 2^{-m} \sum_{z=0}^{2^m-1} f(z) \qquad (i = 0, 1, \ldots, Q(f)-1), \tag{2.6.5}$$

where S_f is the Walsh spectrum of f and $\overleftarrow{l_i} = \sum_{s=0}^{m-1} l_i^{(s)} 2^s$.

PROOF. We need only note that $l_i(z)$ is an approximation of f if and only if, for any fixed z^* such that $f(z^*) = 1$, it is always true that $l_i(z^*) = 1 \oplus l_i^{(m)}$, $W_{\overleftarrow{l_i}}(z^*) = (-1)^{1 \oplus l_i^{(m)}}$; consequently, $|S_f(\overleftarrow{l_i})| = 2^{-m} \sum_{z=0}^{2^m-1} f(z)$. ∎

COROLLARY 2.6.1. *The set of linearity points of an arbitrary BF is a group relative to addition modulo 2.*

PROOF. We first note that $e(z) = 0$ is a linearity point for any BF, since $1 \oplus e(z) = 1$ is an approximation of any BF. Furthermore, if $f(z) = l_i(z) f_{l_i}(z) = l_j(z) f_{l_j}(z)$, then for any fixed z^* it follows from $f(z^*) = 1$ that

$$W_{\overleftarrow{l_i \oplus l_j}}(z^*) = W_{\overleftarrow{l_i} \oplus \overleftarrow{l_j}}(z^*) = W_{\overleftarrow{l_i}}(z^*) \cdot W_{\overleftarrow{l_j}}(z^*) = (-1)^{l_i^{(m)} + l_j^{(m)}} \qquad (\text{mod } 2),$$

and so $|S_f(\overleftarrow{l_i \oplus l_j})| = 2^{-m} \sum_z f(z)$. Hence it follows by Theorem 2.6.1 that $l_i \oplus l_j$ is a linearity point for f. ∎

COROLLARY 2.6.2. *The number of points at which the Walsh spectrum of an arbitrary Boolean function $f(z^{(0)}, \ldots, z^{(m-1)})$ assumes its maximum absolute value $2^{-m} \sum_z f(z)$ is always a power of 2.*

This corollary, which follows from Theorem 2.6.1 and Corollary 2.6.1, yields yet another simple method for checking the validity of a calculation of spectra.

Comparing Theorem 2.6.1 and Corollary 2.6.1 with Theorem 2.1.3 and Corollary 2.1.2, we see that while the maximum absolute value of the spectrum of a BF determines its linearity group, the maximum values of the autocorrelation function define its inertia (anti-self-duality) group.

Now let $l_0, l_1, \ldots, l_{q(f)-1}$ be an arbitrary basis (maximal linearly independent set of vectors over $GF(2)$) of the group of linearity points of f. Denote

$$\log_2 \xi(\tilde{f}_{P_\xi}) = \xi_n(f). \tag{2.6.6}$$

THEOREM 2.6.2.

$$P_\xi(z) = \prod_{i=0}^{q(f)-1} \left(\bigoplus_{s=0}^{m-1} l_i^{(s)} z^{(s)} \oplus \operatorname{sign} S_f(\overleftarrow{l_i}) \right) \quad (\text{mod } 2), \tag{2.6.7}$$

where

$$\operatorname{sign} x = \begin{cases} 1 & \text{if} \quad x > 0, \\ 0 & \text{if} \quad x \leqq 0, \end{cases}$$

and

$$\xi_n(f) \leqq m - q(f). \tag{2.6.8}$$

PROOF. The fact that

$$P_\xi(z) = \prod_{i=0}^{q(f)-1} \left(\bigoplus_{s=0}^{m-1} l_i^{(s)} z^{(s)} \oplus \operatorname{sign} S_f(\overleftarrow{l_i}) \right) \quad (\text{mod } 2)$$

is an approximation of f follows from the definition of $l_0, \ldots, l_{q(f)-1}$ and from Theorem 2.6.1 and Corollary 2.6.1, in view of the fact that if $\operatorname{sign} S_f(\overleftarrow{l_i}) = 0$ ($\operatorname{sign} S_f(\overleftarrow{l_i}) = 1$) then $f(z^*) = 1$ always implies $W_{\overleftarrow{l_i}}(z^*) = -1$ ($W_{\overleftarrow{l_i}}(z^*) = 1$) and

$$l_i(z) = \bigoplus_{i=0}^{m-1} l_i^{(s)} z^{(s)} \qquad (1 \oplus l_i(z)) \qquad (\text{mod } 2)$$

is an approximation of f.

Now, if $\hat{P}(z) \in \pi_f$ and

$$\hat{P}(z) = \prod_{i=0}^{\hat{q}-1} \left(\bigoplus_{s=0}^{m-1} \hat{l}_i^{(s)} z^{(s)} \oplus \hat{l}_i^{(m)} \right) \qquad (\text{mod } 2),$$

where $\hat{l}_i^{(s)} \in \{0, 1\}$, $i = 0, 1, \ldots, \hat{q} - 1$, $s = 0, 1, \ldots, m$, then the vectors $\hat{l}_i = (\hat{l}_i^{(0)}, \ldots, \hat{l}_i^{(m-1)})$ $(i = 0, 1, \ldots, \hat{q} - 1)$ lie in some subgroup of the group of linearity points. Since $l_0, l_1, \ldots, l_{q(f)-1}$ form a basis for this group, it follows that for any z^*, if $P_\xi(z^*) = 1$, then necessarily $\hat{P}(z^*) = 1$ and so $\xi(\tilde{f}_{P_\xi}) \leqq \xi(\tilde{f}_P)$. Thus the polynomial $P_\xi(z)$ defined by (2.6.6) is an optimal approximation.

We now prove (2.6.8). Consider a system of $q(f)$ linearly independent equations in the m unknowns $z^{(0)}, \ldots, z^{(m-1)}$:

$$\bigoplus_{s=0}^{m-1} l_i^{(s)} z^{(s)} \oplus \operatorname{sign} S_f(\overleftarrow{l_i}) = 1. \tag{2.6.9}$$

There always exist $s_0, s_1, \ldots, s_{q(f)-1}$ such that $\sum_{i=0}^{q(f)-1} l_i^{(s_r)} > 0$, since $l_0, \ldots, l_{q(f)-1}$ are linearly independent; then, by (2.6.9),

$$z^{(s)} = g_{s_r}(z^{(0)}, \ldots, z^{(m-1)}) \qquad (r = 0, 1, \ldots, q(f) - 1), \tag{2.6.10}$$

where the g_{s_r} are linear functions each of which is independent of $z^{(s_0)}, \ldots, z^{(s_{q(f)-1})}$. Now, if $f(z) = P_\xi(z) f_{P_\xi}(z)$, substitution of g_{s_r} for $z^{(s_r)}$ $(r = 0, \ldots, q(f) - 1)$ in $f_{P_\xi}(z)$ yields a function $f'_{P_\xi}(z)$ which does not depend on $z^{(s_r)}$ and satisfies the following condition: if $P(z^*) = 1$ for some z^* (i.e., all the equations of system (2.6.9) are satisfied), then $f'_{P_\xi}(z^*) = f_{P_\xi}(z^*)$. Hence $f'_{P_\xi} \in \pi_f$ and

$$\xi_n(f) = \min_{P(z) \in \pi_f} \log_2 \xi(\tilde{f}_P) \leqq \log_2 \xi(f'_{P_\xi}) \leqq m - q(f). \quad \blacksquare$$

Theorems 2.6.1 and 2.6.2 yield a simple procedure for determining the optimal polynomial approximation $P_\xi(z)$ and estimating the complexity $\xi_n(f)$ of the nonlinear part: calculate the spectrum S_f of the function f, construct the group of linearity points l_i for which $|S_f(\overset{\leftarrow}{l_i})| = 2^{-m} \sum_z f(z)$, and select an arbitrary basis of this group. An upper bound for the complexity $\xi_n(f)$ of the nonlinear part of f may even be established without finding a basis, as $\xi_n(f)$ depends only on the number m of arguments of f and on the number $q(f)$, which is equal to the binary logarithm of the number of elements in the group of linearity points of f.

EXAMPLE 2.6.1. Table 2.6.1 defines a BF $f(z)$ and its spectrum $S_f(\omega)$ $(m = 4)$. The table shows that $|S_f(0)| = |S_f(2)| = |S_f(8)| = |S_f(10)| = 2^{-m} \sum_z f(z) = 3/16$. The group of linearity points contains the vectors (0, 0, 0, 0), (0, 1, 0, 0), (0, 0, 0, 1), (0, 1, 0, 1) $(q(f) = 2)$. As basis we take $l_0 = (0, 0, 0, 1)$, $l_1 = (0, 1, 0, 1)$. Since $\operatorname{sign} S_f(\overset{\leftarrow}{l_0}) = \operatorname{sign} S_f(8) = 0$ and $\operatorname{sign} S_f(\overset{\leftarrow}{l_1}) = \operatorname{sign} S_f(10) = 0$, the optimal approximation is $P_\xi(z) = (z^{(1)} \oplus z^{(3)}) z^{(3)} = z^{(3)} \oplus z^{(1)} z^{(3)}$, and by (2.6.8) we have $\xi_n(f) \leqq 2$. The nonlinear part $\tilde{f}_{P_\xi}(z)$ (where $f(z) = P_\xi(z) \tilde{f}_{P_\xi}(z)$) in this example is $\tilde{f}_{P_\xi}(z) = z^{(0)} \vee z^{(2)}$. Finally:

$$f(z) = (z^{(3)} \oplus z^{(1)} z^{(3)})(z^{(0)} \vee z^{(2)}).$$

2.6.3. We now generalize the approximation procedure of the preceding subsection. Hitherto we have been studying multiplicative approximations:

$$f(z) = f_P(z) \prod_{i=0}^{q-1} l_i(z). \tag{2.6.12}$$

We now consider "additive" approximations, with logical product replaced throughout by logical sum:

$$f(z) = f_P(z) \vee \bigvee_{i=0}^{q-1} l_i(z) \quad \text{and} \quad P(z) = \bigvee_{i=0}^{q-1} l_i(z). \tag{2.6.13}$$

TABLE 2.6.1

z, ω	$f(z)$	$16S_f(\omega)$
0	0	3
1	0	−1
2	0	3
3	1	−1
4	0	−1
5	0	−1
6	0	−1
7	0	−1
8	0	−3
9	1	1
10	0	−3
11	1	1
12	0	1
13	0	1
14	0	1
15	0	1

As before, we shall say that an approximation $P(z) = P_\xi(z)$ is optimal if it minimizes $\xi_n(f)$ (see (2.6.4), (2.6.6)).

The physical realization of an additive approximation is described by the same block-diagram as in Figure 2.6.1, except that the logical multiplication elements should be replaced throughout by logical adders.

The optimal additive approximation may be determined with the help of De Morgan's laws [70]:

$$1 \oplus \prod_{i=0}^{q-1} a_i = \bigvee_{i=0}^{q-1} (1 \oplus a_i) \qquad (\text{mod } 2), \qquad a_i \in \{0, 1\}. \tag{2.6.14}$$

Applying (2.6.14) to (2.6.12), we obtain a simple rule for the additive approximation: if $\prod_{i=0}^{q-1} l_i(z)$ is the optimal multiplicative approximation to f, then $\bigvee_{i=0}^{q-1} (1 \oplus l_i(z))$ is the optimal additive approximation to $\bar{f} = 1 \oplus f$.

Since $f(z) = f(z^{(0)}, \ldots, z^{(m-1)})$ has a nontrivial multiplicative approximation only if $\sum_z f(z) \leqq 2^{m-1}$ and a nontrivial additive approximation exists only if $\sum_z f(z) \geqq 2^{m-1}$, it is clear that these two types of approximation complement each other in a natural way.

The above approximation procedures may also be generalized to systems of BF. Namely, call $P(z)$ an *approximation of the system* $f^{(i)}(z)$ $(i = 0, 1, \ldots, k-1)$ if each $f^{(i)}(z)$ may be expressed as

$$f^{(i)}(z) = f_P^{(i)}(z)\, P(z) \qquad (i = 0, 1, \ldots, k-1). \tag{2.6.12a}$$

An *optimal approximation* is one which minimizes the quantity

$$\xi_n(f) = \sum_{i=0}^{k-1} \xi_n(f^{(i)}).$$

The solution of the optimal approximation problem for a system of BF is entirely analogous to the case of a single BF. The only difference is that the linearity group is the set of vectors which are linearity points simultaneously for all Boolean functions in the system.

2.6.4. We conclude this section with a discussion of the relationship between the optimal approximation problems just considered and the problem of minimizing the complexity of devices realizing systems of BF.

Let f be a system of k BF of m arguments and P_ξ a ξ-optimal approximation of f. Let $L(f)$ denote the complexity of a minimal network realizing the system f (i.e., the minimal number of one- and two-input logic elements necessary for the purpose).

THEOREM 2.6.3. *As* $m \to \infty$,

$$L(f) \lesssim \frac{m^2}{\log_2 m} + \frac{k \cdot 2^{\xi_n(f)}}{\xi_n(f)}. \tag{2.6.15}$$

PROOF. By Theorem 2.6.2, $P_\xi(z)$ is a product of $q(f)$ linear functions of m arguments, and so it can be realized by the superposition of a linear network computing these functions and a network computing the product of its outputs. The complexity of the multiplication network is at most $q(f) - 1$ (two-input elements). Since $q(f) \leqq m$, it follows, in view of the complexity estimate $m^2/\log_2 m$ ($m \to \infty$) for the linear network (see § 3.2), that the complexity $L(P_\xi)$ of the block realizing P_ξ satisfies the estimate

$$L(P_\xi) \lesssim \frac{m^2}{\log_2 m}. \tag{2.6.16}$$

We can now use the Shannon-Lupanov estimate (2.3.4) for the complexity $L(\tilde{f}_{P_\xi})$ of the system $\tilde{f}_{P_\xi}$ of k Boolean functions of $\xi_n(f) = \log_2 \xi(\tilde{f}_{P_\xi})$ arguments:

$$L(\tilde{f}_{P_\xi}) \lesssim \frac{k \cdot 2^{\xi_n(f)}}{\xi_n(f)}. \tag{2.6.17}$$

Using the relation

$$L(f) \leqq L(P_\xi) + L(\hat{f}_{P_\xi}), \tag{2.6.18}$$

we now deduce the desired estimate from (2.6.16) and (1.6.17). ∎

Theorem 2.6.3 is of course valid for both multiplicative and additive approximations. It provides a good illustration of how important it is to know the complexity

$\xi_n(f)$ of the nonlinear part in order to estimate the complexity in terms of the number of logic elements necessary for a minimal realization of f. For example, if $k = 1$, $\xi_n(f) = 0$ (f is a product or logical sum of linear functions), it follows from Theorem 2.6.3 that $L(f) \lesssim m^2/\log_2 m$; if $\xi_n(f) = m$ (i.e., $q(f) = 0$, $P_\xi(z) = 1$, and there is no linear part), the estimate of Theorem 2.6.3 coincides with the Shannon-Lupanov estimate (2.3.4), and if $q(f) \geqq 1$ it is sharper.

2.6.5. The above polynomial approximation methods will now be generalized to p-valued logical functions.

As before, our criterion for the complexity of a completely specified p-valued logical function $f(z^{(0)}, z^{(1)}, \ldots, z^{(n-1)})$, depending essentially on all its m arguments, will be $\xi(f) = p^m$.

Let $l_r(z)$ be a linear p-valued logical function,

$$l_r(z) = \bigoplus_{s=0}^{m-1} l_r^{(s)} z^{(s)} \qquad (\text{mod } p) \qquad (l_r^{(s)} \in \{0, 1, \ldots, p-1\}) \tag{2.6.19}$$

Let $d_t(l_r)$ denote the t-th characteristic function of $l_r(z)$:

$$d_t(l_r(z)) = \begin{cases} 1 & \text{if} \quad l_r(z) = t, \\ 0 & \text{otherwise} \end{cases} \tag{2.6.20}$$

$$(t = 0, 1, \ldots, p-1).$$

In the case $p = 2$, we have $t \in \{0, 1\}$ in (2.6.20), and

$$d_0(l_r(z)) = 1 \oplus l_r(z), \qquad d_1(l_r(z)) = l_r(z) \qquad (\text{mod } 2),$$

so that the characteristic functions thus defined generalize those defined by (2.6.1) for $p = 2$.

We shall say that f is *approximated* by a product of q linear functions $l_0, \ldots, l_{q-1}$ if and only if there exist $t_0, \ldots, t_{q-1} \in \{0, 1, \ldots, p-1\}$ such that

$$f(z) = f_P(z) \prod_{r=0}^{q-1} d_{t_r}(l_i(z)). \tag{2.6.21}$$

We then call $P(z) = \prod_{r=0}^{q-1} d_{t_r}(l_r(z))$ a *polynomial approximation* for f.

Let $\mathscr{F}_P$ denote the set of functions f_P satisfying (2.6.21) for given f and $P(z)$, and set

$$\min_{f_P \in \mathscr{F}_P} \xi(f_P) = \xi(\tilde{f}_P). \tag{2.6.22}$$

An approximation $P(z) = P_\xi(z)$ is said to be *optimal* if

$$\min_{P(z) \in \pi_f} \xi(\tilde{f}_P) = \xi(\tilde{f}_{P_\xi}), \tag{2.6.23}$$

where π_f is the set of all polynomial approximations of f; in other words, as before, an optimal approximation minimizes the ξ-complexity of the nonlinear part f_P.

A vector $l_r = (l_r^{(0)}, \ldots, l_r^{(m-1)})$ $(l_r^{(s)} \in \{0, 1, \ldots, p-1\}; s = 0, \ldots, m-1)$ will be called a *linearity point* for a p-valued logical function $f(z) = f(z^{(0)}, \ldots, z^{(m-1)})$ if $l_r(z) = \bigoplus_{s=0}^{m-1} l_r^{(s)} z^{(s)} \pmod p$ approximates $f(z)$, i.e., there exist $t_r \in \{0, 1, \ldots, p-1\}$ such that

$$f(z) = d_{t_r}(l_r(z)) \cdot f_P(z).$$

THEOREM 2.6.4. *A p-valued logical function $f(z) = f(z^{(0)}, \ldots, z^{(m-1)})$ may be expressed as*

$$f(z) = d_{t_r}(l_r(z)) \cdot f_P(z), \tag{2.6.24}$$

where $l_r(z) = \bigoplus_{s=0}^{m-1} l_r^{(s)} z^{(s)} \pmod p$, if and only if

$$S_f(\overleftarrow{l_r}) = p^{-m} \bar{e}_{t_r} \sum_{z=0}^{p^m-1} f(z), \tag{2.6.25}$$

where $S_f(\overleftarrow{l_r})$ is the spectral coefficient for $l_r = \sum_{s=0}^{m-1} l_r^{(s)} 2^s$ in the expansion of $f(z)$ in Chrestenson functions $\chi_\omega^{(p)}(z)$ (see § 1.3) and $\bar{e}_{t_r} = \exp(-2\pi i t_r/p)$ $(i = \sqrt{-1})$.

PROOF. We first note that (2.6.24) holds if and only if, for any fixed z^*, the truth of $f(z^*) \neq 0$ always implies $l_r(z^*) = t_r$. Furthermore, by virtue of the isomorphism of the linear p-valued logical functions and the Chrestenson functions (Theorem 1.3.5),

$$\chi_{\overleftarrow{l_r}}(z^*) = \exp\left(\frac{2\pi}{p} i t_r\right) = e_{t_r},$$

and therefore

$$S_f(\overleftarrow{l_r}) = p^{-m} \sum_{z=0}^{p^m-1} f(z) \overline{\chi_{\overleftarrow{l_r}}(z)} = p^{-m} \bar{e}_{t_r} \sum_{z=0}^{p^m-1} f(z),$$

proving (2.6.25). ∎

Setting $p = 2$ in Theorem 2.6.4, we obtain Theorem 2.6.1 (since $|\bar{e}_{t_r}| = 1$ for any $t_r \in \{0, 1, \ldots, p-1\}$).

Theorem 2.6.4 thus states that l_r is a linearity point of $f(z)$ if and only if

$$|S_f(\overleftarrow{l_r})| = p^{-m} \sum_{z=0}^{p^m-1} f(z). \tag{2.6.26}$$

Here, too, in analogy with Corollary 2.6.1, one sees that the set of linearity points of any p-valued logical function is a group with respect to componentwise addition modulo p (this also follows from Theorem 2.6.4. Hence:

COROLLARY 2.6.3. *The number of points at which the Chrestenson spectrum of an arbitrary p-valued logical function* $f(z^{(0)}, \ldots, z^{(m-1)})$ *(where p is a prime) is of absolute value* $p^{-m} \sum_z f(z)$ *is always an integral power of p.*

Corollary 2.6.3 is a generalization of Corollary 2.6.2 and provides a check on calculations of Chrestenson spectra.

When $p > 2$ the calculation of spectra may be simplified by using the identity

$$S_f(\bar{\omega}) = \overline{S_f(\omega)}, \qquad \omega \oplus \bar{\omega} = 0 \qquad (\text{mod } p), \tag{2.6.27}$$

which follows from $\chi_{\bar{\omega}}^{(p)}(z) = \overline{\chi_{\omega}^{(p)}(z)}$ (mod p). This halves the computational labor involved in calculating Chrestenson spectra.

We can now prove our main result concerning polynomial approximation of p-valued logical functions, generalizing Theorem 2.6.2.

THEOREM 2.6.5. *Let* $f(z) = f(z^{(0)}, \ldots, z^{(m-1)})$ *be a p-valued logical function, and* $l_0, l_1, \ldots, l_{q(f)-1}$ *a basis of its linearity group. Let*

$$S_f(\overleftarrow{l_r}) = p^{-m}\bar{e}_{t_r} \sum_{z=0}^{p^m-1} f(z)$$

(where S_f *is the Chrestenson spectrum of f and* $\bar{e}_{t_r} = \exp(-2\pi i t_r/p)$*).*

Then an optimal polynomial approximation $P_\xi(z)$ *and the complexity* $\xi_n(f) = \log_p \xi(\tilde{f}_{P_\xi})$ *of the nonlinear part of f satisfy the conditions*

$$P_\xi(z) = \prod_{r=0}^{q(f)-1} d_{t_r}\left(\bigoplus_{s=0}^{m-1} l_r^{(s)} z^{(s)}\right) \qquad (\text{mod } p), \tag{2.6.27}$$

$$\xi_n(f) \leqq m - q(f). \tag{2.6.28}$$

PROOF. Analogous to that of Theorem 2.6.2, except that in proving (2.6.28) one considers instead of (2.6.9) the system

$$\bigoplus_{s=0}^{m-1} l_r^{(s)} z^{(s)} = t_r \qquad (r = 0, 1, \ldots, q(f) - 1) \qquad (\text{mod } p) \tag{2.6.29}$$

of $q(f)$ linearly independent linear equations over $GF(p)$ in the m unknowns $z^{(0)}, \ldots, z^{(m-1)}$. ∎

EXAMPLE 2.6.2. Table 2.6.2 defines a ternary logical function of two arguments ($p = 3$, $m = 2$) and its Chrestenson spectrum (using the notation $e_1 = \exp(2\pi i/3)$, $e_2 = \bar{e}_1 = \exp(4\pi i/3)$). The linearity group consists of the points $\{(0,0), (1,2), (2,1)\}$, so that a suitable basis will include the single vector $(1, 2)$ and $q(f) = 1$. By Theorem

2.6.4, in view of the relation

$$S_f(\overleftarrow{5}) = S_f(7) = 3^{-2}\bar{e}_2 \cdot 5,$$

we find that the optimal approximation is

$$P_\xi(z) = d_2(z^{(0)} \oplus 2z^{(1)}) \qquad (\text{mod } 3)$$

and $\xi_n(f) \leqq 1$. The function $d_2(z^{(0)} \oplus 2z^{(1)})$ (mod 3) is also shown in Table 2.6.2.

TABLE 2.6.2.

z, ω	$z^{(0)}$	$z^{(1)}$	$f(z)$	$9S_f(\omega)$	$d_2(z)$
0	0	0	0	5	0
1	0	1	2	$-e_2$	1
2	0	2	0	$-e_1$	0
3	1	0	0	$-e_1$	0
4	1	1	0	-1	0
5	1	2	1	$5e_2$	1
6	2	0	2	$-e_2$	1
7	2	1	0	$5e_1$	0
8	2	2	0	-1	0

TABLE 2.6.3.

$z^{(0)}$	$\tilde{f}_{P_\xi}(z^{(0)})$
0	2
1	1
2	2

It is evident from Table 2.6.2 that a suitable function $\tilde{f}_{P_\xi}(z)$ such that $f(z) = d_2(z^{(0)} \oplus 2z^{(1)}) \cdot \tilde{f}_{P_\xi}(z)$ is the function of one argument $z^{(0)}$ ($\xi_n(f) = 1$) defined by Table 2.6.3.

REMARK. Our procedure for the optimization of polynomial approximations may be generalized to systems of p-valued logical functions; in that case a linearity point of the system is defined as a vector which is a linearity point for each function in the system. Further generalizations, to systems of functions defined on arbitrary finite commutative groups, may be found in [90].

2.7. BOUNDS FOR THE SPECTRAL COMPLEXITY OF SOME CLASSES OF BOOLEAN FUNCTIONS

2.7.1. In this section we shall consider bounds for the spectral complexity of certain familiar classes of completely specified BF. As we shall show later (Chapter VI), these bounds enable the complexity of networks realizing functions of these classes to be estimated by spectral methods. In addition, a comparison of the bounds for Walsh and Haar bases is frequently an aid to more rational choice of a basis system.

The *spectral complexity* of a BF f relative to the Walsh (Haar) basis, denoted by $L^W(f)$ ($L^H(f)$), is defined as the number of nonvanishing coefficients in the Walsh (Haar) expansion of f. As will be shown later (see §§6.1–6.4), the spectral complexity of a BF determines the complexity of a block-diagram realizing the function, or, equivalently, the necessary storage capacity of the computer if the function is being computed by a special program.

In the Appendix the reader may find the Walsh transforms, autocorrelation functions and bounds for the Walsh and Haar spectral complexities for 21 important classes of BF. This section is in fact a commentary on the Appendix.

The Appendix employs the following notation. For any $z \in \{0, 1, \dots, 2^m - 1\}$, $z = \sum_{i=0}^{m-1} z^{(i)} 2^{m-1-i}$ ($z^{(i)} \in \{0, 1\}$), we put

$$\overleftarrow{z} = \sum_{i=0}^{m-1} z^{(i)} 2^i, \qquad \|z\| = \sum_{i=0}^{m-1} z^{(i)}.$$

2.7.2. We begin with a few considerations which enable us to extend the applicability of the bounds for spectral complexity given in the Appendix.

THEOREM 2.7.1. *Let $f(z)$ be a BF of m arguments, $\tau \in \{0, 1, \dots, 2^m - 1\}$ and σ a nonsingular matrix over $GF(2)$. Then*

$$L^W(\overline{f(z)}) = L^W(1 \oplus f(z)) = L^W(f(z)) \pmod 2, \tag{2.7.1}$$

$$L^H(\overline{f(z)}) = L^H(1 \oplus f(z) = L^H(f(z)) \pmod 2, \tag{2.7.2}$$

$$L^W(f(z \oplus \tau)) = L^W(f(z)) \pmod 2, \tag{2.7.3}$$

$$L^H(f(z \oplus \tau)) = L^H(f(z)) \pmod 2, \tag{2.7.4}$$

$$L^W(f(\sigma \otimes z)) = L^W(f(z)) \pmod 2. \tag{2.7.5}$$

Theorem 2.7.1 in effect states that the bounds for the Walsh and Haar spectral complexities are invariant under inversion of functions and translation (inversion) of arguments. The Walsh complexity is also invariant under linear transformations of the argument, but the Haar complexity is not; in particular, it depends on the order of the arguments of the function. The bounds given in the Appendix for the Haar spectral complexity refer to the optimal order of arguments (i.e., that minimizing the complexity).

Theorem 2.7.1 implies a certain duality principle for the spectral complexities L^W and L^H. We first note that any BF may be built up from its arguments by superposition of the following operations: disjunction "$\vee$" (logical addition), conjunction "$\cdot$" (logical multiplication), negation "$^-$" and addition modulo 2 "$\oplus$". (This set of operations is of course redundant—it is sufficient, for example, to take disjunction or conjunction together with negation.)

COROLLARY 2.7.1. *Let $f^*(z)$ be the Boolean function of m arguments derived from a Boolean function $f(z)$ by interchanging the disjunction and conjunction operations in any representation of $f(z)$ in terms of disjunction, conjunction, negation and addition modulo 2. Then*

$$L^W(f^*(z)) = L^W(f(z)), \qquad L^H(f^*(z)) = L^H(f(z)). \tag{2.7.6}$$

The proof of (2.7.6) employs De Morgan's laws (2.6.14) and formulas (2.7.1), (2.7.2), (2.7.4), (2.7.5) with $\tau = 2^m - 1$.

By virtue of the duality principle (2.7.6) one can considerably expand the range of application of the spectral complexity bounds listed in the Appendix. For example, according to the principle the Boolean function $\bigoplus_{i=0}^{k-1} (z^{(i)} \vee z^{(i+k)})$ has the same spectral complexity as the nonrepetitive quadratic form $\bigoplus_{i=0}^{k-1} z^{(i)}z^{(i+k)}$ (mod 2), that is (see Appendix, 13)

$$L^W\left(\bigoplus_{i=1}^{k-1} (z^{(i)} \vee z^{(i+k)})\right) = 2^{2k} \qquad (\text{mod } 2),$$

$$L^H\left(\bigoplus_{i=0}^{k-1} (z^{(i)} \vee z^{(i+k)})\right) = \tfrac{1}{3}(2^{2k} - 1) + k + 1 \qquad (\text{mod } 2). \tag{2.7.7}$$

An analysis of rows 2 through 5 of the Appendix shows that the spectral complexity of conjunction or disjunction increases with the number of arguments, exponentially for the Walsh basis and linearly for the Haar basis; for these functions, therefore, the Haar basis is far more efficient.

On the other hand, for linear functions (Appendix, 6), quadratic forms $\bigoplus_{i,q=0}^{m-1} c^{(i)}d^{(q)}z^{(i)}z^{(q)}$ $(c^{(i)}, d^{(q)} \in \{0, 1\})$ which are products of two linear functions (Appendix, 4), and k-ary forms (i.e., products of k linear functions of m arguments, $k \ll m$; see Appendix, 5), the use of Walsh functions is more efficient, since the Walsh complexity is an exponential function of k and the Haar complexity an exponential function of m.

We now look more closely into the question of complexity bounds of k-ary forms.

Let A be a binary matrix, $A = (a_{is})$ $(i = 1, 2, \ldots, k;\ s = 0, 1, \ldots, m - 1)$ whose rows are linearly independent over $GF(2)$. A function $f(z^{(0)}, \ldots, z^{(m-1)})$ is called a *k-ary form* of m arguments if

$$f(z^{(0)}, \ldots, z^{(m-1)}) = \prod_{i=0}^{k} \bigoplus_{s=0}^{m-1} a_{is}z^{(s)} = \bigoplus_{s_1,\ldots,s_k=0}^{m-1} a_{is_i}z^{(s_i)} \qquad (\text{mod } 2). \tag{2.7.8}$$

EXAMPLE 2.7.1. Let

$$A = \begin{pmatrix} 1 & 1 & 0 & 0 & 0 & 0 \\ 0 & 0 & 1 & 1 & 0 & 0 \\ 0 & 1 & 0 & 0 & 1 & 1 \end{pmatrix}$$

($k = 3, m = 6$). Then

$$f(z) = f(z^{(0)}, \ldots, z^{(5)}) = (z^{(0)} \oplus z^{(1)})(z^{(2)} \oplus z^{(3)})(z^{(1)} \oplus z^{(4)} \oplus z^{(5)}) \qquad (\text{mod } 2).$$

For the function $f(z)$ defined by (2.7.8), we set

$$f_r(z) = \frac{f(z)}{\bigoplus_{s=0}^{m-1} a_{rs} z^{(s)}} = \prod_{\substack{i=1 \\ i \neq r}}^{k} \bigoplus_{s=0}^{m-1} a_{is} z^{(s)} \qquad (\text{mod } 2). \tag{2.7.9}$$

In addition, we let $f_r(z^{(s_1)} = \alpha_{s_1}, z^{(s_2)} = \alpha_{s_2}, \ldots, z^{(s_t)} = \alpha_{s_t})$ denote the subfunction of $f_r(z)$ obtained from $f(z)$ by fixing $z^{(s_1)} = \alpha_{s_1}, z^{(s_2)} = \alpha_{s_2}, \ldots, z^{(s_t)} = \alpha_{s_t}$ $(1 \leqq t \leqq m)$.

In the above example, for instance,

$$f_2(z) = (z^{(0)} \oplus z^{(1)})(z^{(1)} \oplus z^{(4)} \oplus z^{(5)}),$$
$$f_2(z^{(1)} = 0) = z^{(0)}(z^{(4)} \oplus z^{(5)}) \qquad (\text{mod } 2).$$

THEOREM 2.7.2. *Let* $f(z) = \prod_{i=1}^{k} \bigoplus_{s=0}^{m-1} a_{is} z^{(s)}$ (mod 2) *be a k-ary form,* $\min_{1 \leqq i \leqq k} \sum_{s=0}^{m-1} a_{is} = \sum_{s=0}^{m-1} a_{qs}$ *and* $\{s_1, s_2, \ldots, s_t\}$ *all the column indices of the matrix* (a_{is}) *such that* $a_{qs_1} = a_{qs_2} = \cdots = a_{qs_t} = 1$.

Then

$$L^W(f) = 2^k, \tag{2.7.10}$$

$$L^H(f) \leqq L^H(f_q(z^{(s_1)} = 0, z^{(s_2)} = 0, \ldots, z^{(s_t)} = 0))\, 2^{\sum_{s=0}^{m-1} a_{qs} - 1} + 1. \tag{2.7.11}$$

Formulas (2.7.10) and (2.7.11) may be proved by induction on the number k of factors in the k-ary form; the proof of (2.7.10) also involves the convolution theorem (Theorem 1.4.4).

Formula (2.7.11) generates a $(k - 1)$-step recursive procedure which computes a bound for $L^H(f)$, since it essentially reduces estimation of the complexity of a k-ary form f to the same problem for a $(k - 1)$-ary form f_q (which generally depends on fewer variables than f).

To illustrate the procedure, we estimate $L^H(f)$ for the function f of Example 2.7.1. In this case,

$$\min_{1 \leqq i \leqq 3} \sum_{s=0}^{5} a_{is} = \sum_{s=0}^{5} a_{1s} = 2, q = 1, s_1 = 0, s_2 = 1.$$

By (2.7.11),

$$L^H(f) \leqq L^H(f_1(z^{(0)} = 0, z^{(1)} = 0))\, 2^1 + 1,$$

where

$$f_1(z^{(0)} = 0, z^{(1)} = 0) = (z^{(2)} \oplus z^{(3)})(z^{(4)} \oplus z^{(5)}) \qquad (\text{mod } 2).$$

Proceeding in the same way for $L^H(f_1(z^{(0)} = 0, z^{(1)} = 0))$, we get

$$L^H(f_1(z^{(0)} = 0, z^{(1)} = 0)) \leqq L^H(f_{1,2}(z^{(0)} = 0, z^{(1)} = 0, z^{(2)} = 0, z^{(3)} = 0)) \cdot 2 + 1,$$

where

$$f_{1,2}(z^{(0)} = 0, z^{(1)} = 0, z^{(2)} = 0, z^{(3)} = 0) = z^{(4)} \oplus z^{(5)} \quad (\text{mod } 2), \quad L^H(z^{(4)} \oplus z^{(5)}) = 3.$$

Finally, therefore,

$$L^H(f) \leqq 2(2 \cdot 3 + 1) + 1 = 15.$$

Rows 16 to 18 of the Appendix give the Walsh transforms, autocorrelation functions and spectral Walsh and Haar complexities for the signs of the elementary trigonometric functions sign sin $2^{-k}\pi z$, sign cos $2^{-k}\pi z$, sign tan $2^{-k}\pi z$ = sign cot $2^{-k}\pi z$ $(z \in [0, 2^m))$. The sign functions are viewed as step functions representing Boolean functions, and (see (1.1.4)) are assumed to be right continuous, that is to say, $f(z+0) = f(z)$ or $\lim_{\varepsilon \to 0} f(z+\varepsilon) = f(z)$ $(\varepsilon > 0)$.

Rows 16 to 18 illustrate the relationship of the trigonometric functions to the Walsh functions.

Rows 19 to 21 list the transforms, autocorrelation functions and complexities of the unit step and impulse functions, widely used in automatic control, and also of the sign functions for the logarithm. These are again step functions representing Boolean functions. In all these cases the Haar basis is usually more convenient than the Walsh basis, since the Haar complexity is a linear function of the number of arguments and the Walsh complexity an exponential function.

The notation for the transforms in rows 20 and 21 employs the integral Walsh functions $I_\omega(y) = \sum_{z=0}^{y-1} W_\omega(z)$ $(\omega = 0, 1, \ldots, 2^m - 1)$. These may be viewed as piecewise-linear functions; they are linearly independent and therefore, after orthogonalization, may be utilized as a basis for expansion of systems of LF or continuous functions. In that case it is convenient to represent a system of LF not by a step function $\Phi(z)$ but by a piecewise-linear function. Similar considerations apply to the integral Haar functions or Schauder functions $I_i^{(j)}(y) = \sum_{z=0}^{y-1} H_i^{(j)}(z)$ [135].

Chapter Three

SPECTRAL METHODS FOR SYNTHESIS OF NETWORKS REALIZING LOGICAL FUNCTIONS

The present chapter is devoted to methods for the design of digital devices (DD) realizing systems of logical functions (LF). These methods are based on the orthogonal-series expansions of systems of LF described in Chapter I, with the various functional transforms and correlation functions as our main tool.

All the methods described in this chapter for the design of DD realizing systems of LF are also applicable to the construction of algorithms for computer evaluation of LF; in other words, our spectral methods may be interpreted from both the network and the programming standpoint. The synthesis methods are illustrated by network interpretations.

3.1. BLOCK DIAGRAM OF DEVICE REALIZING A SYSTEM OF LOGICAL FUNCTIONS BY SPECTRAL METHODS

3.1.1. Before describing the design procedure, we briefly review the finite orthogonal-series-representations of systems of LF described in Chapter I.

Given a system of LF

$$y^{(s)} = y^{(s)}(z^{(0)}, \dots, z^{(m-1)}) \qquad (s = 0, \dots, k-1; y^{(s)}(z), z^{(s)} \in \{0, \dots, p-1\}), \tag{3.1.1}$$

we define the discrete function $y = f(z)$ corresponding to the system by

$$f(z) = \sum_{s=0}^{k-1} p^{k-1-s} y^{(s)}(z^{(0)}, \dots, z^{(m-1)}), \tag{3.1.2}$$

where

$$z = \sum_{s=0}^{m-1} z^{(s)} p^{m-1-s}. \tag{3.1.3}$$

The function $f(z)$ is defined at all integer points of the half-open interval $[0, p^m)$; we complete $f(z)$ to a step function $\Phi(z)$ by setting

$$\Phi(z) = f(\delta) \quad \text{for} \quad z \in [\delta, \delta + 1). \tag{3.1.4}$$

This step function *represents* system (3.1.1).

Let $\{\Psi_\omega(z)\}$ be one of the complete systems of orthogonal functions described in Chap. I. Expand $\Phi(z)$ in series in terms of $\{\Psi_\omega(z)\}$:

$$\Phi(z) = \sum_{\omega=0}^{p^m-1} S(\omega)\,\Psi_\omega(z), \tag{3.1.5}$$

where the Fourier coefficients $S(\omega)$ are defined by

$$S(\omega) = \left(\sum_{z=0}^{p^m-1} \Psi_\omega(z)\,\overline{\Psi_\omega(z)}\right)^{-1} \sum_{z=0}^{p^m-1} \Phi(z)\,\overline{\Psi_\omega(z)}. \tag{3.1.6}$$

Basing ourselves on formula (3.1.5), we can set up a block-diagram of a device realizing system (3.1.1), as illustrated in Figure 3.1.1. The main blocks are as follows. First we have a basis-function generator, whose input is the argument code z; its output consists of the values $\Psi_\omega(z)$ ($\omega = 0, 1, 2, \ldots, p^m - 1$). The memory stores the values of $S(\omega)$. These data are fed to a multiplier which computes the products of $\Psi_\omega(z)$ and $S(\omega)$, and the sum of the series (3.1.5) is computed by an adder-accumulator. At the end of the summation process, the adder outputs the value of the step function $\Phi(z)$ representing the original system of LF.

The block-diagram of Figure 3.1.1 will be used constantly in what follows, in connection with all spectral methods of synthesis to be described; the complexity and computing time of the physical realizations of the blocks may of course vary from method to method.

The reader should note that according to this method of synthesis the networks realizing the individual functions of the system are minimized jointly; indeed, each component of the device represented by the block-diagram participates in the computation of each function, since the step function $\Phi(z)$ is defined by all the functions in the system. This points to the expediency of spectral methods of synthesis for systems consisting of large numbers of LF.

Although we have included a multiplier in the block-diagram, in the case of most practical importance, namely, realization of systems of Boolean functions using the Walsh and Haar bases, the multiplier becomes superfluous, since the basis functions then take on values -1, $+1$, or 0, ± 1, so that they merely play the role of signs to be ascribed to the coefficients $S(\omega)$ in the adder.

3.1.2. The spectral methods with which we are concerned may be classified according to the following three parameters:

a) The basis $\{\Psi_\omega(z)\}$ used (Walsh or Haar when $p = 2$, Chrestenson or generalized Haar when $p > 2$).

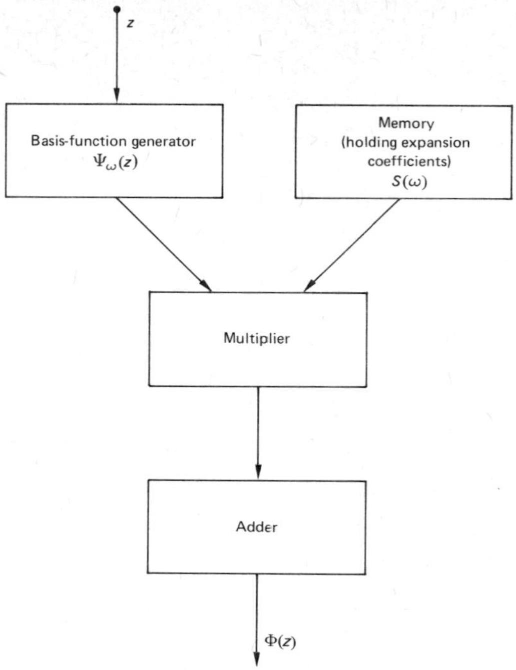

FIGURE 3.1.1.

b) The field over which the arithmetical operations involved in calculating $\Phi(z)$ by formulas (3.1.5) and (3.1.6) are implemented (complex numbers or residues modulo Q; for expansions over the field of residues modulo Q one uses the discrete transforms over Galois fields described in §1.6, and the adder and multiplier must of course implement the appropriate operations modulo Q).

c) Parallel or serial summation of the terms in series (3.1.5).

The question of how to choose the mode of realization according to parameters (b) and (c) will considered in Chapter VI, where appropriate bounds for complexity will also be established.

Our main topic here is the choice of the basis $\{\Psi_\omega(z)\}$. The question as to which basis is more effective in regard to some given system of LF must be settled in each individual case. Our aim here is to provide some signposts indicating the comparative efficiency of one basis or another for certain classes of systems.

3.1.3. When one is using the Walsh or Chrestenson bases, each coefficient in (3.1.5) depends on the values of the system at all points of specification, i.e., on the "global" behavior of $\Phi(z)$. As we know, this is not the case for the Haar bases. For example, in the case of the Haar basis (and the situation for the generalized Haar basis is similar) only the two "leading" coefficients ($c_0^{(0)}, c_0^{(1)}$) (see (1.2.24)) depend on the "global" behavior of $\Phi(z)$. Each of the next two coefficients ($c_1^{(1)}, c_1^{(2)}$) depends on the values of $\Phi(z)$ on one half of the interval; the relevant interval is further contracted as one proceeds along the series, until finally each of the 2^{m-1} lowest-order coefficients ($c_{m-1}^{(1)}, \ldots, c_{m-1}^{(2m-1)}$) (see (1.2.22)) is determined by the values of $\Phi(z)$ at the corresponding two "neighboring" readings. Thus the coefficients of the Haar expansion of $\Phi(z)$ depend essentially only on the "local" behavior of $\Phi(z)$.

Consequently, the use of the Walsh or Chrestenson basis should be recommended when the original system of LF is of a class whose defining properties relate to the behavior of each function at all points of specification. An illustration of this position is the following

THEOREM 3.1.1. *Let $f(z^{(0)}, \ldots, z^{(m-1)})$ be a Boolean function with 2^q anti-self-duality points.* Then*

$$L^W(f) \leqq 2^{m-q}, \tag{3.1.7}$$

where $L^W(f)$ is the number of nonvanishing coefficients in the Walsh expansion of f.

PROOF. Let $\beta_1, \ldots, \beta_q$ be anti-self-duality points which are linearly independent over $GF(2)$. Then

$$f(z) = f(z \oplus \beta_i) = \cdots = f(z \oplus \beta_q) \qquad (\text{mod } 2),$$

and, if $S(\omega)$ is the spectrum of f, it follows from the translation theorem (Theorem 1.4.2) that

$$S(\omega) = W_{\beta_1}(\omega)\, S(\omega) = \cdots = W_{\beta_q}(\omega)\, S(\omega).$$

Hence, if $S(\omega) \neq 0$, then $W_{\beta_1}(\omega) = \cdots = W_{\beta_q}(\omega) = 1$. Using the linear independence of $\beta_1, \ldots, \beta_q$, we see that the number of distinct ω-values satisfying this relation is 2^{m-q}, proving (3.1.7). ∎

According to Theorem 3.1.1, the use of the Walsh basis is advisable when the number of anti-self-duality points of the functions exceeds unity.

By contrast, Haar functions are more convenient when the defining properties of the relevant class of LF are "local," i.e., depend on the values of the function only at a few points. An example is the class of systems of LF represented by step functions $\Phi(z)$ with a small number $V(\Phi)$ of discontinuities. We have

$$V(\Phi) = \sum_{z=0}^{p^m-1} \operatorname{sign} |\Phi(z+1) - \Phi(z)|, \tag{3.1.8}$$

* For the definition of anti-self-duality points and the proof that the number of such points is a power of 2, see §2.1.

where, as we recall,

$$\operatorname{sign}(x) = \begin{cases} 1 & \text{if} \quad x > 0, \\ 0 & \text{if} \quad x \leqq 0. \end{cases}$$

The important factor here is that, as shown by the following theorem, $V(\Phi)$ determines an upper bound for the number $L^H(\Phi)$ of nonvanishing coefficients in the generalized Haar expansion of $\Phi(z)$.

THEOREM 3.1.2. *Let $\Phi(z)$ be a step function representing a system of p-valued logical functions of m arguments. Then*

$$L^H(\Phi) \leqq V(\Phi)(p-1)(m-1) + p. \tag{3.1.9}$$

PROOF. The proof is by induction on $V(\Phi)$: a unit increase in the number of discontinuities implies an increase of at most $(p-1)$ $(m-1)$ in the number of nonvanishing coefficients of functions that take nonzero values only on part of the interval of specification; the second term in (3.1.9) is due to the p coefficients of the basis functions which do not vanish at any point of the interval. ▮

For example, if $p = 2$ and

$$\Phi(z) = \begin{cases} 1 & \text{if} \quad z \geqq a, \\ 0 & \text{otherwise} \end{cases}$$

($\Phi(z)$ represents a single-threshold function; see §2.2), then $V(\Phi) = 1$ and by Theorem 3.1.2 we have $L^H(\Phi) \leqq m + 1$ for any a. By contrast, the number L^W of nonvanishing coefficients of the Walsh expansion of this function $\Phi(z)$ depends on a and may reach as much as 2^m, so that the Haar basis is clearly preferable.

In §3.5.7 we shall describe a method for minimization of $V(\Phi)$ by means of a special linear transformation of the arguments, thus minimizing the upper bound for the complexity $L^H(\Phi)$ of the Haar expansion.

The Appendix lists founds for L^W and L^H for 22 important classes of BF (see the comments on the Appendix in §2.7). The results presented in the Appendix and in §2.7 may also be used when selecting bases for various classes of BF.

3.1.4. We now consider another important feature of Haar expansions, related to their behavior under linear transformations of the arguments in the original system of LF.

Let σ be some permutation of the argument vector $z = (z^{(0)}, \ldots, z^{(m-1)})$. The image of z under the permutation σ will be denoted by

$$\sigma \otimes z = (z^{(\sigma(0))}, z^{(\sigma(1))}, \ldots, z^{(\sigma(m-1))}).$$

The discrete function $y = f_\sigma(z)$ corresponding to the "permuted" system of LF is (see (3.1.1)–(3.1.3))

$$y = f_\sigma(z) = \sum_{s=0}^{k-1} p^{k-1-s} y^{(s)}(z^{(0)}, \ldots, z^{(m-1)}), \tag{3.1.10}$$

where

$$z = \sum_{s=0}^{m-1} z^{(\sigma(s))} 2^{m-1-s}. \tag{3.1.11}$$

The function $f(z)$ defined by (3.1.2), (3.1.3) represents the original system (3.1.1) for the identity permutation $\sigma(s) = s$ $(s = 0, 1, \ldots, m-1)$.

The step-function completion of $f_\sigma(z)$ (see (3.1.4)) will be denoted by $\Phi_\sigma(z)$; we shall say that $\Phi_0(z)$ *represents system* (3.1.1) *for the permutation* σ.

Letting σ vary over different permutations and expanding the corresponding functions $\Phi_\sigma(z)$ in series as in (3.1.5), one generally obtains different expansion coefficients and accordingly different complexities of realization. Since the realization of a permutation requires no special hardware, one is thus presented with the problem of finding an optimal permutation, minimizing the number L^H of nonvanishing coefficients, since L^H determines the "size" of the most complex component of the block diagram in Figure 3.1.1—the coefficient-storage block.

It follows from Theorem 1.4.5 that when the Walsh and Chrestenson bases are used the number of nonvanishing coefficients is independent of permutations of the arguments. The situation is quite different for the Haar bases.

EXAMPLE 3.1.1. Consider the function $f(z^{(0)}, z^{(1)}, z^{(2)})$ of Table 3.1.1. Also shown in the table are the discrete functions $f(z)$ and $f_\sigma(z)$ for $\sigma \otimes z = (z^{(0)}, z^{(2)}, z^{(1)})$. One sees that $L^H = 5$ for $\Phi(z)$, but $L^H = 3$ for $\Phi_\sigma(z)$.

TABLE 3.1.1.

$z^{(0)}$	$z^{(1)}$	$z^{(2)}$	z	$f(z^{(0)}, z^{(1)}, z^{(2)})$	$f(z)$	$f_\sigma(z)$
0	0	0	0	0	0	0
0	0	1	1	1	1	0
0	1	0	2	0	0	1
0	1	1	3	1	1	1
1	0	0	4	1	1	1
1	0	1	5	0	0	1
1	1	0	6	1	1	0
1	1	1	7	0	0	0

This illustrates the dependence of L^H on permutations of the argument vector; a similar situation obtains with regard to the generalized Haar functions.

In §§3.2 and 3.4, in a more general setting, we shall present an analytical solution of the problem of the optimal permutation of arguments (i.e., minimizing L^H) for the Haar bases, in the case of completely specified LF. The case of incompletely specified LF will be treated in §§3.3 and 3.4.

Permutations of the argument vector thus considerably enhance the advantage of Haar bases over Walsh and Chrestenson bases as tools for the synthesis of network realizations of systems of LF.

3.1.5. The Haar bases possess other advantages in regard to the physical realization of Figure 3.1.1.

It was shown in §§1.2.4 and 1.3.4 that when one is calculating $\Phi(z^*)$ for any fixed z^* the number of nonvanishing terms of the series is at most $(p-1)m+1$ for the generalized Haar functions ($m+1$ for the ordinary Haar functions). Thus, if the adder is of serial type, the number of additions performed in summation of the series is at most $(p-1)m+1$ for the generalized Haar functions, whereas the corresponding figure for the Chrestenson functions may reach p^m. (There is also an advantage in parallel summation: for the generalized Haar functions the adder need not have more than $(p-1)m+1$ inputs, while for the Chrestenson functions one may need as many as p^m inputs.)

The advantages of using Haar bases in the synthesis of logical networks are thus evident. In the sequel, therefore, our discussions of spectral methods for synthesis will be concerned for the most part with the Haar bases and the generalized Haar functions; Walsh and Chrestenson expansions will usually be used only as an intermediate but important auxiliary tool.

3.2. OPTIMAL ORDERING OF ARGUMENTS FOR HAAR EXPANSIONS OF SYSTEMS OF BOOLEAN FUNCTIONS

3.2.1. We showed in §3.1.4 that the number L^H of nonvanishing coefficients, and hence also the complexity of the realizing network, depends on the order of the arguments when Haar bases are employed, whereas the corresponding figure for the Walsh and Chrestenson bases is independent of the order. Our immediate goal is therefore to find the optimal order of the arguments in a system of completely specified BF, minimizing L^H.

Any order of arguments in a system of BF of m arguments may be described by an $m \times m$ matrix $\sigma = (\sigma_{is})$, where $\sigma_{is} \in \{0, 1\}$ and $\sum_{i=1}^{m} \sigma_{is} = \sum_{s=1}^{m} \sigma_{is} = 1$, i.e., each row (column) of σ contains exactly one unit. Then the argument vector z_σ defined by the matrix is obtained by multiplying $z = (z^{(0)}, \ldots, z^{(m-1)})$ by σ, i.e.,

$$z_\sigma = \sigma \otimes z \qquad (\text{mod } 2). \qquad (3.2.1)$$

For example, if

$$\sigma = \begin{pmatrix} 1 & 0 & 0 \\ 0 & 0 & 1 \\ 0 & 1 & 0 \end{pmatrix} \qquad (m = 3),$$

then $z_\sigma = \sigma \otimes z = (z^{(0)}, z^{(2)}, z^{(1)})$. To indicate the dependence of L^H on σ we shall write $L^H(\sigma)$.

We may thus state the problem of the optimal order of arguments for the Haar basis as follows: find a matrix σ minimizing $L^H(\sigma)$. In §§3.2.3–3.2.5 we shall solve a more general problem, in which σ may be any nonsingular matrix of ones and zeros. For example, if

$$\sigma = \sigma^* = \begin{pmatrix} 0 & 1 & 1 & 0 \\ 1 & 0 & 0 & 1 \\ 0 & 0 & 1 & 1 \\ 1 & 1 & 1 & 0 \end{pmatrix} \qquad (m = 4),$$

then

$$z_{\sigma^*} = (z^{(1)} \oplus z^{(2)}, z^{(0)} \oplus z^{(3)}, z^{(2)} \oplus z^{(3)}, z^{(0)} \oplus z^{(1)} \oplus z^{(2)}) \qquad (\text{mod } 2).$$

Recall that σ is *nonsingular* if its determinant $|\sigma|_2$ over the field $GF(2)$ does not vanish. Let Ξ denote the class of all nonsingular matrices. Our generalized problem may be formulated as follows:

Find a nonsingular matrix $\sigma_{\min} \in \Xi$ such that

$$\min_{\sigma \in \Xi} L^H(\sigma) = L^H(\sigma_{\min}). \qquad (3.2.2)$$

3.2.2. When using a linear transformation of the arguments, we may represent the block diagram of a device realizing the relevant system of BF as a series connection of two blocks, the first (σ) implementing the linear transformation of the argument vector, the second of the type shown in Figure 3.1.1, realizing the function $f_\sigma(z)$, where

$$f_\sigma(\sigma \otimes z) = f(z) \qquad (\text{mod } 2).$$

The σ-block may be based on mod 2 adders.

Before proceeding to methods for determination of $\sigma_{\min}$, we estimate the complexity of the block implementing multiplication by an arbitrary matrix σ, in terms of the number $N^{(2)}(\sigma)$ of two-input mod 2 adders. It is obvious that if $\sum_{i=1}^{m} \sigma_{is} = \sum_{s=1}^{m} \sigma_{is} = 1$ then $N^{(2)}(\sigma) = 0$, since then multiplication by σ is simply permutation of the components of z. An upper bound for $N^{(2)}(\sigma)$ in the general case is given by the next theorem.

THEOREM 3.2.1. *For any* $\sigma = (\sigma_{is})$ $(\sigma_{is} \in \{0, 1\}, s = 0, 1, \ldots, m-1)$,

$$N^{(2)}(\sigma) \leqq \min_{a \in \{1,2,\ldots,m\}} \left(\left(2^{]m/a[} - \left]\frac{m}{a}\right[- 1 \right) \mathrm{res}_a m \right.$$

$$\left. + \left(2^{[m/a]} - \left[\frac{m}{a}\right] - 1 \right) (a - \mathrm{res}_a m) + (a-1)\, m \right), \quad (3.2.3)$$

where $\mathrm{res}_a m$ *denotes the remainder upon division of m by a, and* $[x]$ $(]x[)$ *is the largest (smallest) integer* $\leqq x$ $(\geqq x)$.

For the proof, we shall need the following lemma, which is also of independent interest.

Lemma 3.2.1. *Let* $L_{\oplus}^{(2)}(n)$ *be the minimal number of two-input* mod 2 *adders necessary to implement all linear Boolean functions of n arguments, i.e., functions of the form* $\bigoplus_{i=0}^{n-1} c^i z^{(i)}$ $(c_i \in \{0, 1\})$ (mod 2). *Then*

$$L_{\oplus}^{(2)}(n) = 2^n - n - 1. \quad (3.2.4)$$

PROOF. Realization of the functions $0, z^{(0)}, \ldots, z^{(n-1)}$ requires no hardware. Let $R^{(2)}(n)$ be the number of linear BF of n arguments that depend essentially on more than one argument. Then $R^{(2)}(n) \leqq L_{\oplus}^{(2)}(n)$ and, since $R^{(2)}(n) = 2^n - n - 1$,

$$L_{\oplus}^{(2)}(n) \geqq 2^n - n - 1. \quad (3.2.5)$$

We now prove by induction on n that

$$L_{\oplus}^{(2)}(n) \leqq 2^n - n - 1. \quad (3.2.6)$$

Suppose we have a network realizing all linear BF of $n - 1$ arguments. We can obtain all functions depending on the n-th argument, with the exception of the function $z^{(n-1)}$, by connecting one more adder to each output of the appropriate network. Since the number of such functions is $2^{n-1} - 1$, we have

$$L_{\oplus}^{(2)}(n) \leqq L_{\oplus}^{(2)}(n-1) + 2^{n-1} - 1. \quad (3.2.7)$$

Hence (3.2.6) is true for all n. The assertion now follows from (3.2.5) and (3.2.6). ▌

Lemma 3.2.1 is readily generalized to linear p-valued logical functions $(p \geqq 2)$. If $L_{\oplus}^{(p)}(n)$ is the minimal number of two-input mod p adders necessary to implement all linear p-valued logical functions, then

$$L_{\oplus}^{(p)}(n) = p^n - n - 1. \quad (3.2.8)$$

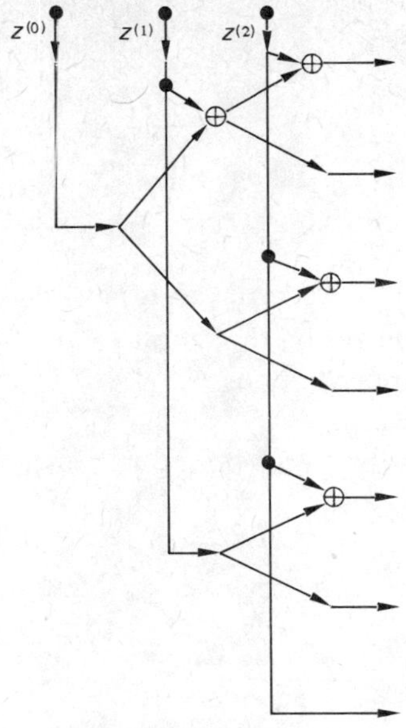

FIGURE 3.2.1.

Figure 3.2.1 illustrates a minimal network realizing all linear BF of 3 arguments (except the identically zero function) in accordance with the proof of Lemma 3.2.1.

The method thus provided for synthesis of minimal linear networks enables us to synthesize minimal generators for the Walsh or Chrestenson functions, in view of the isomorphism between the linear LF and the Walsh or Chrestenson functions.

PROOF OF THEOREM 3.2.1. We first observe that for any natural numbers a and m, $a \leqq m$,

$$m = \left[\frac{m}{a}\right](a - \mathrm{res}_a m) + \left]\frac{m}{a}\right[\mathrm{res}_a m. \tag{3.2.9}$$

In accordance with this formula, we partition the $m \times m$ matrix σ into a submatrices, of which each of the first $a - \mathrm{res}_a m$ containing $[m/a]$ columns of σ, and each of the remaining $\mathrm{res}_a m$ containing $]m/a[$ columns of σ. A network realizing multiplication by σ may now be constructed from a networks multiplying by each of the submatrices, in series with a network summing their outputs modulo 2. Each of the submatrix networks implements linear BF of $[m/a]$ or $]m/a[$ arguments, respectively, so that the corresponding complexities are $L_{\oplus}^{(2)}([m/a])$ or $L_{\oplus}^{(2)}(]m/a[)$. Summation of the outputs of these networks requires at most $(a - 1)\,m$ mod 2 adders. Thus, by Lemma 3.2.1, for any matrix σ,

$$N^{(2)}(\sigma) \leqq L_{\oplus}^{(2)}\left(\left[\frac{m}{a}\right]\right)(a - \operatorname{res}_a m) + L_{\oplus}^{(2)}\left(\left]\frac{m}{a}\right[\right)\operatorname{res}_a m + (a-1)m$$

$$= \left(2^{[m/a]} - \left[\frac{m}{a}\right] - 1\right)(a - \operatorname{res}_a m) + \left(2^{]m/a[} - \left]\frac{m}{a}\right[- 1\right)\operatorname{res}_a m + (a-1)m. \tag{3.2.10}$$

Since a may be selected arbitrarily, this implies the conclusion of our theorem. ▮

Table 3.2.1 lists lowest upper bounds $\tilde{N}^{(2)}$ for $N^{(2)}(\sigma)$ as given by (3.2.3), and the corresponding values of $a = a_{\text{opt}}$.

A fairly sharp approximation to the bound $\tilde{N}^{(2)}$ is given by

$$\tilde{N}^{(2)} \cong \frac{2m^2}{\log_2 m} - 2m - \frac{m}{\log_2 m}. \tag{3.2.11}$$

This estimate follows from Theorem 3.2.1 by setting $a = \,]m/\log_2 m[$.

It can be shown that the bound (3.2.3) is asymptotically optimal ($m \to \infty$) for almost all matrices σ (i.e., the relative abundance of matrices for which this statement is false tends to zero as $m \to \infty$). The asymptotically optimal value of the parameter a is

$$Q_{\text{opt}} = \left[\frac{m}{\log_2 m - \log_2 \log_2 m}\right], \tag{3.2.12}$$

and for any σ

$$N^{(2)}(\sigma) \sim \frac{m^2}{\log_2 m}; \tag{3.2.13}$$

in other words, the quotient of the number of two-input mod 2 adders needed for a best realization of any $m \times m$ matrix σ by the quantity $m^2/\log_2 m$ tends to unity as $m \to \infty$, for almost all matrices σ.

TABLE 3.2.1.

m	1	2	3	4	5	6	7	8	9	10	11	12	13	14	15	16	17	18	19	20
a_{opt}	1	1	1	2	2	2	3	3	3	3	4	4	4	4	5	5	5	5	5	5
$\tilde{N}^{(2)}$	0	1	4	6	10	14	20	25	30	39	46	52	62	72	80	91	102	113	124	135

Figure 3.2.2 illustrates a network implementing multiplication by

$$\sigma = \begin{pmatrix} 0 & 1 & 1 & 0 \\ 1 & 0 & 0 & 1 \\ 0 & 0 & 1 & 1 \\ 1 & 1 & 1 & 0 \end{pmatrix}$$

in accordance with the proof of Theorem 3.2.1. Here $m = 4$ and, as is evident from Table 3.2.1, $a_{\mathrm{opt}} = 2$.

Theorem 3.2.1 may also be generalized to matrices σ with elements in $\{0, 1, \ldots, p-1\}$, and also to the case of rectangular matrices σ. In these cases again the resulting estimate is asymptotically optimal for almost all matrices σ.

3.2.3. We can now tackle our main problem—to find a nonsingular matrix $\sigma_{\min}$ that minimizes the number $L^H(\sigma)$ of Haar expansion coefficients. Since the upper bound for the complexity of the block realizing σ is of the order of $m^2/\log_2 m$, while the complexity of the entire network is in general an exponential function of m, minimization of the whole network should be approached via minimization of only the nonlinear part of the network, in terms of the number of nonvanishing coefficients $L^H(\sigma)$.

The definition and properties of the Haar basis were considered in detail in § 1.2. Recall that the expansion coefficients of a discrete function $f_\sigma(z)$ in terms of the Haar basis are defined by

$$c_{m-1}^{(q)} = 2^{-1}(f_\sigma(2q-2) - f_\sigma(2q-1)) \qquad (q = 1, \ldots, 2^{m-1}), \tag{3.2.14}$$

. . . .

$$c_{m-l}^{(q)} = 2^{-l}\left(\sum_{i=2^{l-1}+1}^{2^l} f_\sigma(2^l q - i) - \sum_{i=1}^{2^{l-1}} f_\sigma(2^l q - i)\right) \quad (q = 1, \ldots, 2^{m-l}), \tag{3.2.15}$$

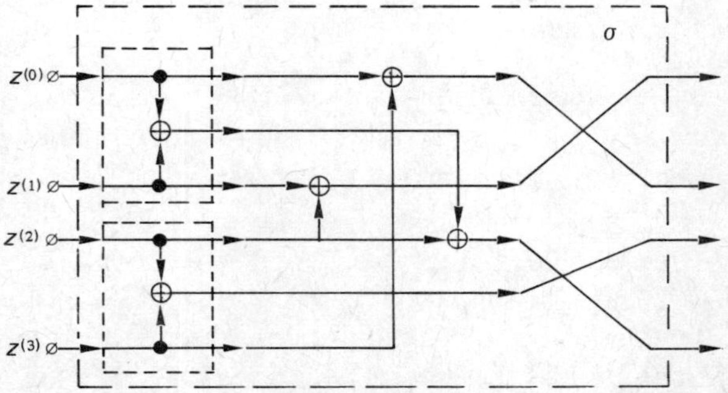

FIGURE 3.2.2.

where

$$f_\sigma(z_\sigma) = f(z) \quad \text{and} \quad z_\sigma = \sigma \otimes z \qquad (\text{mod } 2). \tag{3.2.16}$$

Let $L^H_{m-l}(\sigma)$ denote the number of nonvanishing coefficients with subscript $m - l$. Then it follows from (3.2.14), (3.2.15) that for any σ

$$\begin{gathered} 0 \leqq L^H_{m-1}(\sigma) \leqq 2^{m-1}, \\ \cdots\cdots\cdots\cdots\cdots\cdots \\ 0 \leqq L^H_{m-l}(\sigma) \leqq 2^{m-l} \qquad (l = 1, \ldots, m-1), \end{gathered} \tag{3.2.17}$$

and

$$L^H(\sigma) = \sum_{l=1}^{m-1} L^H_{m-l}(\sigma) \leqq 2^m. \tag{3.2.18}$$

Thus our search for $\sigma_{\min}$ may be organized recursively along the following lines. Let $\Xi_m = \Xi$ be the class of all nonsingular matrices over $GF(2)$.

1) Find the class $\Xi_{m-1} \subseteqq \Xi_m$ of matrices σ such that

$$L^H_{m-1}(\sigma_{m-1}) = \min_{\sigma \in \Xi_m} L^H_{m-1}(\sigma), \tag{3.2.19}$$

where σ_{m-1} is an arbitrary matrix in Ξ_{m-1}.

2) Find the class $\Xi_{m-2} \subseteqq \Xi_{m-1}$ such that

$$L^H_{m-2}(\sigma_{m-2}) = \min_{\sigma \in \Xi_{m-1}} L^H_{m-2}(\sigma), \tag{3.2.20}$$

where σ_{m-2} is an arbitrary matrix in Ξ_{m-2}.

....

l) Find the class $\Xi_{m-l} \subseteqq \Xi_{m-l+1}$ ($l = 1, 2, \ldots, m-1$) such that

$$L^H_{m-l}(\sigma_{m-l}) = \min_{\sigma \in \Xi_{m-l+1}} L^H_{m-l}(\sigma), \tag{3.2.21}$$

where σ_{m-l} is an arbitrary matrix in Ξ_{m-l}.

We may now take any matrix in Ξ_1 as $\sigma_{\min}$. This matrix makes $L^H_{m-1}(\sigma)$ assume its absolute minimum, while the numbers $L^H_{m-l}(\sigma)$ ($l = 2, 3, \ldots, m-1$) take certain local minima.

3.2.4. We now consider how to find the class of matrices Ξ_{m-1} minimizing $L^H_{m-1}(\sigma)$.

First let $\sigma = E_m$ be the identity matrix of order m, and denote the discrete function $f_{E_m}(z) = f(z)$ corresponding to the prescribed system of k Boolean functions of m arguments by $f^{(m-1)}(z)$; define the characteristic functions $f_i^{(m-1)}(z)$ ($i = 0, 1, \ldots, 2^k - 1$)

by

$$f_i^{(m-1)}(z) = \begin{cases} 1 & \text{if } f^{(m-1)}(z) = i, \\ 0 & \text{if } f^{(m-1)}(z) \neq i. \end{cases} \tag{3.2.22}$$

EXAMPLE 3.2.1. Table 3.2.2 defines a system of two BF $y^{(0)}$, $y^{(1)}$ of four arguments, showing also the discrete function $f^{(m-1)}(z)$ and the system of characteristic functions $\{f_i^{(m-1)}(z)\}$ $(i = 0, 1, 2, 3)$.

For each characteristic function $f_i^{(m-1)}(z)$ we have an autocorrelation function $B_i^{(m-1)}(\tau)$:

$$B_i^{(m-1)}(\tau) = \sum_{z=0}^{2^m-1} f_i^{(m-1)}(z)\, f_i^{(m-1)}(z \oplus \tau) \qquad (\text{mod } 2). \tag{3.2.23}$$

The corresponding functions for the example are shown in Table 3.2.2. (The properties of the autocorrelation functions and their relationship to double Walsh transforms were considered in § 1.5.)

The desired class Ξ_{m-1} is determined by the following theorem.

THEOREM 3.2.2. *Let*

$$B^{(m-1)}(\tau) = \sum_{i=0}^{2^k-1} B_i^{(m-1)}(\tau).$$

TABLE 3.2.2.

$z^{(0)}$	$z^{(1)}$	$z^{(2)}$	$z^{(3)}$	$y^{(0)}$	$y^{(1)}$	z, τ	$f^{(3)}(z)$	$f_i^{(3)}(z)$ $i=0$	1	2	3	$B_i^{(3)}(z)$ $i=0$	1	2	3	$B^{(3)}(\tau) = \sum_{i=0}^{3} B_i^{(3)}(\tau)$	$f_{\sigma_3}^{(3)}(z)$
0	0	0	0	0	0	0	0	1	0	0	0	4	4	4	4	16	0
0	0	0	1	1	0	1	2	0	0	1	0	2	0	0	2	4	0
0	0	1	0	1	1	2	3	0	0	0	1	2	2	0	0	4	0
0	0	1	1	1	1	3	3	0	0	0	1	2	2	0	0	4	3
0	1	0	0	0	1	4	1	0	1	0	0	0	0	2	2	4	1
0	1	0	1	1	0	5	2	0	0	1	0	0	0	2	2	4	1
0	1	1	0	0	1	6	1	0	1	0	0	0	0	0	0	0	1
0	1	1	1	1	1	7	3	0	0	0	1	0	0	0	0	0	2
1	0	0	0	1	1	8	3	0	0	0	1	0	0	0	0	0	2
1	0	0	1	0	1	9	1	0	1	0	0	0	0	0	0	0	1
1	0	1	0	0	1	10	1	0	1	0	0	0	0	2	2	4	3
1	0	1	1	1	0	11	2	0	0	1	0	0	0	2	2	4	3
1	1	0	0	0	0	12	0	1	0	0	0	2	2	0	0	4	0
1	1	0	1	0	0	13	0	1	0	0	0	2	2	0	0	4	3
1	1	1	0	1	0	14	2	0	0	1	0	0	2	2	0	4	2
1	1	1	1	0	0	15	0	1	0	0	0	2	2	2	2	8	2

Then Ξ_{m-1} is the set of all matrices σ_{m-1} such that

$$\sigma_{m-1} \otimes \tau_{m-1} = \begin{pmatrix} 0 \\ \vdots \\ 0 \\ 1 \end{pmatrix} \pmod 2, \tag{3.2.24}$$

where

$$\max_{\tau \neq 0} B^{(m-1)}(\tau) = B^{(m-1)}(\tau_{m-1}). \tag{3.2.25}$$

PROOF. Let $\bar{L}^H_{m-1}(\sigma)$ denote the number of *vanishing* coefficients with subscript $m-1$ $(\bar{L}^H_{m-1}(\sigma) = 2^{m-1} - L^H_{m-1}(\sigma))$. Setting $f^{(m-1)}_\sigma(\sigma \otimes z) = f^{(m-1)}(z)$, we denote the characteristic functions of $f^{(m-1)}_\sigma(z)$ by $f^{(m-1)}_{i,\sigma}(z)$, and the corresponding autocorrelation functions by $B^{(m-1)}_{i,\sigma}(\tau)$. Then by (3.2.14) $\bar{L}^H_{m-1}(\sigma)$ is equal to the number of pairs of identical values of $f^{(m-1)}_\sigma(z)$ at points of the type $z = 2s$, $z = 2s+1$ $(s = 0, 1, \ldots, 2^{m-1} - 1)$, i.e., the number of pairs of units for all functions $f^{(m-1)}_{i,\sigma}(z)$ at points $z = 2s$, $z = 2s + 1$ $(i = 0, 1, \ldots, 2^k - 1)$. Thus the required number is $\frac{1}{2}\sum_{i=0}^{2k-1} B^{(m-1)}_{i,\sigma}(1)$. In view of Theorem 1.5.4, this implies the desired assertion. ∎

It follows from Theorem 3.2.2 that, in order to determine Ξ_{m-1}, it is sufficient to find the maximum of the total autocorrelation function $B^{(m-1)}(\tau) = \sum_{i=0}^{2k-1} B^{(m-1)}_i(\tau)$.

Going back to Example 3.2.1, we find the function $B^{(3)}(\tau) = \sum_{i=0}^{3} B^{(3)}_i(\tau)$ as specified in Table 3.2.2; we have $\tau_{m-1} = \tau_3 = 15$, and by Theorem 3.2.2 we may put

$$\sigma_{m-1} = \sigma_3 = \begin{pmatrix} 1 & 1 & 1 & 1 \\ 0 & 0 & 1 & 1 \\ 1 & 0 & 0 & 1 \\ 1 & 1 & 1 & 0 \end{pmatrix}.$$

The function $f^{(3)}_{\sigma_3}(z)$ is also shown in Table 3.2.2. For this example, we have $L^H_3(E_4) = 6$, $L^H_3(\sigma_3) = 4$.

3.2.5. We now determine the classes $\Xi_{m-2} \supseteqq \Xi_{m-3} \supseteqq \cdots \supseteqq \Xi_1$. To determine Ξ_{m-2}, we construct the function $f^{(m-1)}_{\sigma_{m-1}}(z)$ for $f^{(m-1)}(z)$ (see (3.2.16)), hence obtaining the "contracted" function $f^{(m-2)}(z)$ defined at 2^{m-1} points:

$$f^{(m-2)}(z) = f^{(m-1)}_{\sigma_{m-1}}(2z) + f^{(m-1)}_{\sigma_{m-1}}(2z+1). \tag{3.2.26}$$

We now apply the algorithmic procedure of subsection 3.2.4 to $f^{(m-2)}(z)$, and denote the resulting $(m-1) \times (m-1)$ matrix by $\sigma^{(m-2)}$.

THEOREM 3.2.3. *Ξ_{m-2} is the set of all matrices σ_{m-2} such that*

$$\sigma_{m-2} = \left(\begin{array}{c|c} \sigma^{(m-2)} & 0 \\ \hline 0 & 1 \end{array}\right) \otimes \sigma_{m-1} \qquad (\text{mod } 2) \tag{3.2.27}$$

where $\sigma_{m-1} \in \Xi_{m-1}$.

PROOF. It follows from (3.2.27) that the last $((m-1)\text{-th})$ components of the vectors $\sigma_{m-2} \otimes z$ and $\sigma_{m-1} \otimes z$ (mod 2) are the same for any z, and so

$$L^H_{m-1}(\sigma_{m-2}) = L^H_{m-1}(\sigma^{(q)}_{m-1}).$$

In addition, by (3.2.15) and (3.2.26) all the coefficients $c^{(q)}_{m-l}(l > 1)$ of the functions $f^{(m-1)}_{\sigma_{m-1}}$ and $f^{(m-2)}$ coincide. ∎

The construction of the classes $\Xi_{m-3}, \Xi_{m-4}, \ldots, \Xi_1$ is analogous; the final result is

$$\sigma_{\min} = \left(\begin{array}{c|c} \sigma^{(1)} & 0 \\ \hline 0 & E_{m-2} \end{array}\right) \otimes \cdots \otimes \left(\begin{array}{c|c} \sigma^{(m-2)} & 0 \\ \hline 0 & E_1 \end{array}\right) \otimes \sigma_{m-1} \qquad (\text{mod } 2). \tag{3.2.28}$$

Continuing our treatment of Example 3.2.1, we construct the classes $\Xi_{m-2} = \Xi_2, \Xi_1$. The "contracted" function $f^{(m-2)}(z) = f^{(2)}(z)$, the total autocorrelation function $B^{(2)}(\tau)$ for $f^{(2)}(z)$, and the transformed function $f^{(2)}_{\sigma^{(2)}}(z)$ are given in Table 3.2.3; we have

$$\sigma^{(m-2)} = \sigma^{(2)} = \begin{pmatrix} 1 & 0 & 0 \\ 0 & 0 & 1 \\ 0 & 1 & 0 \end{pmatrix}.$$

The functions $f^{(1)}(z)$, $B^{(1)}(\tau)$ and $f^{(1)}_{\sigma^{(1)}}(z)$ are also shown in Table 3.2.3; we have

$$\sigma^{(m-3)} = \sigma^{(1)} = \begin{pmatrix} 1 & 1 \\ 0 & 1 \end{pmatrix}.$$

Finally, using (3.2.28), we see that the minimizing matrix is

$$\sigma_{\min} = \begin{pmatrix} 1 & 1 & 0 & 0 \\ 0 & 1 & 0 & 0 \\ 0 & 0 & 1 & 0 \\ 0 & 0 & 0 & 1 \end{pmatrix} \otimes \begin{pmatrix} 1 & 0 & 0 & 0 \\ 0 & 0 & 1 & 0 \\ 0 & 1 & 0 & 0 \\ 0 & 0 & 0 & 1 \end{pmatrix} \otimes \begin{pmatrix} 1 & 1 & 1 & 1 \\ 0 & 0 & 1 & 1 \\ 1 & 0 & 0 & 1 \\ 1 & 1 & 1 & 0 \end{pmatrix} = \begin{pmatrix} 0 & 1 & 1 & 0 \\ 1 & 0 & 0 & 1 \\ 0 & 0 & 1 & 1 \\ 1 & 1 & 1 & 0 \end{pmatrix} \qquad (\text{mod } 2).$$

A network realization of $\sigma_{\min}$ is shown in Figure 3.2.2.

TABLE 3.2.3.

z, τ	$f^{(2)}(z)$	$B^{(2)}(\tau)$	$f^{(2)}_{\sigma^{(2)}}(z)$	$f^{(1)}(z)$	$B^{(1)}(\tau)$	$f^{(1)}_{\sigma^{(1)}}(z)$
0	0	8	0	2	4	2
1	3	0	2	6	0	10
2	2	4	3	6	0	6
3	3	0	3	10	2	6
4	3	0	3			
5	6	4	3			
6	3	0	6			
7	4	4	4			

TABLE 3.2.4.

z, t	$f(z)$	$16c(t)$	$f_{\sigma_{\min}}(z)$	$16c_{\sigma_{\min}}(t)$
0	0	24	0	24
1	2	6	0	0
2	3	2	1	−16
3	3	10	1	0
4	1	−16	3	−8
5	2	−4	3	8
6	1	4	2	0
7	3	−8	2	0
8	3	−16	2	0
9	1	0	1	0
10	1	−8	0	0
11	2	−16	3	0
12	0	16	0	8
13	0	−8	3	−24
14	2	0	1	−24
15	0	16	2	−8

Table 3.2.4 gives the values of the original function $f(z)$, its Haar spectrum $c(t) = c_l^{(q)}$, where $t = 2^l + q - 1$ (note that q and l are uniquely determined by this condition for any given t), and also $f_{\sigma_{\min}}(z)$ and the corresponding Haar spectrum $c_{\sigma_{\min}}(t)$. It is evident from the table that in this example the linear transformation $\sigma_{\min}$ yields a savings of approximately 45% in the number of nonvanishing coefficients.

Figure 3.2.3 is a block-diagram of our algorithm for determination of the optimal linear transformation $\sigma_{\min}$, relative to the Haar basis and a system of completely specified Boolean functions. The notation $a := b$ used in the figure signifies assignment of the value b to the variable a.

The above algorithm for computation of $\sigma_{\min}$ involves $m - 1$ iterations of the

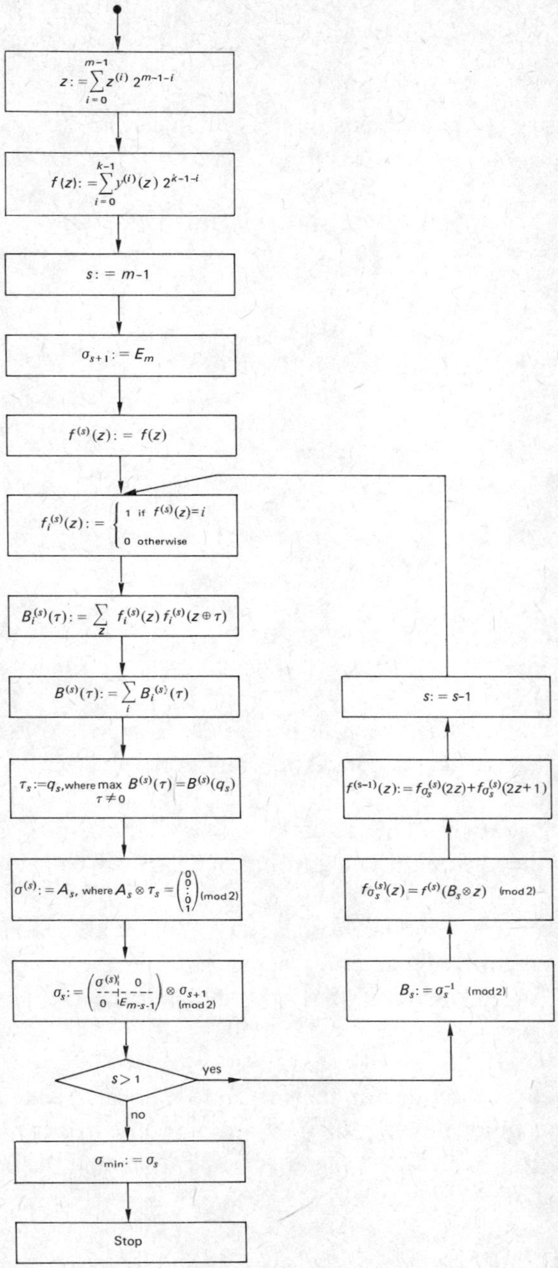
$z := \sum_{i=0}^{m-1} z^{(i)} 2^{m-1-i}$
$f(z) := \sum_{i=0}^{k-1} y^{(i)}(z) 2^{k-1-i}$
$s := m-1$
$\sigma_{s+1} := E_m$
$f^{(s)}(z) := f(z)$
$f_i^{(s)}(z) := \begin{cases} 1 \text{ if } f^{(s)}(z)=i \\ 0 \text{ otherwise} \end{cases}$
$B_i^{(s)}(\tau) := \sum_z f_i^{(s)}(z) f_i^{(s)}(z \oplus \tau)$
$B^{(s)}(\tau) := \sum_i B_i^{(s)}(\tau)$
$\tau_s := q_s$, where $\max_{\tau \neq 0} B^{(s)}(\tau) = B^{(s)}(q_s)$
$\sigma^{(s)} := A_s$, where $A_s \otimes \tau_s = \begin{pmatrix} 0 \\ 0 \\ \vdots \\ 0 \\ 1 \end{pmatrix}$ (mod 2)
$\sigma_s := \begin{pmatrix} \sigma^{(s)} & 0 \\ 0 & E_{m-s-1} \end{pmatrix} \otimes \sigma_{s+1}$ (mod 2)
$s > 1$
yes
no
$\sigma_{\min} := \sigma_s$
Stop
$B_s := \sigma_s^{-1}$ (mod 2)
$f\sigma_s^{(s)}(z) = f^{(s)}(B_s \otimes z)$ (mod 2)
$f^{(s-1)}(z) := f\sigma_s^{(s)}(2z) + f\sigma_s^{(s)}(2z+1)$
$s := s-1$

FIGURE 3.2.3.

maximization procedure for the total autocorrelation function

$$B^{(m-l)}(\tau) = \sum_i \sum_z f_i^{(m-l)}(z)\, f_i^{(m-l)}(z \oplus \tau) \qquad (\text{mod } 2) \qquad (l = 1, \ldots, m-1).$$

Instead, one can use $m - 1$ iterations of the minimization procedure for the *total cross-correlation functions*

$$A^{(m-l)}(\tau) = \sum_{i \neq s} \sum_z f_i^{(m-l)}(z)\, f_s^{(m-l)}(z \oplus \tau) \qquad (\text{mod } 2) \qquad (l = 1, \ldots, m-1).$$

This follows from the following theorem.

THEOREM 3.2.4. *Let* $\{f_i(z)\}$ *be a system of discrete functions,* $f_i(z) \in \{0, 1\}$, $z \in \{0, 1, \ldots, p^m - 1\}$, $i = 0, 1, \ldots, p^k - 1$, *and*

$$B(\tau) = \sum_i \sum_z f_i(z)\, f_i(z \ominus \tau) \qquad (\text{mod } p),$$

$$A(\tau) = \sum_{i \neq s} \sum_z f_i(z)\, f_s(z \ominus \tau) \qquad (\text{mod } p).$$

Then

$$B(\tau) + A(\tau) = p^m. \tag{3.2.29}$$

The only reason that we have been using the autocorrelation functions $B^{(m-l)}(\tau)$ throughout, rather than the cross-correlation functions $A^{(m-l)}(\tau)$, is that the former are easier to calculate.

It is worth noting that the same remark as to possible use of cross-correlation functions applies to the linearization algorithm of §2.4.3, which also made use of the total autocorrelation function.

To end this subsection, we emphasize once again that the autocorrelation functions $B_i(\tau)$ are most conveniently calculated by using their representation in terms of the double Walsh transform (Theorem 1.5.1) and applying the fast Hadamard–Walsh transform (see §1.2).

3.2.6. A good efficiency criterion for the method of linear transformations of the arguments is the maximal number of nonvanishing coefficients for an arbitrary system of k completely specified BF of m arguments, assuming the best choice of the linear transformation σ.

THEOREM 3.2.5. *Let* $\{f_{m,k}\}$ *be the set of all discrete functions corresponding to systems of k completely specified BF of m arguments and* $L^H_{m-1}(f, \sigma)$ *the number of nonvanishing*

Haar coefficients with subscript $m-l$ for a discrete function $f \in \{f_{m,k}\}$ with arguments transformed by σ. Then

$$\max_{\{f_{m,k}\}} \min_{\{\sigma\}} L_0^H(f,\sigma) \leqq 2, \tag{3.2.30}$$

$$\max_{\{f_{m,k}\}} \min_{\{\sigma\}} L_{m-l}^H(f,\sigma) \leqq 2^{m-l} - \left] \frac{[d]\,[d-1]}{2d} \right[, \tag{3.2.31}$$

where

$$d = \frac{2^{m-l+1}}{2^{k+l-1} - 2^{l-1} + 1} \qquad (l = 1, \ldots, m-1). \tag{3.2.32}$$

PROOF. Formula (3.2.30) follows from the definition of $L_0^H(f,\sigma)$. The proof of (3.2.31) will require more effort.

Let $f^{(m-l)}$ be the function produced at the l-th iteration of the algorithm (see Figure 3.2.3), $\{f_i^{(m-l)}\}$ its system of characteristic functions, and $\{B_i^{(m-l)}\}$ the corresponding system of autocorrelation functions. Then, by Theorems 3.2.2, 3.2.3,

$$\max_{\{f_{m,k}\}} \min_{\{\sigma\}} L_{m-l}^H(f,\sigma) = 2^{m-l} - \min_{\{f_{m,k}\}} \max_{\tau \neq 0} \frac{1}{2} \sum_{i=0}^{2^{k+l-1}-2^{l-1}} B_i^{(m-l)}(\tau). \tag{3.2.33}$$

We proceed to estimate the min-max term on the right of (3.2.33). Denote

$$\sum_{z=0}^{2^{m-l+1}-1} f_i^{(m-l)}(z) = N_i. \tag{3.2.34}$$

Then

$$\sum_{i=0}^{2^{k+l-1}-2^{l-1}} N_i = 2^{m-l+1}$$

and

$$\frac{1}{2} \sum_{\tau=1}^{2^{m-l+1}-1} B_i^{(m-l)}(\tau) = \binom{N_i}{2} \qquad (i = 0, \ldots, 2^{k-l-1} - 2^{l-1}). \tag{3.2.35}$$

Since $f^{(m-l)}$ and $f_i^{(m-l)}$ are defined at 2^{m-l+1} points and the number of characteristic functions $f_i^{(m-l)}$ is $2^{k+l-1} - 2^{l-1}$, it follows that

$$\min_{\{f_{m,k}\}} \max_{\tau \neq 0} \frac{1}{2} \sum_{i=0}^{2^{k+l-1}-2^{l-1}} B_i^{(m-l)}(\tau) \geqq \min_{\{f_{m,k}\}} \left] \frac{\sum_{i=0}^{2^{k+l-1}-2^{l-1}} \binom{N_i}{2}}{2^{m-l+1}} \right[$$

$$\geqq \left] \frac{1}{d} \binom{[d]}{2} \right[= \left] \frac{[d]\,[d-1]}{2d} \right[, \tag{3.2.36}$$

where

$$d = \frac{2^{m-l+1}}{2^{k+l-1} - 2^{l-1} + 1}.$$

The estimate (3.2.31) now follows from (3.2.36) and (3.2.33). ▮

In practical computations, the bound (3.2.31) may be simplified as follows, provided $k \ll m$ (so that $[d] \gg 1$):

$$\left]\frac{[d]\,[d-1]}{2d}\right[\geqq \left]\frac{([d]-1)\,[d]}{2([d]+1)}\right[= \left]\frac{[d]}{2} + \sum_{i=0}^{\infty} (-1)^{i+1} \frac{1}{[d]^i}\right[\tag{3.2.37}$$

$$\geqq \left]\frac{[d]}{2} - 1\right[\geqq \left]\frac{d-3}{2}\right[.$$

This inequality yields the coarser estimate

$$\max_{\{f_{m,k}\}} \min_{\{\sigma\}} L^H_{m-l}(f, \sigma) \leqq 2^{m-l} - \left]\frac{d-3}{2}\right[. \tag{3.2.38}$$

We now estimate the number $L^H(\sigma)$ of nonvanishing coefficients for the case of a single BF ($k = 1$).

COROLLARY 3.2.1.

$$\max_{\{f_{m,1}\}} \min_{\{\sigma\}} L^H(f, \sigma) \leqq 2^m -]0.36 \cdot 2^m[+]1.5(m-1)[. \tag{3.2.39}$$

PROOF. It follows from (3.2.38), in view of (3.2.18), that

$$\max_{\{f_{m,1}\}} \min_{\{\sigma\}} \sum_{l=1}^{m} L^H_{m-l}(\sigma) \leqq 2^m - \sum_{l=1}^{m-1} \left]\frac{1}{2}\left(\frac{2^{m-l+1}}{2^{l-1}+1} - 3\right)\right[$$

$$\leqq 2^m +]\tfrac{3}{2}(m-1)[- (]\tfrac{1}{4} \cdot 2^m[+]\tfrac{1}{12} \cdot 2^m[+]\tfrac{1}{40} \cdot 2^m[+]\tfrac{1}{144} \cdot 2^m[)$$

$$\leqq 2^m -]0.36 \cdot 2^m[+]1.5(m-1)[.$$ ▮

Table 3.2.5 lists upper bounds for $\max_{\{f_{m,1}\}} \min_{\{\sigma\}} L^H_{m-l}(f, \sigma)$ $(l = 1, 2, \ldots, m)$ in the case of a single BF, $m = 2, 3, \ldots, 10$; also given in the table is the upper bound $\tilde{L}^H$ of $\max_{\{f_{m,1}\}} \min_{\{\sigma\}} \sum_{l=1}^{m} L^H_{m-1}(f, \sigma)$ $(l = 1, 2, \ldots, m)$.

It follows from Theorem 3.2.5 that in the case $k = 1$ the use of linear transformations powers the upper bound for the number of nonvanishing coefficients with subscript $m - 1$ by about one half, while by (3.2.39) the total number of nonvanishing coefficients is reduced by about one third.

TABLE 3.2.5

l \ m	2	3	4	5	6	7	8	9	10
1	1	2	4	8	16	32	64	128	256
2	2	2	3	6	11	22	43	86	171
3	0	2	2	4	7	13	26	52	103
4	0	0	2	2	4	8	15	29	57
5	0	0	0	2	2	4	8	16	31
6	0	0	0	0	2	2	4	8	16
7	0	0	0	0	0	2	2	4	8
8	0	0	0	0	0	0	2	2	4
9	0	0	0	0	0	0	0	2	2
10	0	0	0	0	0	0	0	0	2
$\tilde{L}^H$	3	6	11	22	42	83	164	327	650

In this case, then, introduction of a linear block σ, requiring a number of two-input mod 2 adders of the order of $m^2/\log_2 m$ (see §3.2.2), produces a savings of at least $]0.36 \cdot 2^m[-]1.5(m-1)[$ multibit storage cells for the coefficients.

3.3. SPECTRAL METHODS FOR REALIZATION OF SYSTEMS OF INCOMPLETELY SPECIFIED BOOLEAN FUNCTIONS

3.3.1. The design of networks realizing systems of incompletely specified Boolean functions is one of the most important and difficult problems of modern digital network design theory. The added difficulty, in comparison with the analogous problem for completely specified functions, is that here one has to look for the optimal completions of the functions involved. In this section we shall show how to determine an optimal completion of a system and an optimal linear transformation of its arguments, in the sense that the number of nonvanishing expansion coefficients is minimized.

Before proceeding to the solution of these problems, we shall estimate the degree to which networks may be simplified if the function to be realized is allowed to be incompletely specified.

THEOREM 3.3.1. *Let* $\{\Psi_\omega(z)\}$, $z \in [0, p^m)$, *be a complete orthogonal system of step functions, and* $f(z)$ *a function defined at* N *points* $z_0, \ldots, z_{N-1} \in \{0, 1, \ldots, p^m - 1\}$. *Then there exists a completion of* $f(z)$ *to* $[0, p^m)$ *such that the number of nonvanishing coefficients in its expansion in terms of the basis* $\{\Psi_\omega(z)\}$ *is at most* N.

In the case of the Walsh basis, this is simply Lemma 2.2.1. In regard to the Haar basis, we shall develop an algorithm computing the desired completion in §3.3.2 below.

3.3.2. We consider the optimal completion problem for a system of BF, in the case of the Haar basis. If L^H_{m-l} is the number of nonvanishing coefficients $c^{(q)}_{m-l}$ ($q = 1, 2, \dots, 2^{m-l}$), then $0 \leqq L^H_{m-l} \leqq 2^{m-l}$ (where m, as usual, is the number of arguments). The optimal completion may therefore be sought by the following recursive procedure. The procedure will produce completion classes $K_{m-1} \supseteqq K_{m-2} \supseteqq \cdots \supseteqq K_0$, where K_{m-l} ($l = 1, 2, \dots, m$) is the set of all completions in K_{m-l+1} which minimize L^H_{m-l} (K_m is the set of all completions of the system).

Suppose we have a system of k incompletely specified BF of m arguments:

$$y^{(s)} = y^{(s)}(z^{(0)}, \dots, z^{(m-1)}) \qquad (s = 0, 1, \dots, k-1). \tag{3.3.1}$$

Note that all the functions in this system are assumed to be undefined at exactly the same points. We represent system (3.3.1) by a discrete function $y = f(z)$, where

$$z = \sum_{s=0}^{m-1} z^{(s)} 2^{m-1-s} \quad \text{and} \quad y = \sum_{s=0}^{k-1} y^{(s)} 2^{k-1-s}.$$

Then K_{m-1} will be the class of all completions such that the number of distinct z for which

$$f(2z-2) = f(2z-1) \tag{3.3.2}$$

is a maximum.

The class $K_{m-2} \subseteqq K_{m-1}$ is the set of all completions in K_{m-1} such that the number of distinct z for which

$$f(4z-4) + f(4z-3) = f(4z-2) + f(4z-1) \tag{3.3.3}$$

is a maximum.

The definition of $K_{m-3} \supseteqq K_{m-4} \supseteqq \cdots \supseteqq K_1 \supseteqq K_0$ is similar. Then any completion in class K_0 will meet our requirements. The essential point here is that determination of a completion in K_0 does not require exhaustive examination. Thus, given a system of incompletely specified BF, with their arguments in a selected order, we can construct a completion corresponding to an absolute minimum of L^H_{m-1} and certain relative minima of L^H_{m-l} ($l = 2, 3, \dots, m$).

EXAMPLE 3.3.1. Consider the BF of 4 variables defined in Table 3.3.1, where the asterisk $*$ stands for undefined values. The classes $K_3, K_2, K_1 = K_0$ are shown in the table. Here $a_1, a_2, a_3, a_4, a_5 \in \{0, 1\}$; $a_6 \in \{a_2, a_3\}$; $a_7 \in \{a_4, a_5\}$; $a_8 \in \{a_6, a_7\}$. There are two functions in K_0, one of which (φ) is shown in Table 3.3.1.

3.3.3. Our next task is to look for an optimal linear transformation of arguments $\sigma_{\min}$. For any σ, we retain the same method of completion as before.

To determine $\sigma_{\min}$ we use the algorithm of §3.2 (see Figure 3.2.3), except that the characteristic functions $f_i^{(m-1)}(z)$ will now be defined not by (3.2.22) but by

$$f_i^{(m-1)}(z) = \begin{cases} 1 & \text{if } f^{(m-1)}(z) = i \text{ or } f^{(m-1)}(z) = *, \\ 0 & \text{otherwise,} \end{cases} \tag{3.3.4a}$$

TABLE 3.3.1.

$z^{(0)}$	$z^{(1)}$	$z^{(2)}$	$z^{(3)}$	f	K_3	K_2	$K_1 = K_0$	$\varphi \in K_0$
0	0	0	0	1	1	1	1	1
0	0	0	1	0	0	0	0	0
0	0	1	0	*	0	0	0	0
0	0	1	1	0	0	0	0	0
0	1	0	0	*	a_1	0	0	0
0	1	0	1	*	a_1	0	0	0
0	1	1	0	0	0	0	0	0
0	1	1	1	1	1	1	1	1
1	0	0	0	*	a_2	a_6	a_8	0
1	0	0	1	*	a_2	a_6	a_8	0
1	0	1	0	*	a_3	a_6	a_8	0
1	0	1	1	*	a_3	a_6	a_8	0
1	1	0	0	*	a_4	a_7	a_8	0
1	1	0	1	*	a_4	a_7	a_8	0
1	1	1	0	*	a_5	a_7	a_8	0
1	1	1	1	*	a_5	a_7	a_8	0

and

$$f_*^{(m-1)}(z) = \begin{cases} 1 & \text{if } f^{(m-1)}(z) = *, \\ 0 & \text{otherwise;} \end{cases} \tag{3.3.4b}$$

the function $\sum_{i=0}^{2k-1} B_i^{(m-1)}(\tau)$ in (3.2.25) is replaced by

$$\sum_{i=0}^{2k-1} (B_i^{(m-1)}(\tau) - B_*^{(m-1)}(\tau)) + B_*^{(m-1)}(\tau) = \sum_{i=0}^{2k-1} B_i^{(m-1)}(\tau) - (2^k - 1)\, B_*^{(m-1)}(\tau),$$

where $B_*^{(m-1)}(\tau)$ is the autocorrelation function of $f_*^{(m-1)}(z)$, and in (3.2.26) we set

$$a + * = 2a \quad \text{and} \quad * + * = *. \tag{3.3.5}$$

We note that for weakly specified functions it follows from (3.3.4) that

$$\sum_{z=0}^{2^m-1} f_i^{(m-1)}(z) \gg 2^{m-1},$$

and therefore it is convenient to calculate $B_i^{(m-1)}(\tau)$ with the aid of Corollary 1.5.1, according to which one can use the inverse function

$$\overline{f_i^{(m-1)}(z)} = 1 - f_i^{(m-1)}(z) \qquad (i = 0, 1, \ldots, 2^k - 1, *).$$

EXAMPLE 3.3.2. Let us find the classes $\Xi_{m-1} = \Xi_3, \Xi_2, \Xi_1$ and the matrix $\sigma_{\min}$ for the BF of Example 3.3.1. Table 3.3.2 presents the original incompletely specified BF $f^{(3)}(z)$, its inverse characteristic functions $\overline{f_*^{(3)}(z)}$, $\overline{f_0^{(3)}(z)}$ and $\overline{f_1^{(3)}(z)}$, the autocorrelation functions $B_{*}^{(3)}(\tau)$, $B_0^{(3)}(\tau)$ and $B_1^{(3)}(\tau)$ (constructed using Corollary 1.5.1), and the function $B_0^{(3)}(\tau) + B_1^{(3)}(\tau) - B_{*}^{(3)}(\tau)$.

According to Table 3.3.2, we can set $\tau_{m-1} = \tau_3 = 2$, and by Theorem 3.2.2,

$$\sigma_3 = \begin{pmatrix} 1 & 0 & 0 & 0 \\ 0 & 1 & 0 & 0 \\ 0 & 0 & 0 & 1 \\ 0 & 0 & 1 & 0 \end{pmatrix}.$$

The function $f_{\sigma_3}^{(3)}(z)$ defined by

$$f_{\sigma_3}^{(3)}(\sigma_3 \otimes z) = f^{(3)}(z) \qquad (\text{mod } 2)$$

is shown in Table 3.3.2. With the completion defined as above, this gives $L_3^H(\sigma_3) = 0$, whereas for the original function $L_3^H(E_4) = 2$.

Table 3.3.3 shows the "contracted" function $f^{(2)}(z)$ constructed by (3.2.26) and (3.3.5), and the function

$$B^{(2)}(\tau) = \sum_{i=0}^{3} B_i^{(2)}(\tau) - 3B_{*}^{(2)}(\tau) = B_0^{(2)}(\tau) + B_2^{(2)}(\tau) - B_{*}^{(2)}(\tau)$$

since $B_1^{(2)}(\tau) = B_3^{(2)}(\tau) = B_{*}^{(2)}(\tau)$).

TABLE 3.3.2.

$z^{(0)}$	$z^{(1)}$	$z^{(2)}$	$z^{(3)}$	z, τ	$f^{(3)}(z)$	$\overline{f_i^{(3)}(z)}$ $i = *$	0	1	$B_i^{(3)}(\tau)$ $i = *$	0	1	$B_0^{(3)}(\tau) + B_1^{(3)}(\tau) - B_{*}^{(3)}(\tau)$	$f_{\sigma_3}^{(3)}(z)$
0	0	0	0	0	1	1	1	0	11	14	13	16	1
0	0	0	1	1	0	1	0	1	10	12	10	12	*
0	0	1	0	2	*	0	0	0	8	12	12	16	0
0	0	1	1	3	0	1	0	1	8	12	10	14	0
0	1	0	0	4	*	0	0	0	8	12	10	14	*
0	1	0	1	5	*	0	0	0	8	12	12	16	0
0	1	1	0	6	0	1	0	1	10	12	10	12	*
0	1	1	1	7	1	1	1	0	10	14	12	16	1
1	0	0	0	8	*	0	0	0	6	12	10	16	*
1	0	0	1	9	*	0	0	0	6	12	10	16	*
1	0	1	0	10	*	0	0	0	6	12	10	16	*
1	0	1	1	11	*	0	0	0	6	12	10	16	*
1	1	0	0	12	*	0	0	0	6	12	10	16	*
1	1	0	1	13	*	0	0	0	6	12	10	16	*
1	1	1	0	14	*	0	0	0	6	12	10	16	*
1	1	1	1	15	*	0	0	0	6	12	10	16	*

TABLE 3.3.3.

z, τ	$f^{(2)}(z)$	$B^{(2)}(\tau)$	$f^{(2)}_{\sigma(2)}(\tau)$	$f^{(1)}(z)$	$B^{(1)}(\tau)$	$f^{(1)}_{\sigma(1)}(z)$
0	2	8	2	4	4	4
1	0	4	*	0	0	4
2	0	4	0	0	0	0
3	2	8	*	4	4	0
4	*	8	0			
5	*	8	*			
6	*	8	2			
7	*	8	*			

Table 3.3.3 yields $\tau_2 = 4$, and by Theorem 3.2.2,

$$\sigma^{(2)} = \begin{pmatrix} 0 & 0 & 1 \\ 0 & 1 & 0 \\ 1 & 0 & 0 \end{pmatrix}.$$

The function $f^{(2)}_{\sigma(2)}(z)$, the "contracted" function $f^{(1)}(z)$, the corresponding $B^{(1)}(\tau) = B^{(1)}_0(\tau) + B^{(1)}_4(\tau)$ and $f^{(1)}_{\sigma(1)}(z)$ $\left(\tau_1 = 3 \text{ and } \sigma^{(1)} = \begin{pmatrix} 1 & 1 \\ 0 & 1 \end{pmatrix}\right)$ are also shown in Table 3.3.3. The final result, using (3.2.21), is:

$$\sigma_{\min} = \begin{pmatrix} 1 & 1 & 0 & 0 \\ 0 & 1 & 0 & 0 \\ 0 & 0 & 1 & 0 \\ 0 & 0 & 0 & 1 \end{pmatrix} \otimes \begin{pmatrix} 0 & 0 & 1 & 0 \\ 0 & 1 & 0 & 0 \\ 1 & 0 & 0 & 0 \\ 0 & 0 & 0 & 1 \end{pmatrix} \otimes \begin{pmatrix} 1 & 0 & 0 & 0 \\ 0 & 1 & 0 & 0 \\ 0 & 0 & 0 & 1 \\ 0 & 0 & 1 & 0 \end{pmatrix} = \begin{pmatrix} 0 & 1 & 0 & 1 \\ 0 & 1 & 0 & 0 \\ 1 & 0 & 0 & 0 \\ 0 & 0 & 1 & 0 \end{pmatrix} \pmod 2.$$

Realization of the $\sigma_{\min}$-block requires one two-input mod 2 adder. The results for this example are summarized in Table 3.3.4, which presents the original function $f(z)$. the function $f_{\sigma_{\min}}(z)$, and the Haar spectra of their optimal completions $\varphi(z)$ and $\varphi_{\sigma_{\min}}(z)$, with $c(t) = c_l^{(q)}$ $(t = 2^l + q - 1)$.

The data in Table 3.3.4 show that whereas the optimal completion of $f(z)$ gives $L^H = L^H(E_4) = 6$, the result after optimal rearrangement of the arguments is $L^H(\sigma_{\min}) = 2$.

3.4. SPECTRAL METHODS OF SYNTHESIS FOR NETWORKS OF MANY-VALUED ELEMENTS

3.4.1. Recent years have seen considerable progress in the development of economical multistable elements [8, 17, 51, 91, 156]. As a result, one is now faced with the need to

TABLE 3.3.4.

z, t				f	φ	$8c(t)$	$f_{\sigma_{\min}}$	$\varphi_{\sigma_{\min}}$	$8c_{\sigma_{\min}}(t)$
0	0	0	0	1	1	1	1	1	4
0	0	0	1	0	0	1	*	1	4
0	0	1	0	*	0	0	*	1	0
0	0	1	1	0	0	0	*	1	0
0	1	0	0	*	0	2	*	1	0
0	1	0	1	*	0	−2	1	1	0
0	1	1	0	0	0	0	*	1	0
0	1	1	1	1	1	0	*	1	0
1	0	0	0	*	0	4	0	0	0
1	0	0	1	*	0	0	0	0	0
1	0	1	0	*	0	0	*	0	0
1	0	1	1	*	0	−4	*	0	0
1	1	0	0	*	0	0	*	0	0
1	1	0	1	*	0	0	0	0	0
1	1	1	0	*	0	0	*	0	0
1	1	1	1	*	0	0	*	0	0

formalize the problem of network synthesis for elements of this type. The problem turns out to be far more complex than its analog for binary elements, in view of the fact that the methods of many-valued switching logic [51, 91, 156] are noneffective from the practical standpoint, and indeed sometimes quite unsuitable for the synthesis of networks from multistable elements.

In this section we shall study spectral methods for the solution of this problem; they yield fairly simple but approximate results.

3.4.2. Networks based on p-stable elements are described by systems of p-valued logical functions. In the framework of spectral methods, we proceed as in the two-valued case, expanding the step function representing a given system in a series of generalized Haar functions $\{M^{(p,q)}_{r,m-l}(z)\}$. These are step functions, defined on $[0, p^m)$ and taking on at most $p+1$ complex values from the set $\{0, \exp(2\pi iq/p)\}$ ($q = 0, 1, \ldots, p-1$; $i = \sqrt{-1}$). The properties of these functions and methods for calculating the generalized Haar spectrum were considered in §1.3. As before, we shall measure the complexity by the number L^M of nonvanishing series coefficients; if the original system depends on m arguments, then $L^M \leqq p^m$. We shall assume throughout that p is a prime.

The use of spectral methods in many-valued logic raises two problems:

1) the optimal completion of the original system of p-valued logical functions (see §3.4.3);

2) choice of the optimal order of arguments or, in more general terms, the optimal linear transformation of the arguments (see §§3.4.4, 3.4.5).

3.4.3. Let L_{m-l}^M denote the number of nonvanishing coefficients $c_{r,m-l}^{(p,q)}$ ($r = 0, 1, \ldots, p-1; q = 1, 2, \ldots, p^{m-l}$). Then

$$0 \leqq L_{m-l}^M < (p-1)\, p^{m-l}. \tag{3.4.1}$$

We shall assume that the functions of the original system are undefined at the same points. In view of (3.4.1), we shall use the following m-step procedure to find an optimal completion. We first determine the class K_{m-1} of completions minimizing L_{m-1}^M, the subclass $K_{m-2} \subseteqq K_{m-1}$ of completions minimizing L_{m-2}^M, and so on, obtaining classes $K_{m-3} \supseteqq K_{m-4} \supseteqq \cdots \supseteqq K_0$. Any member of K_0 may be taken as the optimal completion.

It follows from the definition of the system $\{M_{r,m-l}^{(p,q)}(z)\}$ that the expansion coefficients $c_{r,m-1}^{(p,q)}$ ($r = 0, 1, \ldots, p-1$) of a discrete function $f(z)$ vanish if and only if

$$f(pq-p) = f(pq-(p-1)) = \cdots = f(pq-1). \tag{3.4.2}$$

This condition defines the class K_{m-1}, which will consist of all completions for which the number of distinct numbers q satisfying (3.4.2) is a maximum. The class K_{m-2} will then be the set of all completions in K_{m-1} for which the number of distinct q such that

$$\sum_{i=1}^{p} f(p^2q-i) = \sum_{i=1}^{p} f(p^2q-p-i) = \cdots = \sum_{i=1}^{p} f(p^2q-p(p-1)-i) \tag{3.4.3}$$

is a maximum. The construction of the remaining classes $K_{m-3} \supseteqq K_{m-4} \supseteqq \cdots \supseteqq K_0$ is analogous.

This completion method is a generalization of the method of §3.3 for Boolean functions to p-valued logic. In view of (3.4.1), the method guarantees an absolute minimum for L_{m-1}^M and certain local minima for L_{m-1}^M ($l > 1$).

EXAMPLE 3.4.1. We start with an incompletely specified ternary logical function of two variables (Table 3.4.1). The classes K_1, K_0 are shown in Table 3.4.1 ($a_1, a_2 \in \{0, 1, 2\}$). The optimal completion lowers the number of nonvanishing coefficients to 3.

3.4.4. We now turn to the problem of optimal choice of the order of arguments. As in §§3.2, 3.3, we shall deal with a more general problem—the optimal linear transformation of the arguments. Any linear transformation of a p-ary argument vector $z = (z^{(0)}, \ldots, z^{(m-1)})$ is defined by a nonsingular matrix $\sigma = (\sigma_{ij})$, where $\sigma_{ij} \in \{0, 1, \ldots, p-1\}$ and $|\sigma|_p \neq 0$ ($|\sigma|_p$ is the determinant of σ over the field of residues modulo p).

TABLE 3.4.1.

$z^{(0)}$	$z^{(1)}$	f	K_1	K_0
0	0	1	1	1
0	1	*	a_1	0
0	2	2	2	2
1	0	1	1	1
1	1	1	1	1
1	2	*	1	1
2	0	*	a_2	1
2	1	*	a_2	1
2	2	*	a_2	1

For example, if $p = 3$ and

$$\sigma = \begin{pmatrix} 1 & 0 & 0 \\ 2 & 1 & 0 \\ 1 & 1 & 1 \end{pmatrix},$$

then

$$z_\sigma = \sigma \otimes z = (z^{(0)}, 2z^{(0)} \oplus z^{(1)}, z^{(0)} \oplus z^{(1)} \oplus z^{(2)}) \qquad (\text{mod } 3).$$

Multiplication by the matrix σ requires a special block, which may be based on mod p adders and networks realizing multiplication mod p by the constants $1, 2, \ldots, p - 1$. A network implementing multiplication by the matrix shown above is illustrated in Figure 3.4.1; $\rightarrow ⓐ \rightarrow$ signifies a network for multiplication by a and $\rightarrow \oplus \rightarrow$ (with $\uparrow$ input) a mod p adder.

Let $N^{(p)}(\sigma)$ denote the minimum number of two-input adders and multipliers by constants modulo p necessary to design a network realizing the matrix σ.

THEOREM 3.4.1. *For any* $\sigma = (\sigma_{ij})$ *where* $\sigma_{ij} \in \{0, 1, \ldots, p - 1\}, i, j = 1, 2, \ldots, m,$

$$N^{(p)}(\sigma) \leq \min_{a \in \{1, \ldots, m\}} \left(\left(p^{[m/a]} - \left[\frac{m}{a} \right] - 1 \right) (a - \text{res}_a m) \right.$$
$$\left. + \left(p^{]m/a[} - \left] \frac{m}{a} \right[- 1 \right) \text{res}_a m + (a - 1) m \right), \qquad (3.4.4)$$

where $\text{res}_a m$ *denotes the remainder upon division of* m *by* a.

This theorem is a generalization of Theorem 3.2.1 and its proof employs an analogous construction.

The asymptotic behavior of $N^{(p)}(\sigma)$ as $m \to \infty$ is described for almost all σ by

$$N^{(p)}(\sigma) \sim \frac{m^2}{\log_p m}. \qquad (3.4.5)$$

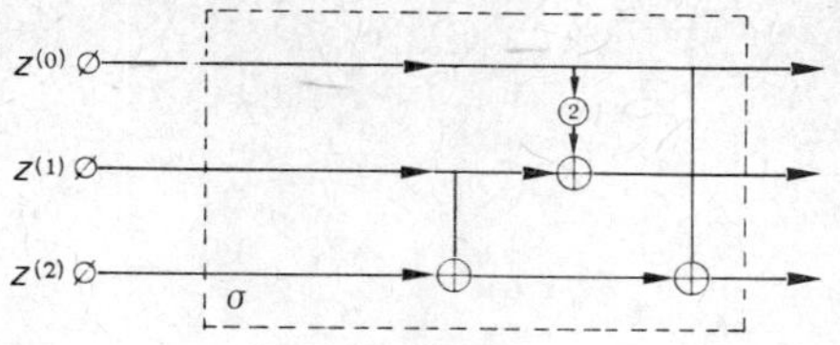

FIGURE 3.4.1.

It follows from Theorem 3.4.1 and formula (3.4.5) that if the network realizing a system of p-valued logical functions is designed as a series connection of two blocks: a linear block implementing the optimal linear transformation $\sigma_{\min}$, and a nonlinear block using spectral methods to realize the appropriate system of functions of the argument $\sigma_{\min} \otimes z \pmod p$, then the complexity of the linear part may be neglected, since it increases with m more slowly than m^2, whereas the complexity of the nonlinear block is generally an exponential function of m.

3.4.5. We now proceed to the construction of an optimal linear transformation $\sigma_{\min}$. This will be done by an $(m-1)$-step procedure, yielding a sequence of classes $\Xi_{m-1} \supseteq \Xi_{m-2} \supseteq \cdots \supseteq \Xi_1$, where Ξ_{m-l} is the set of all matrices in Ξ_{m-l+1} that minimize L^M_{m-l} (Ξ_m is the set of all nonsingular matrices over $GF(p)$).

The construction of Ξ_{m-1} is as follows. Beginning with the system $f(z) = f^{(m-1)}(z)$ of k incompletely specified p-valued logical functions of m arguments we construct the characteristic functions:

$$f_i^{(m-1)}(z) = \begin{cases} 1 & \text{if } f^{(m-1)}(z) = i \quad \text{or} \quad f^{(m-1)}(z) = *, \\ 0 & \text{otherwise} \end{cases}$$

$$(i = 0, 1, \ldots, p^k - 1), \tag{3.4.6}$$

$$f_*^{(m-1)}(z) = \begin{cases} 1 & \text{if } f^{(m-1)}(z) = *, \\ 0 & \text{otherwise.} \end{cases}$$

Now define the function $B_i^{(m-1)}(\tau)$ by

$$B_i^{(m-1)}(\tau) = \sum_{z=0}^{p^m-1} f_i^{(m-1)}(z)\, f_i^{(m-1)}(z \ominus \tau) \ldots f_i^{(m-1)}(z \ominus \overbrace{\tau \ominus \cdots \ominus \tau}^{p-1}) \pmod p. \tag{3.4.7}$$

This is the autocorrelation function of p functions obtained by successive translation modulo p of the argument of the characteristic function $f_i^{(m-1)}(z)$; it coincides with our previous autocorrelation function when $p = 2$.

THEOREM 3.4.2. *The class* Ξ_{m-1} *is the set of all nonsingular matrices* σ_{m-1} *over* $GF(p)$ *such that*

$$\sigma_{m-1} \otimes \tau_{m-1} = \begin{pmatrix} 0 \\ 0 \\ \vdots \\ 0 \\ 1 \end{pmatrix} \pmod p, \tag{3.4.8}$$

where

$$\max_{\tau \neq 0} \left(\sum_{i=0}^{p^k - 1} (B_i^{(m-1)}(\tau) - B_*^{(m-1)}(\tau)) + B_*^{(m-1)}(\tau) \right)$$

$$= \sum_{i=0}^{p^k - 1} (B_i^{(m-1)}(\tau_{m-1}) - B_*^{(m-1)}(\tau_{m-1})) + B_*^{(m-1)}(\tau_{m-1}). \tag{3.4.9}$$

This is in effect a generalization of Theorem 3.2.2 to systems of incompletely specified p-valued logical functions.

EXAMPLE 3.4.2. Consider the ternary logical function of two variables defined by Table 3.4.2: $f(z^{(0)}, z^{(1)}) = f^{(1)}(z)$. The table also gives the system of characteristic functions $f_i^{(1)}(z)$, autocorrelation functions $B_i^{(1)}(\tau)$ $(i = 0, 1, 2)$, and the function $\sum_{i=0}^{2} B_i^{(1)}(\tau)$. According to these data, we have $\tau_1 = \begin{pmatrix} 2 \\ 1 \end{pmatrix}$. By Theorem 3.4.2,

$$\sigma_{\min} = \sigma_1 = \begin{pmatrix} 1 & 1 \\ 1 & 2 \end{pmatrix}.$$

The function $f_{\sigma_1}^{(1)}(z)$ defined by $f_{\sigma_1}^{(1)}(\sigma_1 \otimes z) = f^{(1)}(z) \pmod 3$ is shown in the table. It follows that whereas $L^M = 9$ for the original system $f(z)$, the transformed system $f_{\sigma_1}^{(1)}(z)$ gives $L^M = 5$.

3.4.6. We now outline a recursive procedure constructing the classes $\Xi_{m-2} \supseteq \cdots \supseteq \Xi_1$, which is a generalization of the similar procedure for the two-valued case (see §3.2.5).

To construct Ξ_{m-2}, we start with the function $f^{(m-1)}(z)$, defining a function $f_{\sigma_{m-1}}^{(m-1)}(z)$ by

$$f_{\sigma_{m-1}}^{(m-1)}(\sigma_{m-1} \otimes z) = f^{(m-1)}(z) \pmod p.$$

TABLE 3.4.2.

$z^{(0)}$	$z^{(1)}$	z, τ	$f^{(1)}(z)$	$f_i^{(1)}(z)$ $i=0$	1	2	$B_i^{(1)}(z)$ $i=0$	1	2	$\sum_{i=0}^{2} B_i^{(1)}(\tau)$	$f_{\sigma_1}^{(1)}(z)$
0	0	0	1	0	1	0	2	5	2	9	1
0	1	1	2	0	0	1	0	0	0	0	0
0	2	2	1	0	1	0	0	0	0	0	2
1	0	3	1	0	1	0	0	3	0	3	0
1	1	4	1	0	1	0	0	0	0	0	1
1	2	5	2	0	0	1	0	3	0	3	2
2	0	6	1	0	1	0	0	3	0	3	1
2	1	7	0	1	0	0	0	3	0	3	1
2	2	8	0	1	0	0	0	0	0	0	1

We now define the "contracted" function $f^{(m-2)}(z)$ at p^{m-1} points by

$$f^{(m-2)}(z) = \begin{cases} \sum_{s=0}^{p-1} f_{\sigma_{m-1}}^{(m-1)}(pz+s), & \text{if } f_{\sigma_{m-1}}^{(m-1)}(pz+s)\ (s=0,\dots,p-1) \\ & \text{is completely specified;} \\ pa, & \text{if } f_{\sigma_{m-1}}^{(m-1)}(pz+s) \in \{a, *\} \\ & (a, s = 0, \dots, p-1); \\ *, & \text{otherwise.} \end{cases} \tag{3.4.10}$$

Apply the algorithm of §3.4.5 to the function $f^{(m-2)}(z)$, obtaining an $(m-1) \times (m-1)$ matrix which we denote by $\sigma^{(m-2)}$. The construction of $\sigma^{(m-3)}, \sigma^{(m-4)}, \dots, \sigma^{(1)}$ is analogous.

Finally, we define $\sigma_{\min}$ by

$$\sigma_{\min} = \left(\begin{array}{c|c} \sigma^{(1)} & 0 \\ \hline 0 & E_{m-2} \end{array}\right) \otimes \left(\begin{array}{c|c} \sigma^{(2)} & 0 \\ \hline 0 & E_{m-3} \end{array}\right) \otimes \cdots \otimes \left(\begin{array}{c|c} \sigma^{(m-2)} & 0 \\ \hline 0 & E_1 \end{array}\right) \otimes \sigma_{m-1} \pmod p \tag{3.4.11}$$

REMARK. The algorithms we have described for determination of an optimal linear transformation and an optimal completion in many-valued logic are almost identical with the appropriate algorithms for systems of BF; this is yet another demonstration that the algorithms provided by spectral methods of synthesis are independent of the "arity" of the logic.

3.5. SPECTRAL METHODS OF SYNTHESIS APPLIED TO THE DESIGN OF DIGITAL FUNCTION GENERATORS AND SEQUENCE GENERATORS

3.5.1. When devising complex systems of automatic control, measurement and recording, one is faced with the need to design devices which reproduce given functions. This involves designing digital generators of continuous functions of one or more variables [56, 57, 58]. When the input of such a device is the m-bit binary code of the argument z, the output is the value of the required function $f(z)$, or rather its k-bit code. A *digital function generator* (DFG) may be designed with the aid of any method for the synthesis of an (m, k)-terminal logic network. However, the functions produced by DFG are not continuous, but derived from continuous functions by quantization; thus a more detailed study of DFG design is indispensable. In this section we consider spectral methods to this end and establish a few estimates.

3.5.2. The principal technical parameters of a DFG are the accuracy with which it reproduces the given function, its speed, and the amount of hardware required. The *accuracy* of reproduction is estimated by the distance ε between the object function $f(z)$ and the approximating function $\Phi(z)$ produced at the output. There are various definitions of this "distance"; the *uniform distance* ε_c is defined by

$$\varepsilon_c = \max_{z\in[0,1]} |f(z) - \Phi(z)| \tag{3.5.1}$$

and the *mean-square distance*

$$\varepsilon_2 = \left[\int_0^1 (f(z) - \Phi(z))^2 \, dz \right]^{1/2}. \tag{3.5.2}$$

(Throughout this section we shall assume that $f(z)$, $\Phi(z)$, as well as the Walsh and Haar functions, are defined on the closed interval $[0, 1]$.)

The *speed* of the DFG is estimated by the quantity $1/t_{\text{DFG}}$, where t_{DFG} is the time interval elapsing from application of the argument code at the input to production of the function value at the output.

The *volume of hardware* or *complexity* is measured, as usual, by the number L_{DFG} of logic elements (from some complete basis system) necessary for realization of the DFG.

3.5.3. The known parameters involved in designing a DFG are the required accuracy ε and the admissible value-generation time t_{DFG}. Analyzing the various methods of synthesis, one selects the optimal solution—that minimizing L_{DFG}. The reader should note that the spectral methods considered in this book for the synthesis of combinational networks are more rational than the classical methods. The reasons are as follows. First, the spectral approach greatly increases the likelihood that the synthesis method actually selected will yield a device answering most completely to the demands. To this end the basis systems should be sought not only among the systems studied in our book (Walsh functions, Haar functions, and their many-valued analogs), but also among the systems obtained, say, by integrating them ($\{\int_0^z W_\omega(z)\,dz\}$ [56], Schauder functions [135]).

Second, the DFG designed by spectral methods are multiphase devices—the value of the function is produced only after several arithmetic operations have been performed. The decrease in speed nevertheless makes it possible to reduce the necessary volume of hardware, in other words, to make fuller use of the available time t_{DFG}, which is fixed in advance. Indeed, L_{DGF} is determined largely by the size of the memory device in which the coefficient codes are stored (a single-phase combinational network may be regarded as a memory holding codes of the function values). As we shall show later, when spectral methods are used the number of expansion coefficients to be stored is often much smaller than the number of function values to be reproduced; this is the case for a quite broad range of continuous functions. All this implies a reduction in the complexity of realization.

Finally, spectral methods are applicable in cases where the arguments and functions involve a large number of digits (say, $m = k = 20$), where classical methods for synthesis of combination networks demand a prohibitive amount of labor. This is because the functions to be produced by DFG are generally specified in analytical form, and the techniques for analysis of the errors arising in approximation of these functions by series are independent of the length of the argument or function-value codes (see §§3.5.4–3.5.6).

3.5.4. The volume of hardware L_{DFG} will be measured by the number L of nonvanishing coefficients stored in the memory. This number depends on the form of $f(z)$, the prescribed accuracy of reproduction ε, the type of approximation (uniform or mean-square), and also on the selected basis. We now proceed to estimate L as a function of these parameters.

To approximate $f(z)$ by a finite Walsh or Haar series, we must first approximate $f(z)$ by a step function $\Phi(z)$. (Throughout this section we shall adopt the approximation in the uniform norm (3.5.1), since this is customary in the theory of function generators.) Let $\{\Phi(z)\}^n$ be the set of step functions having n discontinuities of the first kind [77].

THEOREM 3.5.1. *For any differentiable function $f(z)$ defined on $[0, 1]$ and any arbitrarily small $\varepsilon > 0$, there exists $\Phi(z) \in \{\Phi(z)\}^n$ such that*

$$\max_{z\in[0,1]} |\Phi(z) - f(z)| \leqq \varepsilon,$$

provided

$$n \geqq \left] \frac{\max_{z \in [0,1]} |f'(z)|}{2\varepsilon} \right[, \tag{3.5.3}$$

where $f'(z)$ is the derivative of $f(z)$ with respect to z.

PROOF. The required step function $\Phi(z) \in \{\Phi(z)\}^n$ may be constructed as follows. Divide the interval $[0, 1]$ into n subintervals of equal length, where n satisfies (3.5.3), by points $z_0 = 0, z_1, \ldots, z_n = 1$; then set

$$\Phi(z) = \tfrac{1}{2}(f(z_i) + f(z_{i+1})) \tag{3.5.4}$$

for $z_i \leqq z < z_{i+1}$, $i = 0, 1, \ldots, n-1$ and $\max_{z \in [0,1]} |\Phi(z) - f(z)| \leqq \varepsilon$. ∎

COROLLARY 3.5.1. *Denote*

$$m = \left]\log_2 \max_{z \in [0,1]} |f'(z)| - \log_2 \varepsilon - 1\right[. \tag{3.5.5}$$

Then for any differentiable function $f(z)$ $(z \in [0, 1])$ there exists

$$\Phi(z) = \sum_{\omega=0}^{2^m-1} S(\omega)\, W_\omega(z) = c_0^{(0)} + \sum_{l=0}^{m-1} \sum_{q=1}^{2^l} c_l^{(q)} H_l^{(q)}(z)$$

such that $\max_{z \in [0,1]} |f(z) - \Phi(z)| \leqq \varepsilon$, i.e., $L^W, L^H \leqq 2^m$, where $L^W (L^H)$ is the number of nonvanishing coefficients in the Walsh (Haar) expansion of $\Phi(z)$.

PROOF. The assertion follows from Theorem 3.5.1 (with the interval divided into $2^m \geqq n$ parts); one uses the fact that any left-continuous step function $\Phi(z)$ with discontinuities at the points $i/2^m$ $(i = 0, 1, \ldots, 2^m - 1)$ has a Walsh (Haar) expansion containing at most 2^m terms. ∎

EXAMPLE 3.5.1. Suppose we wish to design a function generator for $f(z) = \arctan z$ $(z \in [0, 1])$ with accuracy 1° ($\varepsilon = \pi/180$ radians), with $m = 20$ and $k = 20$. The number $L^W (L^H)$ of nonvanishing coefficients in the Walsh (Haar) expansion of $f(z)$ is calculated as follows.

Calculate m by (3.5.5), noting that

$$\max_{z \in [0,1]} (\arctan z)' = (1 + z^2)^{-1}\big|_{z=0} = 1; \qquad m = \left]1 - \log_2 \frac{\pi}{180} - 1\right[= 6,$$

and so $L^W, L^H \leqq 64$.

The working of this example does not take into consideration that there is an additional error δ in the values of $f(z)$, due to the fact that the number k of binary digits of the function at the generator output is finite. It is clear that this error does not exceed one half of the lowest-order digit; in other words, assuming that the maximum of $f(z)$ is normalized to 1, we have

$$\delta \leqq 2^{-k-1}. \tag{3.5.6}$$

To summarize: the uniform error $\tilde{\varepsilon}$ in reproduction of a function $f(z)$ by a DFG is

$$\tilde{\varepsilon} \leqq \varepsilon + \delta, \tag{3.5.7}$$

where ε is the error in approximation of $f(z)$ by a step function $\Phi(z)$, and δ the error due to the finite accuracy of reproduction of $\Phi(z)$. Thus, when calculating the complexities $L^W(L^H)$ as in Corollary 3.5.1, the error in (3.5.5) should be taken equal to $\varepsilon = \tilde{\varepsilon} - \delta$, where $\tilde{\varepsilon}$ is the prescribed uniform error.

3.5.5. Sharper bounds for L^W may be established if one approximates $f(z)$ by a power series, such as the Taylor series. Recall that the weight $\|\omega\|$ of the index of the Walsh function $W_\omega(z)$ is the number of units in the binary expansion of ω.

THEOREM 3.5.2. *For any function $f(z)$, $z \in [0, 1]$, having a continuous q-th derivative, the coefficients $S(\omega)$ of its expansion in a series of Walsh functions with index weight q ($\|\omega\| = q$) satisfy the inequality*

$$|S(\omega)| \leqq (2^{(q^2+3q)})^{-1/2} \left| \max_{z\in[0,1]} f^{(q)}(z) \right|, \tag{3.5.8}$$

where $f^{(q)}(z)$ denotes the q-th derivative of $f(z)$.

PROOF. By Taylor's formula,

$$f(z) = \sum_{s=0}^{q-1} (s!)^{-1} f^{(s)}\left(\frac{1}{2}\right)\left(z - \frac{1}{2}\right)^s + (q!)^{-1} f^{(q)}(\theta(z))\left(z - \frac{1}{2}\right)^q, \tag{3.5.9}$$

where $0 \leqq \theta(z) \leqq 1$.

The binomial $(z - \frac{1}{2})^s$ may be expanded in Walsh functions [123]:

$$\left(z - \frac{1}{2}\right)^s = \left(-\sum_{i=1}^{\infty} 2^{-i-1} R_i(z)\right)^s. \tag{3.5.10}$$

It follows that the nonvanishing coefficients $D(\omega)$ in the Walsh spectrum of $(z - \frac{1}{2})^q$ have indices of weight at most q, and the maximum absolute value of these coefficients is given by

$$\max_{\|\omega\|=q} |D(\omega)| = 2^{-\sum_{i=1}^{q}(i+1)} \cdot q! = (2^{(q^2+3q)})^{-1/2} \cdot q!. \tag{3.5.11}$$

Substituting (3.5.10) for $(z - \frac{1}{2})^s$ in (3.5.9), we obtain

$$f(z) = \sum_{s=0}^{q-1} (s!)^{-1} f^{(s)}\left(\frac{1}{2}\right)\left(-\sum_{i=1}^{\infty} 2^{-i+1} R_i(z)\right)^s + f^{(q)}(\theta(z))(q!)^{-1}\left(-\sum_{i=1}^{\infty} 2^{-i-1} R_i(z)\right)^q, \tag{3.5.12}$$

where $0 \leqq \theta(z) \leqq 1$. It follows that the nonvanishing coefficients of index weight q in the spectrum of $f(z)$ are determined only by the second term, and by (3.5.11),

$$\max_{\|\omega\| = q} |S(\omega)| = (2^{(q^2+3q)})^{-1/2} \,|\max_{z\in[0,1]} f^{(q)}(z)|. \quad \blacksquare$$

Other estimates based on power series may be found, e.g., in [123].

EXAMPLE 3.5.2. Let us estimate the maximum coefficient of index weight 10 for $f(z) = \sin 2\pi z$ ($z \in [0, 1]$). We have

$$\max_{z\in[0,1]} |\sin^{(10)} 2\pi z| = (2\pi)^{10}.$$

Thus, by (3.5.8),

$$\max_{\|\omega\| = 10} |S(\omega)| \leqq 2^{-65} \cdot (2\pi)^{10} < 2 \cdot 10^{-10}.$$

Theorem 3.5.2 and Example 3.5.2 illustrate the rapid convergence of Walsh series and the advantage of using the Walsh basis in designing DFG for smooth functions.

3.5.6. We now consider the same problem for the number L^H of nonvanishing Haar coefficients.

Let $0 < \alpha \leqq 1$. We shall say that $f(z)$ is in class $H_\alpha(A)$ if, for any $z_1, z_2 \in [0, 1]$,

$$|f(z_1) - f(z_2)| \leqq A(z_1 - z_2)^\alpha, \tag{3.5.13}$$

where $A > 0$. We shall call $H_\alpha(A)$ the *Lipschitz class of order* α *with constant* A.

THEOREM 3.5.3 [146]. If $f(z) \in H_\alpha(A)$, then

$$L^H \leqq 2^{]\log_2((\alpha+1)\varepsilon^{-1}A)^{1/\alpha}[} \tag{3.5.14}$$

These bounds for the number of nonvanishing Haar coefficients may be lowered if the number m of digits in the code of the argument z is fixed in advance.

Let $f(z)$ be approximated with prescribed accuracy ε by a step function $\Phi(z) \in \{\Phi(z)\}^n$, where $\Phi(z)$ has discontinuities at points of the form $i \cdot 2^m$, $i = 0, 1, \ldots, 2^m - 1$. Then, by Theorem 3.1.1,

$$L^H \leqq n(m - 1) + 2. \tag{3.5.15}$$

This bound is useful when $n < 2^m/m$.

The above results demonstrate that in the design of function generators for analytic functions the use of Walsh functions is generally preferable to that of Haar functions, as far as the memory capacity necessary for storing the coefficients is concerned. The underlying reason is that, unlike the Haar coefficients, each of the Walsh coefficients depends on the behavior of $f(z)$ throughout its interval of definition.

Methods for DFG design, estimates of the accuracy of reproduction for some other bases, related to the Walsh and Haar bases, and also methods for the design of DFG computing functions of several variables, were considered in [56, 57, 58].

Structurally speaking, DFG designed on the basis of expansion in orthogonal series are described by the same block-diagrams as devices realizing systems of LF by spectral methods; details will be given in Chapter VI.

3.5.7. We now consider spectral methods for the synthesis of an important special class of DFG, which we call *sequence generators* (SG). An SG generates a function $y = f(z)$ defined at points $\{0, 1, \ldots, M-1\}$; when the codes of the numbers $0, 1, \ldots, M-1$ are applied at its input, the corresponding output is the sequence $f(0), f(1), \ldots, f(M-1)$. Thus, the characteristic feature of SG is that the sequence of argument codes z is fixed in advance. This generally simplifies realization of the SG in comparison with DFG in general. In the majority of SG, the argument z is the time.

All our spectral methods of synthesis may be applied to the design of SG, and the estimates for the convergence of the series representing $f(z)$ remain valid. We proceed now to some special methods for the synthesis of SG which frequently yield substantial reductions in the requisite amount of hardware.

We shall assume that the complexity of a SG is a monotone increasing function of the number L_f of nonzero values of $f(z)$. There are several reasons for this choice of complexity criterion. If the network realization of the SG is an (m, k)-terminal logic network, it is clear that with decreasing L_f the number of zero values of the corresponding Boolean functions will increase, and so, as shown in §2.3, the complexity of a minimal network realizing the system of BF will decrease (compare the criterion $\mu_0(f)$ for the complexity of BF; see §2.3.2). If SG is realized by spectral methods, a decrease in L_f implies a decrease in the absolute values of the expansion coefficients of $f(z)$, thus leading to a reduction in the complexity of the coefficient-storage block.

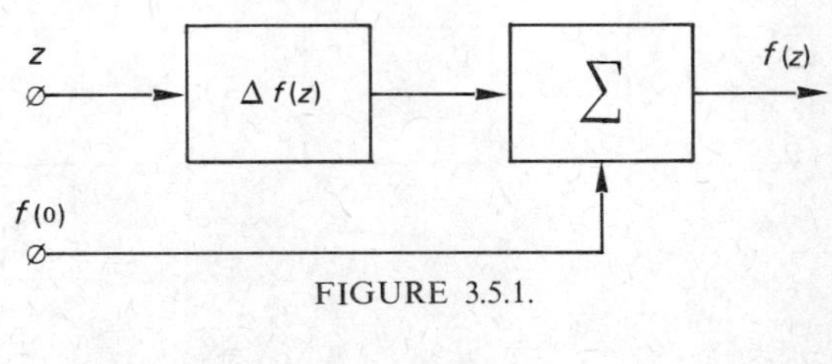

FIGURE 3.5.1.

Denote

$$\Delta f(z) = f(z) - f(z \ominus 1) \qquad (\text{mod } M). \tag{3.5.16}$$

Then

$$f(z) = \sum_{s=1}^{z} \Delta f(z) + f(0) \qquad (z \in \{1, 2, \ldots, M-1\}). \tag{3.5.17}$$

This representation determines a method for designing SG, in terms of the finite differences $\Delta f(z)$ of $f(z)$, as illustrated in Figure 3.5.1.

In the figure, $\Delta f(z)$ is an SG computing the function $\Delta f(z)$, and Σ is an adder-accumulator whose initial state is $f(0)$. For smooth functions, $L_{\Delta f} \ll L_f$, and this method of SG design turns out to be highly effective. An SG implementing the function $\Delta f(z)$ may in turn be designed by the same method, as a result of which $f(z)$ will be represented in terms of higher-order finite differences.

Another method for designing SG is related to the linearization of $f(z)$ with respect to the above criterion. Let $0 < \sigma < M$, and express $f(z)$ as

$$f_\sigma(\sigma \otimes z) = f(z) \qquad (\text{mod } M), \tag{3.5.18}$$

where f_σ is a discrete function defined at the points $0, 1, \ldots, M-1$. Thus, for each $\sigma \in \{1, 2, \ldots, M-1\}$ we have a linear transformation of the arguments modulo M. We confine attention to transformations which are nonsingular modulo M, so that the equation

$$\sigma \otimes z = 1 \qquad (\text{mod } M) \tag{3.5.19}$$

is solvable for any $z \in \{1, 2, \ldots, M-1\}$. The number σ generates a nonsingular transformation if and only if σ and M are relatively prime. Denote the set of nonsingular transformations mod M by Ξ_M. (The number of elements in this set is $\varphi(M)$, where φ is Euler's function [97].) The linearization problem for a sequence generator realizing $f(z)$ is: find a number $\sigma_{\min} \in \Xi_M$ such that

$$L_{\Delta f_{\sigma_{\min}}} = \min_{\sigma \in \Xi_M} L_{\Delta f_\sigma}, \tag{3.5.20}$$

where f_σ, $f_{\sigma_{\min}}$ are defined as in (3.5.18).

An SG designed by linearization of $f(z)$ is illustrated in Figure 3.5.2.

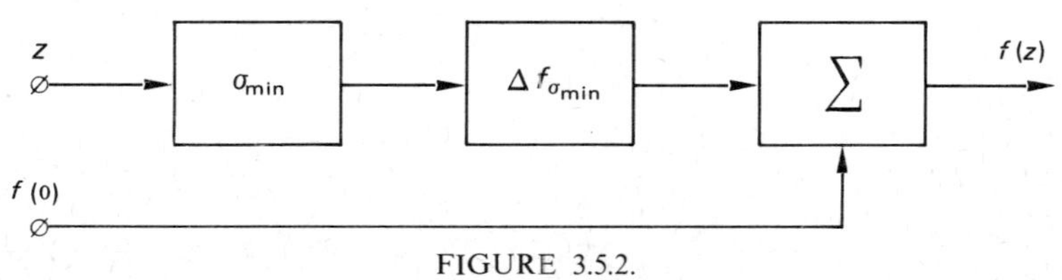

FIGURE 3.5.2.

The transformation $\sigma_{\min}$ itself can be determined by analyzing the appropriate correlation functions.

Let $\{f_i(z)\}$ be the system of characteristic functions of $f(z)$:

$$f_i(z) = \begin{cases} 1 & \text{if } f(z) = i, \\ 0 & \text{otherwise.} \end{cases} \tag{3.5.21}$$

The *cyclic autocorrelation functions* (mod M) of $f_i(z)$ are defined by

$$B_{M,i}(\tau) = \sum_{z=0}^{M-1} f_i(z)\, f_i(z \ominus \tau) \qquad (\text{mod } M). \tag{3.5.22}$$

It is evident from (3.5.22) and Theorem 1.5.1 that if M is a prime the functions $B_{M,i}(\tau)$ may be expressed as double Chrestenson transforms $\chi^{(M)}$ of $f_i(z)$ (see §1.5).

THEOREM 3.5.4. *Let*

$$\max_{\tau \in \Xi_m} \sum_i B_{M,i}(\tau) = \sum_i B_{M,i}(\tau_M). \tag{3.5.23}$$

Then

$$\sigma_{\min} \otimes \tau_M = 1 \qquad (\text{mod } M). \tag{3.5.24}$$

The proof is based on Theorem 1.5.4 for the case $m = 1$, $p = M$.

Thus, to find an optimal nonsingular linear transformation mod M for an SG we need only determine the maximum reading of the total cyclic atuocorrelation function $\sum_i B_{M,i}(\tau)$ over all τ relatively prime to M.

EXAMPLE 3.5.3. Let $M = 11$ and define $f(z)$ as in Table 3.5.1. Also given in the table are $\Delta f(z)$, $L_{\Delta f} = 6$, and the correlations $B_{11,0}(\tau)$, $B_{11,1}(\tau)$, $B_{11,0}(\tau) + B_{11,1}(\tau)$.

TABLE 3.5.1.

z, τ	f	Δf	$B_{11,0}$	$B_{11,1}$	$B_{11,0} + B_{11,1}$	f_8	Δf_8
0	1	0	5	6	11	1	1
1	0	1	2	3	5	1	0
2	1	1	0	1	1	1	0
3	1	0	3	4	7	1	0
4	0	1	4	5	9	1	0
5	0	0	1	2	3	1	0
6	1	1	1	2	3	0	1
7	1	0	4	5	9	0	0
8	0	1	3	4	7	0	0
9	0	0	0	1	1	0	0
10	1	1	2	3	5	0	0

We thus obtain $\tau_M = \tau_{11} = 7$, and by (3.5.24), $\sigma_{\min} = 8$. The functions $f_8(z)$ and $\Delta f_8(z)$ are also shown in the table; the result is $L_{\Delta f_8} = 2$.

REMARK. The linearization method of this section is essentially equivalent to minimization of the number of discontinuities of $f(z)$; it may be used in realization of $f(z)$ by expansion in Haar series, since the number L^H of nonvanishing coefficients is then determined by the number of discontinuities (see (3.5.15)). The method is extremely useful in the realization of logical functions by threshold element networks [30], and in many other cases.

Chapter Four

SPECTRAL METHODS FOR SYNTHESIS OF DIGITAL DEVICES WITH MEMORY

In the previous chapters we discussed the elements of the theory of discrete transforms and methods for their application to the analysis, synthesis and optimization of digital devices implementing logical functions (combinational networks). In this chapter we consider their application to the synthesis of digital devices with memory.

4.1. REALIZATION OF A FINITE AUTOMATON BY EXPANSION OF ITS EXCITATION FUNCTIONS IN ORTHOGONAL SERIES

4.1.1. *A finite automaton* (FA) is a model of the operation of a digital device (DD) with memory. The operation of any device may be modeled at two levels—*abstract* and *structural*. The appropriate mathematical models are known respectively as *abstract* and *structural* FA.

An *abstract* FA is defined as a set of six objects $M = \{X, A, Y, a_0, \varphi(x, a), g(a)\}$, where X is the (finite) set of *input signals*, A the (finite) set of (*internal*) *states*, Y the (finite) set of *output signals*, a_0 the *initial state*, $\varphi(x, a)$ the *next-state function*, and $g(a)$ the *output function* ($x \in X$, $a \in A$, $a_0 \in A$, and $\varphi(x, a) \in A$, $g(a) \in Y$ for any a, x).*

It is clear from this definition that an abstract FA is essentially a formal description of the operational algorithm of the digital device in question.

The synthesis of a DD usually begins with the construction of an abstract FA describing its operation. The abstract FA may be defined by a state table or a state diagram. The *state table* is a matrix whose element $a_{q,r}$ is the state into which the automaton passes from state a_q when the input signal x_r is applied ($a_q, a_{q,r} \in A$; $x_r \in X$); in other words, $a_{qr} = \varphi(x_r, a_q)$. The *state diagram* is a labeled directed graph whose vertices represent the states of the automaton and the corresponding output signals; a vertex a_s is joined to a vertex a_q by a directed arc labeled x_r if and only if $a_s = \varphi(x_r, a_q)$.

EXAMPLE 4.1.1. Table 4.1.1 is the state table of an automaton with $A = \{a_0, a_1, a_2, a_3\}$, $X = \{x_0, x_1, x_2, x_3\}$ (the output signals and states are identified).

* This is the *Moore model* of an abstract automaton [70]. If the *Mealy model* [70] is used instead, no essential changes are needed in the spectral methods developed below.

Once the operation of a DD has been simulated by a model at the abstract level, one proceeds to the next level—structural. Here the operational algorithm of the device is modeled by a structural FA.

A *structural FA* describing the operation of a DD based on p-ary elements is an abstract FA whose input signals, states and output signals are p-ary vectors, so that the next-state and output functions are essentially systems of p-valued logical functions.

4.1.2. Once the transition has been made from an abstract FA to the structural level (this transition is known as a *state-input assignment* or simply *assignment*), synthesis of the FA amounts to realization of the system of logical functions (LF) defined by the next-state and output functions. Throughout this chapter we shall limit the discussion to FA based on binary elements.

EXAMPLE 4.1.2. Suppose that the states and inputs of the abstract FA in Example 4.1.1 are assigned binary codes as shown in Table 4.1.2. Then the truth table for the system of Boolean functions (BF) corresponding to the next-state function is Table 4.1.3. (Here $x^{(s)}$, $a^{(s)}$ denote the s-th components of the input signal and state, respectively.)

The structural FA may be realized* as shown in the block-diagram of Figure 4.1.1, where $x(t)$, $a(t)$ and $y(t)$ denote the input signal, state and output signal, respectively, at time t; the blocks φ and g are combinational networks described by the systems

TABLE 4.1.1.

	x_0	x_1	x_2	x_3
a_0	a_0	a_0	a_3	a_3
a_1	a_2	a_2	a_2	a_0
a_2	a_1	a_3	a_0	a_2
a_3	a_3	a_1	a_1	a_1

TABLE 4.1.2.

a_q	Code of a_q	x_r	Code of x_r
a_0	0 0	x_0	0 0
a_1	0 1	x_1	0 1
a_2	1 0	x_2	1 0
a_3	1 1	x_3	1 1

* By a realization of a finite automaton we mean a digital device whose behavior is simulated by the automaton.

TABLE 4.1.3

$x^{(0)}$	$x^{(1)}$	$a^{(0)}$	$a^{(1)}$	$\varphi(x, a)$
0	0	0	0	0 0
0	0	0	1	1 0
0	0	1	0	0 1
0	0	1	1	1 1
0	1	0	0	0 0
0	1	0	1	1 0
0	1	1	0	1 1
0	1	1	1	0 1
1	0	0	0	1 1
1	0	0	1	1 0
1	0	1	0	0 0
1	0	1	1	0 1
1	1	0	0	1 1
1	1	0	1	0 0
1	1	1	0	1 0
1	1	1	1	0 1

of BF corresponding to the next-state and output functions. Thus, realization of the FA, given some state-input assignment, reduces to realization of the combinational networks φ and g.

Throughout this chapter we shall identify the output signal and the state: $y(t) = g(a(t)) = a(t)$. Whenever this condition is not satisfied, the output-function block g may be implemented by the methods of the previous chapters.

The Boolean functions realized by the next-state function block will be called *excitation functions*.

4.1.3. The methods of § 3.1 may be employed to realize the excitation functions by expanding them in orthogonal series. The adder-accumulator computing the sums of these series will then fulfill the function of memory.

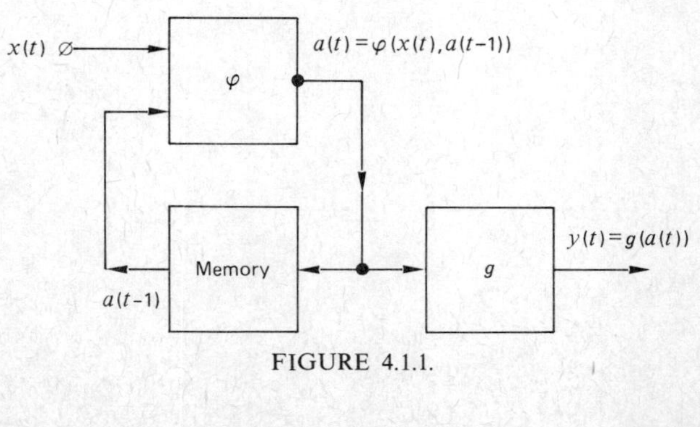

FIGURE 4.1.1.

Once a state-input assignment has been fixed, the realization of a system of excitation functions is our familiar problem of realizing a system of completely or incompletely specified BF (Table 4.1.4). The new notation used in the legend of the table is: n_x—number of input signals ($n_x = 2^c$), n_a = number of states ($n_a = 2^n$).

As usual in our spectral methods, we must first represent the system by a step function $\Phi(z)$. This is done as follows. Set

$$z = \sum_{s=0}^{\log_2 n_x - 1} x^{(s)}(t)\, 2^{\log_2 n_x - 1 - s} + n_x \sum_{s=0}^{\log_2 n_a - 1} a^{(s)}(t-1)\, 2^{\log_2 n_a - 1 - s}, \tag{4.1.1}$$

$$\tilde{a} = \sum_{s=0}^{\log_2 n_a - 1} a^{(s)}(t)\, 2^{\log_2 n_a - 1 - s}. \tag{4.1.2}$$

Then the next-state function $\varphi(x(t), a(t-1))$ defines a discrete function $\tilde{a} = f(z)$. (Table 4.1.5 defines this function for our example.) We complete $f(z)$ to a step function $\Phi(z)$ representing the system of excitation functions:

$$\Phi(z) = f(\delta) \quad \text{for} \quad \delta \leqq z < \delta + 1. \tag{4.1.3}$$

We now expand $\Phi(z)$ in an orthogonal series:

$$\Phi(z) = \sum_{q=0}^{n_a n_x - 1} c_q \Psi_q(z), \tag{4.1.4}$$

where

$$c_q = \left(\sum_{z=0}^{n_a n_x - 1} \Psi_q(z)\, \overline{\Psi_q(z)} \right)^{-1} \cdot \sum_{q=0}^{n_a n_x - 1} \Phi(z)\, \overline{\Psi_q(z)} \tag{4.1.5}$$

and $\{\Psi_q(z)\}$ is some complete system of orthogonal (basis) step functions.

Formula (4.1.4) defines a physical realization of the FA, as a network consisting of a function generator for the basis functions $\Psi_q(z)$, coefficient block c_q and adder.

TABLE 4.1.4

$a^{(0)}(t-1)$	$a^{(1)}(t-1)$	$\cdots$	$a^{(\log_2 n_a - 1)}(t-1)$	$x^{(0)}(t)$	$x^{(1)}(t)$	$\cdots$	$x^{(\log_2 n_x - 1)}(t)$	$a^{(0)}(t)$	$a^{(1)}(t)$	$\cdots$	$a^{(\log_2 n_a - 1)}(t)$

TABLE 4.1.5.

z	0	1	2	3	4	5	6	7	8	9	10	11	12	13	14	15
$f(z)$	0	0	3	3	2	2	2	0	1	3	0	2	3	1	1	1
$f_T(z)$	0	2	3	2	0	2	3	0	1	3	0	1	3	1	2	1

The step function $\Phi(z)$ representing a system of excitation functions may be constructed in various ways, depending on the choice of the order of the arguments $(x^{(0)}, x^{(1)}, \ldots, x^{(\log_2 n_x - 1)}, a^{(0)}, a^{(1)}, \ldots, a^{(\log_2 n_a - 1)})$. Let $T = (T_0, T_1, \ldots, T_{m-1})$, where $m = \log_2 n_a + \log_2 n_x$, be a permutation of the numbers $0, 1, \ldots, m-1$. Set

$$z^{(s)} = \begin{cases} x^{(s)}(t) & \text{for} \quad s = 0, 1, \ldots, \log_2 n_x - 1, \\ a^{(s - \log_2 n_x)}(t-1) & \text{for} \quad s = \log_2 n_x, \log_2 n_x + 1, \ldots, m-1. \end{cases}$$

If $\Phi_T(z)$ is the step function representing the system of excitation functions with order of arguments T, then z is defined by

$$z = \sum_{s=0}^{m-1} z^{(T_s)} 2^{m-1-s}; \tag{4.1.6}$$

$\tilde{a}$ is defined by (4.1.2); in view of (4.1.2), (4.1.6), the next-state function $\varphi(x(t), a(t-1))$ is defined by $\tilde{a} = f_T(z)$, and $\Phi_T(z)$ is the completion of $f_T(z)$ defined as in (4.1.3). (The function $\Phi(z)$ defined by (4.1.1) corresponds to the identity permutation $(0, 1, \ldots, m-1)$.)

EXAMPLE 4.1.3. Consider the assignment of Table 4.1.2 for the FA in Example 4.1.1. Let $T = (2, 1, 0, 3)$ $(m = 4)$. Then $z^{(0)} = x^{(0)}$, $z^{(1)} = x^{(1)}$, $z^{(2)} = a^{(0)}$, $z^{(3)} = z^{(1)}$, and by (4.1.6):

$$z = a^{(0)}2^3 + x^{(1)}2^2 + x^{(0)}2^1 + a^{(1)}2^0.$$

The corresponding function $f_T(z)$ is shown in Table 4.1.5.

The problem of optimizing the order of arguments so as to minimize the number of nonvanishing coefficients in (4.1.4) for a given basis $\{\Psi_q(z)\}$ (and hence also minimize the complexity of the combinational part of the FA) is inextricably bound up with the state-input assignment problem. Both these problems will be attacked in §§4.2 and 4.3.

When the excitation functions are being realized by expansion in orthogonal series, we can utilize all the spectral methods of synthesis described in Chapter III: optimizing the linear transformation (minimizing the number of nonvanishing terms in the series) (see §3.2), optimizing the completion of the excitation functions (see §3.3 and also §4.3), summation of series over finite fields (see §1.6), and so on.

To end this section, we list the main distinguishing features and advantages of spectral methods for the realization of FA:

1) The structure of the digital device is fixed in advance; the only parameter to be minimized is the number of nonvanishing coefficients in the expansion of the step function representing the system of excitation functions (see §§6.1, 6.2).

2) Once the assignment has been fixed, minimization of the number of nonvanishing coefficients and optimization of the completion of the excitation functions may be accomplished by techniques which almost completely avoid the brute-force approach (see §§3.2, 3.3).

3) It is relatively easy to determine asymptotically optimal assignments for abstract FA and asymptotically optimal completions of the next-state function for incompletely specified FA (see §§4.2, 4.3).

4) Automata designed by spectral methods provide for a highly simple organization of error detection and/or correction using arithmetic error-detecting and/or -correcting codes (see §5.4).

A more detailed comparison of spectral and classical methods of synthesis will be carried out in §6.4.

4.2. STATE-INPUT ASSIGNMENT FOR COMPLETELY SPECIFIED AUTOMATA

4.2.1. One of the most complex problems of modern automata theory is optimization of state-input assignments as regards the complexity, reliability, etc. of the projected digital device. It was shown in the last section that, once an assignment has been selected, realization of the FA reduces to realization of systems of BF representing the next-state and output functions of the FA. The disadvantage of classical methods of Boolean algebra in the state-assignment problem for the combinational part of an FA is that the system of BF describing the combinational part under a given assignment does not readily yield information concerning the complexity, reliability, etc. of a corresponding network realization. The situation is further complicated if the FA to be synthesized is incompletely specified, i.e., the next-state function is undefined for certain input signal-state pairs.

Because of these complicating factors, classical solutions of the assignment problem always involve some measure of brute force. The number of alternatives to be checked generally increases exponentially with the number of states or input signals of the FA. This situation is apparently intrinsic in nature, thus limiting application of the classical methods to relatively simple devices.

A complete survey of existing methods for solving the assignment problem by classical methods may be found, e.g., in [62].

By contrast, we shall see presently (§§4.2–4.4) that the spectral approach to realization of the combinational part of an FA yields highly simple assignment algorithms and algorithms computing asymptotically optimal arrangements of arguments in the excitation functions for almost all FA (i.e., the relative abundance of automata for which the state assignment is not optimal tends to zero—quite rapidly—as the number of states tends to infinity). As in Chapter II, we shall measure the complexity of the combinational part by the number of nonvanishing expansion terms;

the basis system will again be the Haar basis (see §1.2). For the sake of convenience, we shall restrict ourselves here and in §4.3 to Haar expansions over the field of real numbers. However, the method described here for optimal assignments, optimal ordering of arguments of the excitation functions and optimal completion of the state table carries over practically unchanged to Haar–Galois expansions (see §6.3).

4.2.2. We proceed to the first two of the last-mentioned problems: optimization of the state-input assignment and optimization of the order of arguments in the excitation functions, for realization of the combinational part of a completely specified FA, using the Haar basis.

We have already discussed the main properties of the Haar system $\{H_l^{(q)}(z)\}$, the reasons for choosing it as a basis, and efficient computational procedures for the expansion coefficients $c_l^{(q)}$ (see §1.2).

Let L_l ($l = 0, 1, \ldots, m - 1$) denote the number of nonvanishing $c_l^{(q)}$ ($q = 1, 2, \ldots, 2^l$) with subscript l in the expansion of $\Phi_T(z)$. Then $0 \leqq L_l \leqq 2^l$, and so our first task is to determine the set of state-input assignments minimizing the number L_{m-1} of nonvanishing coefficients $c_{m-1}^{(q)}$. We shall then look for the subset of assignments minimizing L_{m-2} for the optimal L_{m-1}, and so on.

Arbitrary assignments will be represented by two functions $K^a(q)$ and $K^x(r)$ ($q = 0, 2, \ldots, n_a - 1$; $r = 0, 1, \ldots, n_x - 1$), where $K^a(q)$ $(K^x(r))$ is the number whose binary expansion is the code vector of the state a_q (input signal x_r) in the assignment; in other words.

$$K^a(q) = \sum_{s=0}^{\log_2 n_a - 1} a_q^{(s)} 2^{\log_2 n_a - 1 - s}, \tag{4.2.1}$$

$$K^x(r) = \sum_{s=0}^{\log_2 n_x - 1} x_r^{(s)} 2^{\log_2 n_x - 1 - s}, \tag{4.2.2}$$

where $a_q^{(s)}$ $(x_r^{(s)})$ is the s-th component of the binary code of the state a_q (input signal x_r).

Denote the inverse functions of $K^a(q)$ and $K^x(r)$ by $(K^a)^{-1}(q)$ and $(K^x)^{-1}(r)$, respectively. To be precise: $(K^a)^{-1}(q) = \lambda$ $((K^x)^{-1}(q) = \lambda)$ if and only if the code of the state a_λ (input signal x_λ) is the binary expansion of the number q. For example, if the state a_4 is encoded by 011, then $K^a(4) = 3$ $((K^a)^{-1}(3) = 4)$.

We now consider how to determine the set of assignments minimizing the number L_{m-1}. We shall assume that the original abstract automaton is defined by a state matrix $A_{m-1} = (a_{q,r})$ ($q = 0, 1, \ldots, n_a - 1$; $n_a = 2^n$; $r = 0, 1, \ldots, n_x - 1$; $n_x = 2^c$), where $a_{q,r}$ is the state into which the automaton passes from a_q upon application of the input signal x_r.

We shall need *distance functions* on the set of rows and columns, respectively, of the state matrix. The distance $\rho_{m-1}^a(q; s)$ $(\rho_{m-1}^x(q; s))$ between rows (columns) q

and s of the state matrix is defined as follows:

$$\rho^a_{m-1}(q; s) = \sum_{r=0}^{n_x-1} \rho(a_{q,r}; a_{s,r}), \tag{4.2.3}$$

$$\rho^x_{m-1}(q; s) = \sum_{r=0}^{n_a-1} \rho(a_{r,q}; a_{r,s}), \tag{4.2.4}$$

where

$$\rho(a_{\alpha,\beta}; a_{\gamma,\delta}) = \begin{cases} 1 & \text{if} \quad a_{\alpha,\beta} \neq a_{\gamma,\delta}, \\ 0 & \text{if} \quad a_{\alpha,\beta} = a_{\gamma,\delta}. \end{cases} \tag{4.2.5}$$

Using formulas (4.2.3)–(4.2.5), we construct $n_a \times n_a$ and $n_x \times n_x$ *distance matrices* $(\rho^a(q; s))$ and $(\rho^x(q; s))$, respectively, given the state matrix of the abstract FA.*

THEOREM 4.2.1. *Given an abstract FA* $A_{m-1} = (a_{q,r})$ *and an assignment* $(K^a(q), K^x(r))$. *Then*

$$L_{m-1} = \min(L^a_{m-1}, L^x_{m-1}),$$

where

$$\begin{aligned} L^x_{m-1} &= \sum_{s=1}^{n_x/2} \rho^x_{m-1}((K^x)^{-1}(2s-2); (K^x)^{-1}(2s-1)), \\ L^a_{m-1} &= \sum_{i=1}^{n_a/2} \rho^a_{m-1}((K^a)^{-1}(2i-2); (K^a)^{-1}(2i-1)). \end{aligned} \tag{4.2.6}$$

PROOF. Define a function $\hat{\varphi}(\lambda, \beta)$ by setting $\hat{\varphi}(\lambda, \beta) = \delta$ if and only if there exist $q, t \in \{0, 1, \ldots, n_a - 1\}$ and $r \in \{0, 1, \ldots, n_x - 1\}$ such that

$$\begin{cases} K^a(q) = \lambda, K^x(r) = \beta, K^a(t) = \delta, \\ \varphi(x_r, a_q) = a_t. \end{cases}$$

Then depending on the selected order of arguments of $\hat{\varphi}$, we have

$$L_{m-1} = L^a_{m-1} = \sum_{\beta=0}^{n_x-1} \sum_{i=1}^{n_a/2} \operatorname{sign} |\hat{\varphi}(2i-2, \beta) - \hat{\varphi}(2i-1, \beta)|$$

or

$$L_{m-1} = L^x_{m-1} = \sum_{\lambda=0}^{n_a-1} \sum_{s=1}^{n_x/2} \operatorname{sign} |\hat{\varphi}(\lambda, 2s-2) - \hat{\varphi}(\lambda, 2s-1)|.$$

* The distance thus defined on the state matrix satisfies the usual axioms for a metric; it is a generalization of the Hamming metric [14].

We first consider the case $L_{m-1} = L^a_{m-1}$. Set

$$q' = (K^a)^{-1}(2i-2), \qquad q'' = (K^a)^{-1}(2i-1), \qquad r = (K^x)^{-1}(\beta)$$

$$(i \in \{1, 2, \ldots, n_a/2\};\ q', q'' \in \{0, 1, \ldots, n_a - 1\};\ r, \beta \in \{0, 1, \ldots, n_x - 1).$$

Then

$$\operatorname{sign}|\hat{\varphi}(2i-2, \beta) - \hat{\varphi}(2i-1, \beta)| = 1$$

if and only if

$$\varphi(x_r, a_{q'}) \neq \varphi(x_r, a_{q''}),$$

or, in view of (4.2.5), if and only if

$$\rho(a_{q',r}; a_{q'',r}) = 1.$$

Hence, by (4.2.4),

$$\sum_{\beta=0}^{n_x-1} \operatorname{sign}|\hat{\varphi}(2i-2, \beta) - \hat{\varphi}(2i-1, \beta)| = \sum_{r=0}^{n_x-1} \rho(a_{q',r}; a_{q'',r}) = \rho^a_{m-1}(q'; q'')$$
$$= \rho^a_{m-1}\big((K^a)^{-1}(2i-2); (K^a)^{-1}(2i-1)\big).$$

Consequently,

$$L^a_{m-1} = \sum_{i=1}^{n_a/2} \sum_{\beta=0}^{n_x-1} \operatorname{sign}|\hat{\varphi}(2i-2, \beta) - \hat{\varphi}(2i-1, \beta)|$$
$$= \sum_{i=1}^{n_a/2} \rho^a_{m-1}\big((K^a)^{-1}(2i-2); (K^a)^{-1}(2i-1)\big).$$

The proof of the equality

$$L^x_{m-1} = \sum_{s=1}^{n_x/2} \rho^x_{m-1}\big((K^x)^{-1}(2s-2); (K^x)^{-1}(2s-1)\big)$$

is analogous, completing the proof. ∎

EXAMPLE 4.2.1. Consider the abstract FA with state matrix

$$A_{m-1} = (a_{q,r}) = A_5 = \begin{pmatrix} a_0 & a_4 & a_0 & a_6 & a_4 & a_0 & a_2 & a_0 \\ a_1 & a_0 & a_2 & a_3 & a_0 & a_2 & a_3 & a_2 \\ a_5 & a_1 & a_7 & a_6 & a_7 & a_5 & a_6 & a_6 \\ a_0 & a_3 & a_0 & a_6 & a_3 & a_0 & a_7 & a_0 \\ a_1 & a_0 & a_2 & a_3 & a_4 & a_2 & a_3 & a_2 \\ a_0 & a_3 & a_0 & a_4 & a_3 & a_0 & a_4 & a_0 \\ a_0 & a_6 & a_0 & a_4 & a_6 & a_0 & a_4 & a_0 \\ a_5 & a_2 & a_7 & a_6 & a_2 & a_5 & a_6 & a_6 \end{pmatrix}.$$

Then

$$(\rho_5^a(q;s)) = \begin{pmatrix} 0 & 8 & 7 & \textcircled{2} & 7 & 4 & 4 & 8 \\ 8 & 0 & 8 & 8 & \textcircled{1} & 8 & 8 & 8 \\ 7 & 8 & 0 & 7 & 8 & 8 & 8 & \textcircled{2} \\ \textcircled{2} & 8 & 7 & 0 & 8 & 2 & 4 & 7 \\ 7 & \textcircled{1} & 8 & 8 & 0 & 8 & 8 & 8 \\ 4 & 8 & 8 & 2 & 8 & 0 & \textcircled{2} & 8 \\ 4 & 8 & 8 & 4 & 8 & \textcircled{2} & 0 & 8 \\ 8 & 8 & \textcircled{2} & 7 & 8 & 8 & 8 & 0 \end{pmatrix} \quad \text{and}$$

$$(\rho_5^x(q;s)) = \begin{pmatrix} 0 & 8 & 4 & 8 & 8 & \textcircled{2} & 8 & 4 \\ 8 & 0 & 8 & 8 & \textcircled{1} & 8 & 8 & 8 \\ 4 & 8 & 0 & 8 & 8 & 2 & 8 & \textcircled{2} \\ 8 & 8 & 8 & 0 & 8 & 8 & \textcircled{2} & 6 \\ 8 & \textcircled{1} & 8 & 8 & 0 & 8 & 8 & 8 \\ \textcircled{2} & 8 & 2 & 8 & 8 & 0 & 8 & 2 \\ 8 & 8 & 8 & \textcircled{2} & 8 & 8 & 0 & 6 \\ 4 & 8 & \textcircled{2} & 6 & 8 & 2 & 6 & 0 \end{pmatrix}.$$

By Theorem 4.2.1, the assignment $K^a(q) = q$, $K^x(q) = q$ gives $L_{m-1} = L_5 = \min(31, 30) = 30$. (As we shall show below, the optimal assignment for this example gives $L_{m-1} = 7$.)

It follows from Theorem 4.2.1 that the optimal assignment for L_{m-1} minimizes one of the quantities $\mathcal{L}^a_{m-1}, \mathcal{L}^x_{m-1}$.

To find an optimal assignment, we first determine the class of assignments $K^a_{m-1} = \{K^a(q)\}$ minimizing $\mathcal{L}^a_{m-1}$, and then the class of assignments $K^x_{m-1} = \{K^x(q)\}$ minimizing $\mathcal{L}^x_{m-1}$. The final class of assignments K_{m-1}, minimizing L_{m-1}, is then either K^a_{m-1} or K^x_{m-1}, according as $\mathcal{L}^a_{m-1} \leqq \mathcal{L}^x_{m-1}$ or $\mathcal{L}^a_{m-1} > \mathcal{L}^x_{m-1}$.

If $z^{(T_s)}$ is the s-th argument of the excitation function (see §4.1.3) for the required optimal order T of arguments, we set

$$z^{(T_{m-1})} = \begin{cases} x^{(\log_2 n_x - 1)} & \text{if} \quad \mathcal{L}^x_{m-1} < \mathcal{L}^a_{m-1}, \\ a^{(\log_2 n_a - 1)} & \text{if} \quad \mathcal{L}^a_{m-1} \leqq \mathcal{L}^x_{m-1}. \end{cases} \tag{4.2.7}$$

4.2.3. In order to determine the class $K^a_{m-1}(K^x_{m-1})$ minimizing

$$\mathcal{L}^a_{m-1} = \sum_{i=1}^{n_a/2} \rho^a_{m-1}\big((K^a)^{-1}(2i-2); (K^a)^{-1}(2i-1)\big)$$

$$\left(L^x_{m-1} = \sum_{q=1}^{n_x/2} \rho^x_{m-1}\big((K^x)^{-1}(2q-2); (K^x)^{-1}(2q-1)\big) \right),$$

it suffices to partition the set of rows (columns) of the state matrix into disjoint pairs in such a way that the sum of distances (4.2.3) ((4.2.4)) is minimized. If these pairs are $\{i_0; i_1\}, \ldots, \{i_{n_a-2}; i_{n_a-1}\}$ ($\{g_0; g_1\}, \ldots, \{g_{n_x-2}; g_{n_x-1}\}$) then the required class of assignments K^a_{m-1} (K^x_{m-1}) is the set of all assignments such that

$$\begin{cases} (K^a)^{-1}(2\alpha) = i_{2q} \\ (K^a)^{-1}(2\alpha+1) = i_{2q+1} \end{cases} \text{or} \quad \begin{cases} (K^a)^{-1}(2\alpha) = i_{2q+1} \\ (K^a)^{-1}(2\alpha+1) = i_{2q} \end{cases}$$

$$\left(\alpha = 0, 1, \ldots, \frac{n_a}{2} - 1, \quad q = 0, 1, \ldots, \frac{n_a}{2} - 1 \right) \tag{4.2.8}$$

$$\left(\begin{cases} (K^x)^{-1}(2\alpha) = g_{2s} \\ (K^x)^{-1}(2\alpha+1) = g_{2s+1} \end{cases} \text{or} \quad \begin{cases} (K^x)^{-1}(2\alpha) = g_{2s+1} \\ (K^x)^{-1}(2\alpha+1) = g_{2s} \end{cases} \quad \left(\alpha = 0, 1, \ldots, \frac{n_x}{2} - 1, \quad s = 0, 1, \ldots, \frac{n_x}{2} - 1 \right) \right).$$

It follows from (4.2.8) that $K^a_{m-1}(K^x_{m-1})$ is the set of all assignments for which the states (input signals) in one pair are assigned binary codes differing only in the lowest-order digit.

The above construction of $K^a_{m-1}(K^x_{m-1})$ may be formulated in terms of the matrix $(\rho^a_{m-1}(q; s))$ $((\rho^x_{m-1}(q; s)))$ as follows. We wish to find n_a (n_x) elements of the matrix, say $\rho^a_{m-1}(q_0; s_0), \rho^a_{m-1}(q_1; s_1), \ldots, \rho^a_{m-1}(q_{n_a-1}; s_{n_a-1})$ $(\rho^x_{m-1}(q_0; s_0), \rho^x_{m-1}(q_1; s_1), \ldots, \rho^x_{m-1}(q_{n_x-1}; s_{n_x-1}))$, where $q_i \neq s_i$, such that each row (column) of the matrix contains exactly one of the selected elements; the elements are symmetrically situated about the principal diagonal, and their sum is a minimum. Let us call any set of matrix elements satisfying these conditions a *minimal symmetric matching* of the matrix. It is apparent from Theorem 4.2.1 that half the sum of elements of a minimal symmetric matching for the matrix $(\rho^a_{m-1}(q, s))$ $((\rho^x_{m-1}(q, s)))$ gives $L^a_{m-1}(L^x_{m-1})$. The minimal symmetric matching of a matrix is relatively easy to determine, using the so-called Hungarian algorithm [33, 95, 96, 108, 155].*

Resuming our discussion of Example 4.2.1, let us find the class of assignments $K_{m-1} = K_5$ minimizing L_{m-1}. To determine K^a_5, we find a minimal symmetric matching for $(\rho^a_5(q; s))$, which in this case is

$$\{\rho^a_5(0; 3), \rho^a_5(1; 4), (\rho^a_5(2; 7), \rho^a_5(5; 6)\}.$$

* To ensure that the minimal matching is symmetric, one actually uses a slightly modified version of the Hungarian algorithm.

(In view of the symmetry, we shall henceforth specify matchings in terms of elements above the principal diagonal; elements appearing in the matching are circled in the matrix $(\rho_5^a(q; s))$.) The states are thus partitioned into pairs

$$\{a_0; a_3\}, \{a_1; a_4\}, \{a_2; a_7\}, \{a_5; a_6\}.$$

The class K_5^a is the set of all assignments in which states of the same pair are assigned codes differing only in the lowest-order digit $a^{(\log_2 n_a - 1)}$. For example, the assignment of Table 4.2.1 is in K_5^a.

TABLE 4.2.1.

q	0	1	2	3	4	5	6	7
$K^a(q)$	0	2	4	1	3	6	7	5

Half the sum of the "intrapair" distances is

$$\rho_5^a(0; 3) + \rho_5^a(1; 4) + \rho_5^a(2; 7) + \rho_5^a(5; 6) = 7,$$

and so $L_5^a = 7$.

Similarly, to determine K_5^x we find a minimal symmetric matching for $(\rho_5^x(q; s))$:

$$\{\rho_5^x(0: 5), \rho_5^x(1; 4), \rho_5^x(2; 7), \rho_5^x(3; 6)\};$$

the input signals are divided into pairs $\{x_0; x_5\}$, $\{x_1; x_4\}$, $\{x_2; x_7\}$, $\{x_3; x_6\}$, and

$$L_5^x = \rho_5^x(0; 5) + \rho_5^x(1; 4) + \rho_5^x(2; 7) + \rho_5^x(3; 6) = 7.$$

Since $L_5^a = L_5^x = 7$, we may take, say $K_5 = K_5^a$. Thus, whereas the original assignment for Example 4.2.1 gave $L_{m-1} = L_5 = 30$, any assignment in class K_5 gives $L_5 = 7$.

4.2.4. Continuing the procedure, we must now determine the classes of assignments $K_{m-2} \supseteqq K_{m-3} \supseteqq \cdots \supseteqq K_1 \supseteqq K_0$, where K_{m-s} is the set of all assignments in K_{m-s+1} for which L_{m-s} assumes a local minimum, preserving the previously attained value of L_{m-s+1}. Along with the classes K_{m-s} $(s = 1, 2, \ldots, m)$ we determine the order T (see §4.1.3) of arguments in the excitation functions of the automaton.

We now outline the construction of $K_{m-2} \subseteqq K_{m-1}$, assuming to fix ideas that $L_{m-1}^a \leqq L_{m-1}^x$, so that $K_{m-1} = K_{m-1}^a$. Construct the $n_a/2 \times n_x$ matrix $A_{m-2} = A_{m-2}^a = (b_{q,r})$ whose (q, r)-th element is the (unordered) pair

$$b_{q,r} = \{a_{i_{2q},r}; a_{i_{2q+1},r}\}, \tag{4.2.9}$$

where $q = 0, 1, \ldots, n_a/2 - 1$, $r = 0, 1, \ldots, n_x/2 - 1$, and let

$$\{\{i_0, i_1\}, \ldots, \{i_{n_a-2}; i_{n_a-1}\}\}$$

be the partition into pairs of the rows of A_{m-1}, determined by a minimal matching for the matrix $(\rho^a_{m-1}(q; s))$. Thus, to the q-th row of $A_{m-2} = A^a_{m-2}$ corresponds the pair of rows i_{2q}, i_{2q+1} of the state matrix A_{m-1}.

In the case of Example 4.2.1,

$$A_{m-2} = \begin{pmatrix} \{a_0;a_0\} & \{a_4;a_3\} & \{a_0;a_0\} & \{a_6;a_6\} & \{a_4;a_3\} & \{a_0;a_0\} & \{a_7;a_7\} & \{a_0;a_0\} \\ \{a_1;a_1\} & \{a_0;a_0\} & \{a_2;a_2\} & \{a_3;a_3\} & \{a_0;a_4\} & \{a_2;a_2\} & \{a_3;a_3\} & \{a_2;a_2\} \\ \{a_5;a_5\} & \{a_1;a_2\} & \{a_7;a_7\} & \{a_6;a_6\} & \{a_1;a_2\} & \{a_5;a_5\} & \{a_6;a_6\} & \{a_6;a_6\} \\ \{a_0;a_0\} & \{a_3;a_6\} & \{a_0;a_0\} & \{a_4;a_4\} & \{a_3;a_6\} & \{a_0;a_0\} & \{a_4;a_4\} & \{a_0;a_0\} \end{pmatrix}.$$

As in (4.2.3)–(4.2.5), we define distances $\rho^a_{m-2}(q; s)$ and $\rho^x_{m-2}(q; s)$ on the rows and columns, respectively, of A_{m-2}:

$$\rho^a_{m-2}(q; s) = \sum_{r=0}^{n_x-1} \rho(b_{q,r}; b_{s,r}) \qquad (q, s = 0, 1, \ldots, n_a/2 - 1), \tag{4.2.10}$$

$$\rho^x_{m-2}(q; s) = \sum_{r=0}^{n_a/2-1} \rho(b_{r,q}; b_{s,r}) \qquad (q, s = 0, 1, \ldots, n_x - 1), \tag{4.2.11}$$

where

$$\rho(b_{\alpha,\beta}; b_{\gamma,\delta}) = \begin{cases} 1 & \text{if} \quad b_{\alpha,\beta} \neq b_{\gamma,\delta}, \\ 0 & \text{if} \quad b_{\alpha,\beta} = b_{\gamma,\delta}. \end{cases} \tag{4.2.12}$$

(Note that the equality of elements of A_{m-2} is equality of unordered pairs.) In Example 4.2.1,

$$(\rho^a_4(q; s)) = \begin{pmatrix} 0 & 8 & 7 & \textcircled{4} \\ 8 & 0 & \textcircled{8} & 8 \\ 7 & \textcircled{8} & 0 & 8 \\ \textcircled{4} & 8 & 8 & 0 \end{pmatrix},$$

$$(\rho^x_4(q; s)) = \begin{pmatrix} 0 & 4 & 2 & 4 & 4 & \textcircled{1} & 4 & 2 \\ 4 & 0 & 4 & 4 & \textcircled{1} & 4 & 4 & 4 \\ 2 & 4 & 0 & 4 & 4 & 1 & 4 & \textcircled{1} \\ 4 & 4 & 4 & 0 & 4 & 4 & \textcircled{1} & 3 \\ 4 & \textcircled{1} & 4 & 4 & 0 & 4 & 4 & 4 \\ \textcircled{1} & 4 & 1 & 4 & 4 & 0 & 4 & 1 \\ 4 & 4 & 4 & \textcircled{1} & 4 & 4 & 0 & 4 \\ 2 & 4 & \textcircled{1} & 3 & 4 & 1 & 4 & 0 \end{pmatrix}.$$

If $L^a_{m-1} > L^x_{m-1}$, so that $K_{m-1} = K^x_{m-1}$, we define A_{m-2} as A^x_{m-2} instead of A^a_{m-2} (this is an $n_a \times n_x/2$ matrix, derived from A_{m-1} in the same way as A^a_{m-2}). The procedure is then analogous, except that in formulas (4.2.9)–(4.2.12) the optimal matching $\{i_{2q}; i_{2q+1}\}$ $(q = 0, 1, \ldots, n_a/2 - 1)$ on (ρ^a_{m-1}) is replaced by an optimal matching on (ρ^x_{m-1}): $\{g_{2q}; g_{2q+1}\}$ $(q = 0, 1, \ldots, n_x/2 - 1)$.

Any state (input) assignment in class K^a_{m-1} (K^x_{m-1}) may be defined, as in (4.2.1), (4.2.2), by a function $K^a_{m-2}(q)\,(K^x_{m-2}(q))$:

$$K^a_{m-2}(q) = \sum_{s=0}^{\log_2 n_a - 2} a^{(s)}_{i_{2q}} \cdot 2^{\log_2 n_a - 2 - s}$$

$$= \sum_{s=0}^{\log_2 n_a - 2} a^{(s)}_{i_{2q+1}} 2^{\log_2 n_a - 2 - s} \quad (q = 0, \ldots, \tfrac{1}{2} n_a - 1), \tag{4.2.13}$$

$$\left(K^x_{m-2}(q) = \sum_{s=0}^{\log_2 n_x - 2} x^{(s)}_{g_{2q}} \cdot 2^{\log_2 n_x - 2 - s} \right.$$

$$\left. = \sum_{s=0}^{\log_2 n_x - 2} x^{(s)}_{g_{2q+1}} 2^{\log_2 n_x - 2 - s} \quad (q = 0, \ldots, \tfrac{1}{2} n_x - 1) \right). \tag{4.2.14}$$

It follows from these formulas that for fixed q the value of $K^a_{m-1}(q)\,(K^x_{m-1}(q))$ may be regarded as the number whose binary expansion is determined by the first $\log_2 n_a - 1$ $(\log_2 n_x - 1)$ components of the codes of states $a_{i_{2q}}, a_{i_{2q+1}}$ (input signals $x_{g_{2q}}, x_{g_{2q+1}}$). Since the pair of states $\{a_{i_{2q}}; a_{i_{2q+1}}\}$ (input signals $\{x_{g_{2q}}; x_{g_{2q+1}}\}$) corresponds to the q-th row of A^a_{m-2} (q-th column of A^x_{m-2}), we may assume $K^a_{m-2}(q)$ $(K^x_{m-2}(q))$ defines the number corresponding to the q-th row of A^a_{m-2} (q-th column of A^x_{m-2}) in the given assignment.

The inverse functions of $K^a_{m-2}(q)$ and $K^x_{m-2}(q)$ will be denoted by $(K^a_{m-2})^{-1}(q)$ and $(K^x_{m-2})^{-1}(q)$; these functions define, respectively the row of A^a_{m-2} and column of A^x_{m-2} which the assignment associates with the number q.

For example, if states a_{i_2} and a_{i_3} are assigned codes (110) and (111), then $K^a_{m-2}(1) = 3$, $(K^a_{m-2})^{-1}(3) = 1$.

THEOREM 4.2.2. *For almost all automata with assignment in* K_{m-1},

$$L_{m-2} \sim \min\left(\sum_{q=1}^{n_a/4} \rho^a_{m-2}\big((K^a_{m-2})^{-1}(2q-2); (K^a_{m-2})^{-1}(2q-1)\big), \right.$$

$$\left. \sum_{q=1}^{n_x/4} \rho^x_{m-2}\big((K^x_{m-2})^{-1}(2q-2); (K^x_{m-2})^{-1}(2q-1)\big) \right)$$

$$(m = \log_2 n_a + \log_2 n_x) \tag{4.2.15}$$

as $n = \log_2 n_a \to \infty$ *and* $c = \log_2 n_x \to \infty$.

PROOF. The proof is analogous to that of Theorem 4.2.1. The only new point is the use of the limit relation

$$\rho(b_{\alpha,\beta}, b_{\gamma,\delta}) = \lim_{n_a \to \infty} P\left(\sum_{a_q \in b_{\alpha,\beta}} K^a(q) \neq \sum_{a_q \in b_{\gamma,\delta}} K^a(q) \right),$$

where $\rho(b_{\alpha,\beta}; b_{\gamma,\delta})$ is defined by (4.2.12) and $P(\sum_{\alpha_q \in b_{\alpha,\beta}} K^a(q) \neq \sum_{a_q \in b_{\gamma,\delta}} K^a(q))$ is the relative abundance of assignments $K^a(q)$ for the automaton such that the sum of the numbers corresponding to codes of states in the set $b_{\alpha,\beta}$ is not equal to the analogous sum for the set $b_{\gamma,\delta}$. ∎

Theorem 4.2.2 provides a fairly simple procedure for estimating L_{m-2} in terms of the matrix A_{m-2}, with the right-hand side of (4.2.15) as an upper bound for L_{m-2}. Thus, in Example 4.2.1, for an assignment in $K_5 = K_5^a$, analysis of the matrices $(\rho_4^a(q; s))$ and $(\rho_4^x(q; s))$ by Theorem 4.2.2 with $K_4^a(q) = q$, $K_4^x(q) = q$ gives $L_4 \leqq 16$. (We shall see later that by optimizing the assignment we can lower the bound for L_4 to 4.)

It follows from Theorem 4.2.2 that for almost all automata the asymptotically optimal assignment in class K_{m-2} (minimizing L_{m-2}) minimizes at least one of the quantities

$$L^a_{m-2} = \sum_{q=1}^{n_a/4} \rho^a_{m-2}\left((K^a_{m-2})^{-1}(2q - q); (K^a_{m-2})^{-1}(2q - 1)\right)$$

and

$$L^x_{m-2} = \sum_{q=1}^{n_x/4} \rho^x_{m-2}\left((K^x_{m-2})^{-1}(2q - 2); (K^x_{m-2})^{-1}(2q - 1)\right).$$

To determine K_{m-2}, we first find the class K^a_{m-2} of assignments minimizing L^a_{m-2} and the class K^x_{m-2} of assignments minimizing L^x_{m-2}, and then set

$$K_{m-2} = \begin{cases} K^a_{m-2} & \text{if} \quad L^a_{m-2} \leqq L^x_{m-2} \\ K^x_{m-2} & \text{if} \quad L^x_{m-2} < L^a_{m-2}. \end{cases} \tag{4.2.16}$$

If $K_{m-1} = K^a_{m-1}$, we put

$$z^{(T_{m-2})} = \begin{cases} a^{(\log_2 n_a - 2)} & \text{if} \quad L^a_{m-2} \leqq L^x_{m-2} \\ x^{(\log_2 n_x - 1)} & \text{if} \quad L^x_{m-2} < L^a_{m-2}, \end{cases} \tag{4.2.17a}$$

and if $K_{m-1} = K^x_{m-1}$,

$$z^{(T_{m-2})} = \begin{cases} a^{(\log_2 n_a - 1)} & \text{if} \quad L^a_{m-2} \leqq L^x_{m-2} \\ x^{(\log_2 n_x - 2)} & \text{if} \quad L^x_{m-2} < L^a_{m-2}. \end{cases} \tag{4.2.17b}$$

A comparison of Theorems 4.2.1 and 4.2.2 shows that the procedures for K_{m-2} and K_{m-1} are entirely analogous, amounting to determination of a minimal symmetric matching, except that now this is done for the matrices $(\rho^a_{m-2}(q; s))$ and $(\rho^x_{m-2}(q; s))$. As in (4.2.8), the minimum of these matchings defines the class K_{m-2}. For example, if $\{i_0; i_1\}, \dots, \{i_{n_a/2-2}; i_{n_a/2-1}\}$ is the partition of the rows of A^a_{m-2} into pairs corresponding to the computed matching on $(\rho^a_{m-2}(q; s))$, then K^a_{m-2} is the set of all assignments in K_{m-1} for which

$$\begin{cases} (K^a_{m-2})^{-1}(2\alpha) = i_{2q} \\ (K^a_{m-2})^{-1}(2\alpha + 1) = i_{2q+1} \end{cases} \quad \text{or} \quad \begin{cases} (K^a_{m-2})^{-1}(2\alpha) = i_{2q+1} \\ (K^a_{m-2})^{-1}(2\alpha + 1) = i_{2q}. \end{cases} \tag{4.2.18}$$

Returning to our example, we now compute $K_{m-2} = K_4$. To this end, we need a minimal symmetric matching for $(\rho^a_4(q; s))$; this is found to be $\{\rho^a_4(0; 3), \rho^a_4(1; 2)\}$ (the elements participating in the matching are circled in the matrix $(\rho^a_4(q; s))$). The rows of $(\rho^a_4(q; s))$ are divided into pairs $\{0; 3\}, \{1; 2\}$. The class K^a_4 is the set of assignments in which the codes of states in the same pair may differ only in the second digit (in order of significance). For example, K^a_4 will contain the assignment of K^a_5 shown in Table 4.2.2. We have $L^a_4 = \rho^a_4(0; 3) + \rho^a_4(1; 2) = 12$.

To find K^x_4, we construct a minimal matching for $(\rho^x_4(q; s))$:

$$\{\rho^x_4(0; 5), \rho^x_4(1; 4), \rho^x_4(2; 7), \rho^x_4(3; 6)\}.$$

Then

$$L^x_4 = \rho^x_4(0; 5) + \rho^x_4(1; 4) + \rho^x_4(2; 7) + \rho^x_4(3; 6) = 4.$$

Consequently, $K_{m-2} = K_4 = K^x_4$, $L_4 = L^x_4 = 4$ (instead of 16 in the original assignment) and, since $K_5 = K^a_5$ and $K_4 = K^x_4$, we have $z^{(T_5)} = a^{(2)}$, $z^{(T_4)} = x^{(2)}$. An example of an assignment in the class $K_4 = K^x_4$ is given in Table 4.2.3.

TABLE 4.2.2.

q	0	1	2	3	4	5	6	7
$K^a(q)$	0	4	6	1	5	2	3	7

TABLE 4.2.3.

q	0	1	2	3	4	5	6	7
$K^x(q)$	0	2	4	6	3	1	7	5

4.2.5. The classes $K_{m-3}, K_{m-4}, \ldots, K_0$ may be determined in the same way as K_{m-1} and K_{m-2}, together with the order of arguments for the excitation functions.

A general outline of the procedure is as follows.

The initial data for computation of the class K_{m-t} is the matrix $A_{m-(t-1)}$ determined at the $(t-1)$-th iteration, which defines the class $K^a_{m-(t-1)}$ $(K^x_{m-(t-1)})$ and the minimal symmetric matching for $(\rho^a_{m-(t-1)}(q; s))$ $((\rho^x_{m-(t-1)}(q; s)))$. The t-th iteration of the algorithm consists of the following operations:

1) Using $A_{m-(t-1)}$, construct the matrix $A^a_{m-t}(A^x_{m-t})$ as in (4.2.9).

2) Construct the distance matrices $(\rho^a_{m-t}(q; s))$ and $(\rho^x_{m-t}(q; s))$ for A^a_{m-t} and A^x_{m-t}, as in (4.2.10)–(4.2.12).

3) Employing the "Hungarian" algorithm, determine minimal symmetric matchings for the matrices $(\rho^a_{m-t}(q; s))$ and $(\rho^x_{m-t}(q; s))$ and select a matching with the minimal sum of elements.

4) If the sum of elements of the matching for $(\rho^a_{m-t}(q; s))$ is $\leqq$ $(>)$ the analogous sum for $(\rho^x_{m-t}(q; s))$, set $K_{m-t} = K^a_{m-t}$ $(K_{m-t} = K^x_{m-t})$; the construction of K_{m-t} is analogous to (4.2.8).

5) Suppose that the set of arguments of the excitation functions determined in the previous $t-1$ iterations, i.e., in computation of $K_{m-1}, K_{m-1}, \ldots, K_{m-(t-1)}$,

$$\{z^{(T_{m-1})}, z^{(T_{m-2})}, \ldots, z^{(T_{m-(t-1)})}\},$$

contains the components $a^{(\log_2 n_a - 1)}, a^{(\log_2 n_a - 2)}, \ldots, a^{(\log_2 n_a - \gamma)}$ and $x^{(\log_2 n_x - 1)}, x^{(\log_2 n_x - 2)}, \ldots, x^{(\log_2 n_x - (t-1-\gamma))}$. Then, assuming the conditions of (4) satisfied, we set

$$z^{(T_{m-t})} = a^{(\log_2 n_a - \gamma - 1)}$$

$$(z^{(T_{m-t})} = x^{(\log_2 n_x - (t-1-\gamma)-1)} = x^{(\log_2 n_x - t + \gamma)}).$$

4.2.6. We now construct the classes $K_{m-3} = K_3, K_2, K_1, K_0$ for our example, and find the final state-input assignment and order of arguments in the excitation functions.

We have already seen that $K_5 = K^a_5$, $z^{(T_5)} = a^{(2)}$, $L_5 = 7$; the second iteration gave $K_4 = K^x_4$, $z^{(T_4)} = x^{(2)}$, $L_4 \leqq 4$.

Third iteration. The matrix $A_{m-2} = A_4$ was found in §4.2.4; $K_4 = K^x_4$ and the minimal symmetric matching for $(\rho^x_4(q; s))$ is $\{\rho^x_4(0; 5), \rho^x_4(1; 4), \rho^x_4(2; 7), \rho^x_4(3; 6)\}$.

1. Using A_4 and the above matching, we construct A^x_3:

$$A^x_3 = \begin{pmatrix} \{a_0; a_0; a_0; a_0\} & \{a_4; a_3; a_4; a_3\} & \{a_0; a_0; a_0; a_0\} & \{a_6; a_6; a_7; a_7\} \\ \{a_1; a_1; a_2; a_2\} & \{a_0; a_0; a_0; a_4\} & \{a_2; a_2; a_2; a_2\} & \{a_3; a_3; a_3; a_3\} \\ \{a_5; a_5; a_5; a_5\} & \{a_1; a_2; a_1; a_2\} & \{a_7; a_7; a_6; a_6\} & \{a_6; a_6; a_6; a_6\} \\ \{a_0; a_0; a_0; a_0\} & \{a_3; a_6; a_3; a_6\} & \{a_0; a_0; a_0; a_0\} & \{a_4; a_4; a_4; a_4\} \end{pmatrix}.$$

2. The distance matrices $(\rho_3^a(q;s))$ and $(\rho_3^x(q;s))$ $(q, s = 0, 1, 2, 3)$ are

$$(\rho_3^a(q;s)) = \begin{pmatrix} 0 & 4 & 4 & \textcircled{2} \\ 4 & 0 & \textcircled{4} & 4 \\ 4 & \textcircled{4} & 0 & 4 \\ \textcircled{2} & 4 & 4 & 0 \end{pmatrix}, \qquad (\rho_3^x(q;s)) = \begin{pmatrix} 0 & 4 & \textcircled{2} & 4 \\ 4 & 0 & 4 & \textcircled{4} \\ \textcircled{2} & 4 & 0 & 4 \\ 4 & \textcircled{4} & 4 & 0 \end{pmatrix}.$$

3. The minimal symmetric matchings for $(\rho_3^a(q;s))$ and $(\rho_3^x(q;s))$ are, respectively,

$$\{\rho_3^a(0;3), \rho_3^a(1;2)\} \quad \text{and} \quad \{\rho_3^x(0;2), \rho_3^x(1;3)\}.$$

This gives $L_3 \leqq 6$.

4. Since the sums of elements in the matchings are equal, we set, say, $K_3 = K_3^a$. Since K_5 is the set of assignments in which states in the same pair $\{a_0; a_3\}$, $\{a_1; a_4\}$, $\{a_2; a_7\}$ or $\{a_5; a_6\}$ have codes that differ only in the least significant digit, we see that K_3^a is the set of assignments in which the codes of states in one of the 4-tuples $\{a_0; a_3; a_5; a_6\}$, $\{a_1; a_4; a_2; a_7\}$ differ only in the second digit. For example, K_3^a will contain the state assignment of Table 4.2.4.

TALBE 4.2.4.

q	0	1	2	3	4	5	6	7
$K^a(q)$	0	4	6	1	5	2	3	7

5. Since $K_3 = K_3^a$, $z^{(T_5)} = a^{(2)}$, $z^{(T_4)} = x^{(2)}$, it follows that $z^{(T_3)} = a^{(1)}$.

Fourth iteration. 1. The matrix A_2^a is shown in Figure 4.2.1.

2.

$$(\rho_2^a(q;s)) = \begin{pmatrix} 0 & \textcircled{4} \\ \textcircled{4} & 0 \end{pmatrix}, \qquad (\rho_2^x(q;s)) = \begin{pmatrix} 0 & 2 & \textcircled{1} & 2 \\ 2 & 0 & 2 & \textcircled{2} \\ \textcircled{1} & 2 & 0 & 2 \\ 2 & \textcircled{2} & 2 & 0 \end{pmatrix}.$$

3. Minimal matchings: $\{\rho_2^a(0;1)\}$ and $\{\rho_2^x(0;2), \rho_2^x(1;3)\}$.

4. $K_2 = K_2^x$, $L_2 \leqq 3$. The class $K_4 = K_4^x$ is the set of assignments in which the codes of input signals in each of the pairs $\{x_0; x_5\}$, $\{x_1; x_4\}$, $\{x_2; x_7\}$, $\{x_3; x_6\}$ differ only in the least significant digit, and, since the matching on $(\rho_2^x(q;s))$ is $\{\rho_2^x(0;2), \rho_2^x(1;3)\}$, it follows that K_2^x is the set of assignments in which the codes of input signals in each of the 4-tuples $\{x_0; x_5; x_2; x_7\}$, $\{x_1; x_4; x_3; x_6\}$ differ only in their second digits. For example, K_2 contains the assignment of Table 4.2.5.

5. Since $K_2 = K_2^x$, $z^{(T_5)} = a^{(2)}$, $z^{(T_4)} = x^{(2)}$, $z^{(T_3)} = a^{(1)}$, we have $z^{(T_2)} = x^{(1)}$.

Fifth iteration. 1. The matrix A_1^x is shown in Figure 4.2.2.

$$A_2^a = \begin{pmatrix} \{a_0; a_0; a_0; a_0; a_0; a_0; a_0; a_0\} & \{a_4; a_3; a_4; a_3; a_3; a_6; a_3; a_6\} & \{a_0; a_0; a_0; a_0; a_0; a_0; a_0; a_0\} & \{a_6; a_6; a_7; a_7; a_4; a_4; a_4; a_4\} \\ \{a_1; a_1; a_2; a_2; a_5; a_5; a_5; a_5\} & \{a_0; a_0; a_0; a_4; a_3; a_6; a_3; a_6\} & \{a_2; a_2; a_2; a_2; a_7; a_7; a_6; a_6\} & \{a_3; a_3; a_3; a_3; a_6; a_6; a_6; a_6\} \end{pmatrix}$$

Figure 4.2.1.

$$A_1^x = \begin{pmatrix} \{a_0; a_0; a_0; a_0; a_0; a_0; a_0; a_0; a_0; a_0; a_0; a_0; a_0; a_0; a_0; a_0\} & \{a_4; a_3; a_4; a_3; a_3; a_6; a_3; a_6; a_6; a_6; a_7; a_7; a_4; a_4; a_4; a_4\} \\ \{a_1; a_1; a_2; a_2; a_5; a_5; a_5; a_5; a_2; a_2; a_2; a_2; a_7; a_7; a_6; a_6\} & \{a_0; a_0; a_0; a_4; a_3; a_6; a_3; a_6; a_3; a_3; a_3; a_3; a_6; a_6; a_6; a_6\} \end{pmatrix}$$

Figure 4.2.2.

$$A_0^x = \begin{pmatrix} \{a_0; a_0; a_0; a_0; a_0; a_0; a_0; a_0; a_0; a_0; a_0; a_0; a_0; a_0; a_0; a_0; a_4; a_3; a_4; a_3; a_3; a_6; a_3; a_6; a_6; a_6; a_7; a_7; a_4; a_4; a_4; a_4\} \\ \{a_1; a_1; a_2; a_2; a_5; a_5; a_5; a_5; a_2; a_2; a_2; a_2; a_7; a_7; a_6; a_6; a_0; a_0; a_0; a_4; a_3; a_6; a_3; a_6; a_3; a_3; a_3; a_3; a_6; a_6; a_6; a_6\} \end{pmatrix}$$

Figure 4.2.3.

TABLE 4.2.5

q	0	1	2	3	4	5	6	7
$K^x(q)$	0	4	2	6	5	1	7	3

2.

$$(\rho_1^a(q;s)) = \begin{pmatrix} 0 & \textcircled{2} \\ \textcircled{2} & 0 \end{pmatrix}, \qquad (\rho_1^x(q;s)) = \begin{pmatrix} 0 & \textcircled{2} \\ \textcircled{2} & 0 \end{pmatrix}.$$

3. Minimal matchings: $\{\rho_1^a(0;1)\}$ and $\{\rho_1^x(0;1)\}$.

4. We set $K_1 = K_1^x$, $L_1 \leqq 2$. For example, K_1^x contains the input assignment shown in Table 4.2.6.

5. $z^{(T_1)} = x^{(0)}$.

Sixth iteration. 1. The matrix A_0^x is shown in Figure 4.2.3.

TABLE 4.2.6.

q	0	1	2	3	4	5	6	7
$K^x(q)$	0	4	2	6	5	1	7	3

2.

$$(\rho_0^a(q;s)) = \begin{pmatrix} 0 & \textcircled{1} \\ \textcircled{1} & 0 \end{pmatrix}.$$

3. Minimal matching: $\{\rho_0^a(0;1)\}$.

4. $K_0 = K_0^a$, $L_0 \leqq L_0^a + 1 = 2$.

5. $z^{(T_0)} = a^{(0)}$.

To summarize, the final state-input assignment is that shown in Table 4.2.7.

The order of arguments for the excitation functions is $z = (a^{(0)}, x^{(0)}, x^{(1)}, a^{(1)}, x^{(2)}, a^{(2)})$; the permutation defining this order is $T = (3, 0, 1, 4, 2, 5)$ (see (4.1.6)).

The complexity bounds L (giving the number of nonvanishing coefficients in the initial and optimal assignments) are shown in Table 4.2.8. It is evident that optimal choice of the state-input assignment and optimal ordering of arguments in the excitation functions have reduced the complexity of the automaton to about 2/5 of its original value.

4.2.7. The above method may be generalized to realization of automata based on multistable elements, by using expansions of the excitation functions in series of generalized Haar functions (see §1.3). In addition, if the numbers of states and input signals are not powers of 2, the automaton in question may be treated as an incompletely specified automaton with $2^{]\log_2 n_a[}$ states and $2^{]\log_2 n_x[}$ input signals. The assignment problem for incompletely specified automata will be considered in the next section.

TABLE 4.2.7.

q	0	1	2	3	4	5	6	7
$K^x(q)$	0	4	2	6	5	1	7	3
$K^a(q)$	0	4	6	1	5	2	3	7

TABLE 4.2.8

	l	0	1	2	3	4	5
L_l	Initial assignment	2	2	4	4	14	30
	Optimal assignment	2	2	3	5	4	7

To conclude this section, we note that in order to establish sharper bounds for the expected values of $L_{m-2}, L_{m-3}, \ldots, L_0$ during implementation of the algorithm, one should proceed as follows. To define distances between the rows (columns) of the appropriate matrices $A_{m-2}, A_{m-3}, \ldots, A_0$, put the distance $\rho(c_{\alpha,\beta}, c_{\gamma,\delta})$ between elements $c_{\alpha,\beta}$, $c_{\gamma,\delta}$ of the matrix A_{m-t} $(t = 2, 3, \ldots, m)$ (recall that each element of A_{m-t} is a set of 2^{t-1} elements) equal to

$$\rho(c_{\alpha,\beta}; c_{\gamma,\delta}) = P\left(\sum_{a_q \in c_{\alpha,\beta}} K^a(q) \neq \sum_{a_q \in c_{\gamma,\delta}} K^a(q)\right) \tag{4.2.19}$$

The right-hand side of this expression is the relative abundance of assignments $K^a(q)$ for which the sum of numbers corresponding to codes of states in $c_{\alpha,\beta}$ is not equal to the analogous sum for $c_{\gamma,\delta}$. It can be shown that

$$\lim_{\log_2 n_a \to \infty} P\left(\sum_{a_q \in c_{\alpha,\beta}} K^a(q) \neq \sum_{a_q \in c_{\gamma,\delta}} K^a(q)\right) = \begin{cases} 1 & \text{if} \quad c_{\alpha,\beta} \neq c_{\gamma,\delta}, \\ 0 & \text{if} \quad c_{\alpha,\beta} = c_{\gamma,\delta}. \end{cases} \tag{4.2.20}$$

The quantity on the left of (4.2.20) tends to unity ($c_{\alpha,\beta} \neq c_{\gamma,\delta}$) as $\log_2 n_a \to \infty$, and does so quite rapidly; thus, for as low a value as $\log_2 n_a = 4$ the deviation from unity is at most 0.1. This justifies the use of a formula of type (4.2.12) in the definition of $\rho(c_{\alpha,\beta}, c_{\gamma,\delta})$, instead of the more accurate (4.2.19).

4.3. STATE-INPUT ASSIGNMENT FOR INCOMPLETELY SPECIFIED AUTOMATA

4.3.1. The state-input assignment problem for incompletely specified FA is even more complicated than for completely specified FA. Besides the difficulties enumerated in §4.2.1, one is faced here with the highly complex problem of optimal completion of the state table. At present, there seem to be no really effective assignment procedures for incompletely specified FA, as far as the classical approach is concerned.

When spectral methods are employed, however, one can construct a fairly simple and effective completion and assignment algorithm, producing optimal results for almost all incompletely specified automata. The method of completion, as we shall see, is analogous to the completion of incompletely specified logical functions in the framework of spectral methods (see §3.3).

As before, we shall adopt the Haar functions as our basis system, measuring the complexity of realization in terms of the number L of nonvanishing expansion coefficients.

4.3.2. Consider an incompletely specified FA with state matrix $\tilde{A}_{m-1} = (\tilde{a}_{q,r})$ ($q = 0, 1, \ldots, n_a - 1$; $r = 0, 1, \ldots, n_x - 1$). Using the matrix $\tilde{A}_{m-1}$ of the original automaton, we construct an equivalent* FA with state matrix $A_{m-1} = (a_{q,r})$ ($q = 0, 1, \ldots, 2^{]\log_2 n_a[} - 1$; $r = 0, 1, \ldots, 2^{]\log_2 n_x[} - 1$), where

$$a_{q,r} = \begin{cases} \tilde{a}_{q,r} & \text{if } 0 \leqq q \leqq n_a - 1 \text{ and } 0 \leqq r \leqq n_x - 1, \\ * & \text{otherwise} \end{cases} \tag{4.3.1}$$

(the symbol $*$ stands for undefined elements of the state matrix A_{m-1}).

EXAMPLE 4.3.1. Let

$$\tilde{A}_{m-1} = \tilde{A}_5 = \begin{pmatrix} a_0 & * & a_3 & a_6 & a_4 & * & a_7 \\ a_1 & a_0 & a_2 & a_3 & a_0 & a_2 & * \\ * & a_1 & a_7 & * & a_1 & a_5 & a_6 \\ a_0 & * & a_0 & a_6 & a_3 & * & a_7 \\ a_1 & a_0 & a_2 & a_3 & a_4 & * & * \\ a_0 & a_3 & * & a_4 & a_3 & a_0 & * \end{pmatrix} \quad (n_a = 6, n_x = 7).$$

Then

$$A_{m-1} = A_5 = \begin{pmatrix} a_0 & * & a_3 & a_6 & a_4 & * & a_7 & * \\ a_1 & a_0 & a_2 & a_3 & a_0 & a_2 & * & * \\ * & a_1 & a_7 & * & a_1 & a_5 & a_6 & * \\ a_0 & * & a_0 & a_6 & a_3 & * & a_7 & * \\ a_1 & a_0 & a_2 & a_3 & a_4 & * & * & * \\ a_0 & a_3 & * & a_4 & a_3 & a_0 & * & * \\ * & * & * & * & * & * & * & * \\ * & * & * & * & * & * & * & * \end{pmatrix}.$$

* Two automata A and A' are said to be *equivalent* if they define the same mapping of the set of input words (= strings of input signals) into the set of output words (= strings of output signals).

We now apply to A_{m-1} the method of §4.2 for optimal state-input assignment and optimization of the order T of arguments in the excitation functions. As before, we minimize the numbers $L_{m-1}, L_{m-2}, \ldots, L_0$ consecutively. Now, however, we simultaneously construct a completion of A_{m-1} which is optimal for almost all automata.

The first iteration of the algorithm constructs matrices $(\rho^a_{m-1}(q; s))$ and $(\rho^x_{m-1}(q; s))$ for A_{m-1} by (4.2.3), (4.2.4). Formula (4.2.5) is now replaced by

$$\rho(a_{\alpha,\beta}; a_{\gamma,\delta}) = \begin{cases} 1 & \text{if} \quad a_{\alpha,\beta} \neq a_{\gamma,\delta}, \quad \text{provided } a_{\alpha,\beta} \text{ and } a_{\gamma,\delta} \text{ are defined,} \\ 0 & \text{otherwise.} \end{cases} \tag{4.3.2}$$

For the example given above,

$$(\rho^a_5(q; s)) = \begin{pmatrix} 0 & 4 & 3 & 2 & 3 & 2 & \textcircled{0} & 0 \\ 4 & 0 & 3 & 4 & \textcircled{1} & 5 & 0 & 0 \\ 3 & 3 & 0 & 3 & 3 & 3 & 0 & \textcircled{0} \\ 2 & 4 & 3 & 0 & 4 & \textcircled{1} & 0 & 0 \\ 3 & \textcircled{1} & 3 & 4 & 0 & 4 & 0 & 0 \\ 2 & 5 & 3 & \textcircled{1} & 4 & 0 & 0 & 0 \\ \textcircled{0} & 0 & 0 & 0 & 0 & 0 & 0 & 0 \\ 0 & 0 & \textcircled{0} & 0 & 0 & 0 & 0 & 0 \end{pmatrix}$$

and

$$(\rho^x_5(q; s)) = \begin{pmatrix} 0 & 3 & 3 & 4 & 5 & \textcircled{1} & 2 & 0 \\ 3 & 0 & 3 & 3 & \textcircled{1} & 3 & 1 & 0 \\ 3 & 3 & 0 & 4 & 5 & 1 & 3 & \textcircled{0} \\ 4 & 3 & 4 & 0 & 5 & 2 & \textcircled{2} & 0 \\ 5 & \textcircled{1} & 5 & 5 & 0 & 3 & 3 & 0 \\ \textcircled{1} & 3 & 1 & 2 & 3 & 0 & 1 & 0 \\ 2 & 1 & 3 & \textcircled{2} & 3 & 1 & 0 & 0 \\ 0 & 0 & \textcircled{0} & 0 & 0 & 0 & 0 & 0 \end{pmatrix}$$

We now find minimal symmetric matchings on $(\rho^a_{m-1}(q; s))$ and $(\rho^x_{m-1}(q; s))$. In the example, these are $\{\rho^a_5(0; 6), \rho^a_5(1; 4), \rho^a_5(2; 7), \rho^a_5(3; 5)\}$ and $\{\rho^x_5(0; 5), \rho^x_5(1; 4), \rho^x_5(2; 7), \rho^x_5(3; 6)\}$.

The matchings on $(\rho^a_{m-1}(q; s))$ and $(\rho^x_{m-1}(q; s))$ define partitions of the sets of states and input signals (and also the sets of rows and columns of the matrices $(\rho^a_{m-1}(q; s))$ and $(\rho^x_{m-1}(q; s))$) into disjoint pairs. In our examples, these partitions are

$\{\{a_0;a_6\},\{a_1;a_4\},\{a_2;a_7\},\{a_3;a_5\}\}$ and $\{\{x_0;x_5\},\{x_1;x_4\},\{x_2;x_7\},\{x_3;x_6\}\}$.

As in the completely specified case, the sum of elements of the matching on $(\rho^a_{m-1}(q;s))$ or $(\rho^x_{m-1}(q;s))$ determines L_{m-1}, if the binary codes of states or inputs in each pair differ only in the least significant digit. We thus choose a minimal matching. In our example, the minimal matching is defined on $(\rho^a_5(q;s))$, and so

$$L_5 = L^a_5 = \rho^a_5(0;6) + \rho^a_5(1;4) + \rho^a_5(2;7) + \rho^a_5(3;5) = 2.$$

Since $L_5 = L^a_5$, it follows that

$$z^{(T_{m-1})} = z^{(T_5)} = a^{(\log_2 n_a - 1)} = a^{(2)}.$$

Once a minimal matching has been determined, we construct the class of completions minimizing L_{m-1}. It follows from (4.3.2) that

$$\rho(a_{\alpha,\beta};*) = 0 \tag{4.3.3}$$

for any $a_{\alpha,\beta}$. We therefore complete A_{m-1} to $A'_{m-1} = (a'_{p,r})$ as follows: if $L_{m-1} = L^a_{m-1}$ ($L_{m-1} = L^x_{m-1}$) and $\{\{i_0;i_1\},\ldots,\{i_{n_a-2};i_{n_a-1}\}\}$ is the partition of rows (columns) into pairs corresponding to the matching, we set $a'_{i_{2q},r} = a'_{i_{2q+1},r}$ if $a_{i_{2q},r} = *$ or $a_{i_{2q+1},r} = *$ ($a'_{r,i_{2q}} = a'_{r,i_{2q+1}}$ if $a_{r,i_{2q}} = *$ or $a_{r,i_{2q+1}} = *$).

The matrix A'_{m-1} defines a class of completions minimizing L_{m-1}.

In the example, we have $L_5 = L^a_5$ and

$$A'_5 = \begin{pmatrix} a_0 & *_1 & a_3 & a_6 & a_4 & *_2 & a_7 & *_3 \\ a_1 & a_0 & a_2 & a_3 & a_0 & a_2 & *_4 & *_5 \\ *_6 & a_1 & a_7 & *_7 & a_1 & a_5 & a_6 & *_8 \\ a_0 & \boldsymbol{a_3} & a_0 & a_6 & a_3 & \boldsymbol{a_0} & a_7 & *_9 \\ a_1 & a_0 & a_2 & a_3 & a_4 & \boldsymbol{a_2} & *_4 & *_5 \\ a_0 & a_3 & \boldsymbol{a_0} & a_4 & a_3 & a_0 & \boldsymbol{a_7} & *_9 \\ \boldsymbol{a_0} & *_1 & \boldsymbol{a_3} & \boldsymbol{a_6} & \boldsymbol{a_4} & *_2 & \boldsymbol{a_7} & *_3 \\ *_6 & \boldsymbol{a_1} & \boldsymbol{a_7} & *_7 & \boldsymbol{a_1} & \boldsymbol{a_5} & \boldsymbol{a_6} & *_8 \end{pmatrix}$$

where $*_i$ are undefined values (generally distinct)* and the completed elements of A'_5 are indicated by bold type.

Note that according to this method the fictitious states (input signals) introduced in order to bring the number of states (input signals) up to $2^{]\log_2 n_a[}$ ($2^{]\log_2 n_x[}$) are equivalent** to the states (input signals) paired with them in the selected partition. Thus,

* In other words, in any completion the matrix entries $*_i$ may be replaced by arbitrary states, with the restriction that asterisks with equal subscripts i must be replaced by the same state.

** States a_p and a_q of an automaton A are said to be *equivalent* if $\varphi(x, a_p) = \varphi(x, a_q)$ for any input word x, where φ is the next-state function of A.

the fictitious states a_6 and a_7 in our example are equivalent to a_0 and a_2, respectively.

Thus, the first iteration of our algorithm produces a class of assignments and completions minimizing L_{m-1}.

4.3.3. The second iteration constructs a class of assignments and completions minimizing L_{m-2}. First, using the matrix A'_{m-1} and the optimal matching, we construct a matrix A^a_{m-2} or A^x_{m-2} as in (4.2.9).

In the example, we have

$$A_{m-2} = A_4 = A^a_4 =$$

$$\begin{pmatrix} \{a_0;a_0\} & \{*_1;*_1\} & \{a_3;a_3\} & \{a_6;a_6\} & \{a_4;a_4\} & \{*_2;*_2\} & \{a_7;a_7\} & \{*_3;*_3\} \\ \{a_1;a_1\} & \{a_0;a_0\} & \{a_2;a_2\} & \{a_3;a_3\} & \{a_0;a_4\} & \{a_2;a_2\} & \{*_4;*_4\} & \{*_5;*_5\} \\ \{*_6;*_6\} & \{a_1;a_1\} & \{a_7;a_7\} & \{*_7;*_7\} & \{a_1;a_1\} & \{a_5;a_5\} & \{a_6;a_6\} & \{*_8;*_8\} \\ \{a_0;a_0\} & \{a_3;a_3\} & \{a_0;a_0\} & \{a_6;a_4\} & \{a_3;a_3\} & \{a_0;a_0\} & \{a_7;a_7\} & \{*_9;*_9\} \end{pmatrix}.$$

The next step is to construct the distance matrices $(\rho^x_{m-2}(q;s))$ and $(\rho^a_{m-2}(q;s))$ for A_{m-2}, using formulas (4.2.10)–(4.2.12), the latter formula being modified in the same way as (4.3.2):

$$\rho(b_{\alpha,\beta};b_{\gamma,\delta}) = \begin{cases} 1 & \text{if } b_{\alpha,\beta} \neq b_{\gamma,\delta}, \text{ provided } b_{\alpha,\beta}, b_{\gamma,\delta} \text{ are defined,} \\ 0 & \text{otherwise} \end{cases} \tag{4.3.4}$$

(where $b_{\alpha,\beta}$, $b_{\gamma,\delta}$ are elements of A_{m-2}).

In the example, $A_{m-2} = A^a_4$ and the distance matrices are

$$(\rho^a_4(q;s)) = \begin{pmatrix} 0 & 4 & 3 & \textcircled{3} \\ 4 & 0 & \textcircled{4} & 6 \\ 3 & \textcircled{4} & 0 & 5 \\ \textcircled{3} & 6 & 5 & 0 \end{pmatrix},$$

$$(\rho^x_4(q;s)) = \begin{pmatrix} 0 & 2 & 2 & 3 & 3 & \textcircled{1} & 2 & 0 \\ 2 & 0 & 3 & 2 & \textcircled{1} & 3 & 2 & 0 \\ 2 & 3 & 0 & 3 & 4 & 1 & 3 & \textcircled{0} \\ 3 & 2 & 3 & 0 & 3 & 2 & \textcircled{2} & 0 \\ 3 & \textcircled{1} & 4 & 3 & 0 & 3 & 3 & 0 \\ \textcircled{1} & 3 & 1 & 2 & 3 & 0 & 2 & 0 \\ 2 & 2 & 3 & \textcircled{2} & 3 & 2 & 0 & 0 \\ 0 & 0 & \textcircled{0} & 0 & 0 & 0 & 0 & 0 \end{pmatrix}$$

The minimal matchings are indicated by circled elements, and $L_4 \leqq L_4^x = 4$, $z^{(T_4)} = x^{(2)}$.

We now construct a class of completions minimizing L_{m-2}. The construction of a completion $A'_{m-2} = (b'_{p,q})$ of $A_{m-2} = (b_{p,q})$, using the optimal partition of the set of rows (columns) of A_{m-2} into pairs, is analogous to the construction of A'_{m-1} in the first iteration.

In the example, $L_4 \leqq L_4^x$, the partition of columns is $\{\{0;5\},\{1;4\},\{2;7\},\{3;6\}\}$, and

$$A'_{m-2} = A'_4 =$$

$$\begin{pmatrix} \{a_0;a_0\} & \{a_4;a_4\} & \{a_3;a_3\} & \{a_6;a_6\} & \{a_4;a_4\} & \{a_0;a_0\} & \{a_7;a_7\} & \{a_3;a_3\} \\ \{a_1;a_1\} & \{a_0;a_0\} & \{a_2;a_2\} & \{a_3;a_3\} & \{a_0;a_4\} & \{a_2;a_2\} & \{a_3;a_3\} & \{a_2;a_2\} \\ \{a_5;a_5\} & \{a_1;a_1\} & \{a_7;a_7\} & \{a_6;a_6\} & \{a_1;a_1\} & \{a_5;a_5\} & \{a_6;a_6\} & \{a_7;a_7\} \\ \{a_0;a_0\} & \{a_3;a_3\} & \{a_0;a_0\} & \{a_6;a_4\} & \{a_3;a_3\} & \{a_0;a_0\} & \{a_7;a_7\} & \{a_0;a_0\} \end{pmatrix}$$

In this example, then, all undefined states become defined at the second iteration of the algorithm. The final, completely specified state matrix is

$$A'_5 = \begin{pmatrix} a_0 & \mathbf{a_4} & a_3 & a_6 & a_4 & \mathbf{a_0} & a_7 & \mathbf{a_3} \\ a_1 & a_0 & a_2 & a_3 & a_0 & a_2 & \mathbf{a_3} & \mathbf{a_2} \\ \mathbf{a_5} & a_1 & a_7 & \mathbf{a_6} & a_1 & a_5 & a_6 & \mathbf{a_7} \\ a_0 & a_3 & a_0 & a_6 & a_3 & a_0 & a_7 & \mathbf{a_0} \\ a_1 & a_0 & a_2 & a_3 & a_4 & a_2 & \mathbf{a_3} & \mathbf{a_2} \\ a_0 & a_3 & a_0 & a_4 & a_3 & a_0 & a_7 & \mathbf{a_0} \\ a_0 & \mathbf{a_4} & a_3 & a_6 & a_4 & \mathbf{a_0} & a_7 & \mathbf{a_3} \\ \mathbf{a_5} & a_1 & a_7 & \mathbf{a_6} & a_1 & a_5 & a_6 & \mathbf{a_7} \end{pmatrix}.$$

(The elements defined at the second iteration are in bold type.)

The procedure for assignments minimizing $L_{m-3}, L_{m-4}, \ldots, L_0$ is analogous. We illustrate using the same example as before.

Third iteration.

$$A_3 = A_3^x = \begin{pmatrix} \{a_0;a_0;a_0;a_0\} & \{a_4;a_4;a_4;a_4\} & \{a_3;a_3;a_3;a_3\} & \{a_6;a_6;a_7;a_7\} \\ \{a_1;a_1;a_1;a_1\{ & \{a_0;a_0;a_0;a_4\} & \{a_2;a_2;a_2;a_2\} & \{a_3;a_3;a_3;a_3\} \\ \{a_5;a_5;a_5;a_5\{ & \{a_1;a_1;a_1;a_1\} & \{a_7;a_7;a_7;a_7\} & \{a_6;a_6;a_6;a_6\} \\ \{a_0;a_0;a_0;a_0\} & \{a_3;a_3;a_3;a_3\} & \{a_0;a_0;a_0;a_0\} & \{a_6;a_4;a_7;a_7\} \end{pmatrix},$$

$$(\rho_3^a(q;s)) = \begin{pmatrix} 0 & 4 & 4 & \textcircled{3} \\ 4 & 0 & \textcircled{4} & 4 \\ 4 & \textcircled{4} & 0 & 4 \\ \textcircled{3} & 4 & 4 & 0 \end{pmatrix}, \qquad (\rho_3^x(q;s)) = \begin{pmatrix} 0 & 4 & \textcircled{3} & 4 \\ 4 & 0 & 4 & \textcircled{4} \\ \textcircled{3} & 4 & 0 & 4 \\ 4 & \textcircled{4} & 4 & 0 \end{pmatrix},$$

$$L_3 \leqq L_3^x = 7, \qquad z^{(T_3)} = x^{(1)}.$$

Fourth iteration.

$$A_2 = A_2^x = \begin{pmatrix} \{a_0; a_0; a_0; a_0; a_3; a_3; a_3; a_3\} & \{a_4; a_4; a_4; a_4; a_6; a_6; a_7; a_7\} \\ \{a_1; a_1; a_2; a_2; a_2; a_2; a_2; a_2\} & \{a_0; a_0; a_0; a_4; a_3; a_3; a_3; a_3\} \\ \{a_5; a_5; a_5; a_5; a_7; a_7; a_7; a_7\} & \{a_1; a_1; a_1; a_1; a_6; a_6; a_6; a_6\} \\ \{a_0; a_0; a_0; a_0; a_0; a_0; a_0; a_0\{ & \{a_3; a_3; a_3; a_3; a_6; a_4; a_7; a_7\} \end{pmatrix},$$

$$(\rho_2^a(q; s)) = \begin{pmatrix} 0 & 2 & 2 & \textcircled{2} \\ 2 & 0 & \textcircled{2} & 2 \\ 2 & \textcircled{2} & 0 & 2 \\ \textcircled{2} & 2 & 2 & 0 \end{pmatrix}, \qquad (\rho_2^x(q; s)) = \begin{pmatrix} 0 & \textcircled{4} \\ \textcircled{4} & 0 \end{pmatrix}.$$

$$L_2 \leqq L_2^a = 4, \qquad z^{T_2} = a^1.$$

Fifth iteration. The matrix $A_1 = A_1^a$ is shown in Figure 4.3.1.

$$(\rho_1^a(q; s)) = \begin{pmatrix} 0 & \textcircled{2} \\ \textcircled{2} & 0 \end{pmatrix}, \qquad (\rho_1^x(q; s)) = \begin{matrix} 0 & \textcircled{2} \\ \textcircled{2} & 0 \end{matrix},$$

$$L_1 \leqq L_1^a = 2, \qquad z^{(T_1)} = a^{(0)}.$$

Sixth iteration. The matrix A_0^a is shown in Figure 4.3.2.

$$(\rho_0^x(q; s)) = \begin{pmatrix} 0 & \textcircled{1} \\ \textcircled{1} & 0 \end{pmatrix}, \qquad L_0 \leqq L_0^x + 1 = 2, \qquad z^{(T_0)} = x^{(0)}.$$

TABLE 4.3.1.

q	0	1	2	3	4	5	6	7
$K^a(q)$	0	4	6	2	5	3	1	7
$K^x(q)$	0	4	2	6	5	1	7	3

TABLE 4.3.2.

l	0	1	2	3	4	5
L_l	2	2	4	7	4	2

In our example, therefore, the argument vector of the excitation functions is $z = (x^{(0)}, a^{(0)}, a^{(1)}, x^{(1)}, x^{(2)}, a^{(2)})$. Tables 4.3.1 and 4.3.2 provide the assignments and complexity bounds.

The above method of completion and assignment for incompletely specified FA, like its counterpart (§4.2) for completely specified FA, is asymptotically optimal for almost all automata; indeed, even for comparatively low values of $n = \log_2 n_a$ the

$$A_1^a = \begin{pmatrix} \{a_0; a_0; a_0; a_0; a_3; a_3; a_3; a_3; a_0; a_0; a_0; a_0; a_0; a_0; a_0; a_0\} & \{a_4; a_4; a_4; a_4; a_6; a_6; a_7; a_7; a_3; a_3; a_3; a_3; a_6; a_4; a_7; a_7\} \\ \{a_1; a_1; a_2; a_2; a_2; a_2; a_2; a_2; a_5; a_5; a_5; a_5; a_7; a_7; a_7; a_7\} & \{a_0; a_0; a_0; a_4; a_3; a_3; a_3; a_3; a_1; a_1; a_1; a_1; a_6; a_6; a_6; a_6\} \end{pmatrix}$$

Figure 4.3.1.

$$A_0^a = (\{a_0; a_0; a_0; a_0; a_3; a_3; a_3; a_3; a_0; a_0; a_0; a_0; a_0; a_0; a_0; a_0; a_1; a_1; a_2; a_2; a_2; a_2; a_2; a_2; a_5; a_5; a_5; a_5; a_7; a_7; a_7; a_7\}$$

$$\{a_4; a_4; a_4; a_4; a_6; a_6; a_7; a_7; a_3; a_3; a_3; a_3; a_6; a_4; a_7; a_7; a_0; a_0; a_0; a_4; a_3; a_3; a_3; a_3; a_1; a_1; a_1; a_1; a_6; a_6; a_6; a_6\})$$

Figure 4.3.2.

probability p that the completion and assignment will be optimal is nearly unity (thus, if $n = 4$ we have $p > 0.9$).

4.4. SOME SPECIAL CASES OF THE ASSIGNMENT PROBLEM

4.4.1. The solutions of the assignment problem presented in §§4.2 and 4.3 apply to arbitrary automata. In this section we shall discuss a few important special classes of automata whose specific features yield simplifications of the general algorithms. The special classes considered in this section are autonomous automata and automata for which a fixed state (or input) assignment is given in advance. All the notation of §§4.2, 4.3 will be retained.

4.4.2. An automaton is said to be *autonomous* if it has only one input signal. The next state of an autonomous automaton is determined only by the previous state, and its state matrix $A_{m-1} = (a_{q,r})$ reduces to a single column $(a_{q,1})$. The algorithms of §§4.2 and 4.3 are substantially simplified when applied to autonomous automata. First and foremost, there is no need to construct the function $K^x(r)$, the matrices A^x_{m-t} or $(\rho^x_{m-t}(q;s))$ $(t = 1, \dots, m)$. In addition, $z^{(m-i)} = a^{(m-i)}$. Moreover, for any t we have $\rho^a_{m-1}(q;s) \in \{0, 1\}$ for any q, s, and all the matrices A^a_{m-t} are single columns.

We now establish a bound on the number of nonvanishing expansion coefficients for autonomous automata.

Let $\varphi^{-1}(a_i)$ $(i = 0, 1, \dots, n_a - 1)$, denote the set of states such that if $a_j \in \varphi^{-1}(a_i)$ then $\varphi(a_j) = a_i$, where φ is the next-state function of the automaton; in other words, $a_{j,1} = a_i$ in the matrix $A_{m-1} = (a_{q,1})$. Let $|\varphi^{-1}(a_i)|$ denote the number of elements in $\varphi^{-1}(a_i)$.

THEOREM 4.4.1. *Consider an autonomous automaton with n_a states whose next-state function is undefined at n_* points. Then*

$$L = \sum_{t=1}^{m} L_{m-t} \leqq n_a - n_* - \sum_{t=1}^{]\log_2 n_a[} \sum_{i=0}^{n_a-1} [2^{-t}|\varphi^{-1}(a_i)|] \qquad (m =]\log_2 n_a[). \quad (4.4.1)$$

PROOF. We note that if $\sum_{i=0}^{n_a-1} [2^{-t}|\varphi^{-1}(a_i)|] = \delta$, then the matrix $(\rho^a_{m-t}(q;s))$ contains δ elements $\rho^a_{m-t}(q_i;s_i)$ $(i = 1, 2, \dots, \delta)$ above the principal diagonal such that $\rho^a_{m-t}(q_i;s_i) = 0$, $q_i \neq q_j$, $s_i \neq s_j$ $(i \neq j,\ i, j = 1, 2, \dots, \delta)$. Consequently, the elements $(q_i;s_i)$ $(i = 1, \dots, \delta)$ must be in a minimal symmetric matching for $(\rho^a_{m-t}(q;s))$ (since $\rho^a_{m-t}(q;s) \in \{0, 1\}$ for any $q, s \in \{0, 1, \dots, n_a - 1\}$), and the number of vanishing coefficients $c^{(l)}_{m-t}$ $(l = 1, 2, \dots, 2^{m-t})$ is at least δ. Hence, summing over t and using the fact that the next-state function is defined at $n_a - n_*$ points and Theorem 3.3.1, we obtain (4.4.1). ∎

EXAMPLE 4.4.1. Consider an autonomous automaton with state matrix

$$A_{m-1} = \begin{pmatrix} a_0 \\ a_1 \\ a_1 \\ a_2 \\ * \\ a_1 \\ a_2 \\ * \\ a_1 \end{pmatrix}.$$

(where $*$ stands for undefined value of the next-state function). We have $n_a = 9$, $n_* = 2$, $\varphi^{-1}(a_0) = a_0$, $\varphi^{-1}(a_1) = \{a_1, a_2, a_5, a_8\}$, $\varphi^{-1}(a_2) = \{a_3, a_6\}$, $\varphi^{-1}(a_3) = \varphi^{-1}(a_4) = \varphi^{-1}(a_5) = \varphi^{-1}(a_6) = \varphi^{-1}(a_7) = \varphi^{-1}(a_8) = \varnothing$ (where $\varnothing$ is the symbol for the empty set). Then

$$\sum_{i=0}^{8} [\tfrac{1}{2}|\varphi^{-1}(a^i)|] = 3, \qquad \sum_{i=0}^{8} [\tfrac{1}{4}|\varphi^{-1}(a_i)|] = 1$$

and

$$\sum_{i=0}^{8} [2^{-t}|\varphi^{-1}(a_i)|] = 0$$

for $t > 2$; thus, by Theorem 4.4.1, $L \leqq 3$.

Thus, Theorem 4.4.1 yields a bound on the spectral complexity of autonomous automata, directly in terms of the next-state function; there is no need to construct the matrices $(\rho^a_{m-t}(q; s))$.

The state assignment problem for autonomous automata reduces to determination of minimal symmetric matchings for the matrices $(\rho^a_{m-t}(q; s))$. The construction of these matrices is analogous to that described in §§4.2, 4.3, facilitated by the fact that here all the elements are zeros and ones. The problem of a minimal symmetric matching for a matrix of zeros and ones is known as the matching problem. Efficient algorithms for its solution, based on a modification of the Hungarian algorithm, may be found, e.g., in [32, 33, 71].

4.4.3. We now consider the assignment problem for automata whose input signals or internal states have preassigned codes (fixed before the beginning of the synthesis process).

Let us assume that an assignment $K^x(q)$ of input signals ($K^a(q)$ of states) is fixed. Our problem is to determine the order of arguments in the excitation functions and a state assignment $K^a(q)$ (input assignment $K^x(q)$). The algorithm is again a simplified version of the general procedure. If $K^x(q)$ is fixed, we set

$$K^x(q) = K^x_{m-1}(q) = \sum_{i=0}^{]\log_2 n_x[-1} x_q^{(i)} 2^{]\log_2 n_x[-i-1}, \tag{4.4.2}$$

$$K^x_{m-s}(q) = \sum_{i=0}^{]\log_2 n_x[-s-1} x_q^{(i)} 2^{]\log_2 n_x[-s-i-1}. \tag{4.4.3}$$

If it is the state assignment $K^a(q)$ that is fixed, we construct similar functions $K^a_{m-s}(q)$. The inverse functions to $K^x_{m-s}(q)$ and $K^a_{m-s}(q)$ will be denoted by $(K^x_{m-s})^{-1}(q)$ and $(K^a_{m-s})^{-1}(q)$, respectively.

Now suppose that the t-th iteration has produced a $2^{]\log_2 n_x[-g} \times 2^{]\log_2 n_a[-d}$ matrix A_{m-t} ($g + d = t - 1, m =]\log_2 n_a[+]\log_2 n_x[$). Then, if the assignment $K^x(q)$ (or $K^a(q)$) is fixed, we define a minimal symmetric matching for $(\rho^x_{m-t}(q;s))$ (or $(\rho^a_{m-t}(q;s))$) to be

$$\{\rho^x_{m-t}((K^x_{m-1-d})^{-1}(2q-2); (K^x_{m-1-d})^{-1}(2q-1))\} \quad (q = 1, \ldots, 2^{]\log_2 n_x[-d-1})$$

$$(\{\rho^a_{m-t}((K^a_{m-1-g})^{-1}(2q-2); (K^a_{m-1-g})^{-1}(2q-1))\} \quad (q = 1, \ldots, 2^{]\log_2 n_a[-g-1})).$$

Thus, in this type of automaton there is no need to construct the matrices $(\rho^x_{m-t}(q;s))$ $((\rho^a_{m-t}(q;s)))$ for all $t = 1, 2, \ldots, m$ or to search for minimal symmetric matchings on these matrices.

REMARK. In many cases it is convenient to change the notation for the input signals (states) of the original automaton before actually beginning the assignment procedure, in such a way that the fixed assignment has the form $K^x(q) = q$ (or $K^a(q) = q$). This merely involves a suitable permutation of the columns (rows) of the state matrix A_{m-1} and clearly has no effect on the final result.

If the fixed assignment is $K^a(q)$, we replace the states $a_{q,s}$ in the original matrix A_{m-1} by the values of their codes. The elements $b_{r,s}$ of A_{m-t} were defined as sets of 2^{t-1} elements of A_{m-1} (see, e.g., (4.2.9)). Since $K^a(q)$ is given, we can replace the sets $b_{r,s}$ by the respective numbers $\sum_{a_q \in b_{r,s}} K^a(q)$. Thus the elements of the matrices A_{m-t} are no longer sets but numbers, and this simplifies the construction of the matrices $(\rho^x_{m-t}(q;s))$. In addition, it can be shown that the resulting values of L_{m-t} ($t = 2, 3, \ldots, m$) are not only asymptotically optimal (see Theorem 4.2.2) but exact minima with respect to all assignments which give L_{m-t+1} the previously determined minimal value.

EXAMPLE 4.4.2. Consider the automaton with $K^a(q) = q$ ($q = 0, 1, 2, 3$) defined by the state matrix

$$A_3 = \begin{pmatrix} 0 & 2 & 0 & 3 \\ 0 & 3 & 1 & 3 \\ 2 & 2 & 2 & 0 \\ 3 & 1 & 3 & 1 \end{pmatrix}$$

($m = 4$). Here $n_a = n_x = 4$ and the element a_{ij} of A_3 is s if the input signal x_j passes the automaton from a state assigned the code of the number i into the state assigned the code of s. We determine an assignment $K^x(q)$ and an optimal ordering T of the arguments of the excitation functions.

First iteration. The matrix $(\rho_3^x(q; s))$ is

$$(\rho_3^x(q; s)) = \begin{pmatrix} 0 & 3 & \textcircled{1} & 4 \\ 3 & 0 & 3 & \textcircled{2} \\ \textcircled{1} & 3 & 0 & 4 \\ 4 & \textcircled{2} & 4 & 0 \end{pmatrix}.$$

The minimal matching for $(\rho_3^x(q; s))$ is $\{\rho_3^x(0; 2), \rho_3^x(1; 3)\}$. As a minimal matching for $(\rho_3^a(q; s))$ we take $\{\rho_3^a(0; 1), \rho_3^a(2; 3)\}$. Then $L_3^x = 3$, $L_3^a = 6$, $L_3^x < L_3^a$, $z^{(T_3)} = x^{(1)}$. The input signals x_0, x_2 and x_1, x_3 are assigned codes that differ only in the least significant digit $x^{(1)}$.

Second iteration. The matrix $A_{m-2} = A_2 = A_2^x$ is

$$A_2 = A_2^x = \begin{pmatrix} 0 & 5 \\ 1 & 6 \\ 4 & 2 \\ 6 & 2 \end{pmatrix}.$$

Then

$$(\rho_2^x(q; s)) = \begin{pmatrix} 0 & \textcircled{4} \\ \textcircled{4} & 0 \end{pmatrix}.$$

The minimal matching for $(\rho_2^x(q; s))$ is $\{\rho_2^x(0; 1)\}$. As minimal matching for $(\rho_2^a(q; s))$ we take $\{\rho_2^a(0; 1), \rho_2^a(2; 3)\}$. Then $L_2^x = 4$, $L_2^a = 3$, $L_2^a < L_2^x$, $z^{(T_2)} = a^{(1)}$.

Third iteration. $A_{m-3} = A_1 = A_1^a = \begin{pmatrix} 1 & 11 \\ 10 & 4 \end{pmatrix}$. Then $(\rho_1^x(q; s)) = \begin{pmatrix} 0 & \textcircled{2} \\ \textcircled{2} & 0 \end{pmatrix}$.

The minimal matching for $(\rho_1^a(q; s))$ is $\{\rho_1^a(0; 1)\}$ and that for $(\rho_1^x(q; s))$ is $\{\rho_1^x(0; 1)\}$; $L_1^x = L_1^a = 2$. We set $z^{(T_1)} = a^{(0)}$.

Fourth iteration. $A_0 = A_0^a = (11 \;\; 15)$, $(\rho_0^x(q; s)) = \begin{pmatrix} 0 & \textcircled{1} \\ \textcircled{1} & 0 \end{pmatrix}$, $L_0 = L_0^x + 1 = 2$

$z^{(T_0)} = z^{(0)}$.

In this example, then, the result is $L_3 = 3$, $L_2 = 3$, $L_1 = 2$, $L_0 = 2$, $K^x(0) = 0$, $K^x(1) = 2$, $K^x(2) = 1$, $K^x(3) = 3$. The optimal order of arguments is $(x^{(0)}, a^{(0)}, a^{(1)}, x^{(1)})$.

Chapter Five

SPECTRAL METHODS FOR ANALYSIS OF CORRECTING CAPABILITY OF DIGITAL DEVICES AND SYNTHESIS OF RELIABLE DIGITAL DEVICES

The spectral approach developed over the previous chapters will now be applied to the analysis of the reliability of digital devices (DD) and to the synthesis of reliable DD; these problems will be discussed both for DD including memory elements and for combinational (memoryless) networks.

Our analysis of correcting capability (§§5.1, 5.3) will consider only "nonredundant" DD, that is, DD containing no special "redundant" logic elements or memory elements designed to increase reliability.

Our basic working tools in this chapter will be various correlation characteristics of the functions describing the operation of the DD.

5.1. SPECTRAL METHODS FOR ANALYSIS OF CORRECTING CAPABILITY OF LOGICAL FUNCTIONS

5.1.1. We shall first analyze the types of error that can be corrected by switching (combinational) networks with no memory elements. As we know, the behavior of such networks is described by systems of logical functions (LF).

In this context, the term "error" is reserved for errors at the input of a network realizing the LF (in particular, these may be due to errors in previous logic networks or in communication channels). A knowledge of the types of error that a network is capable of correcting is of paramount importance, both in order to estimate its reliability and in order to use it for synthesis of another network realizing the same system of LF but correcting errors of some prescribed type. We shall confine ourselves to the two most widespread types of error: algebraic errors and arithmetic errors. We begin with a descriptive outline of these types of error and of the appropriate correcting capability of systems of LF. Formal definitions will be given later on in this section.

An *algebraic error* is an error leading to independent distortion of individual components of the input signal. In particular, an *l-fold error* is an error resulting in distortion of exactly l components of the input signal.

An *arithmetic error* is an error which can spread from component to component of the input signal along the carry circuits. In particular, an *l-fold arithmetic error* in m digits of a p-ary input signal is an error which increases (or decreases) the input signal (mod p^m) by any number of the form $\alpha^{(0)}p^{i_0} + \cdots + \alpha^{(l-1)}p^{i_{l-1}}$, where $\alpha^{(0)}, \ldots, \alpha^{(l-1)} \in \{\pm 1, \pm 2, \ldots, \pm(p-1)\}, 0 \leqq i_0, \ldots, i_{l-1} \leqq m-1$. Arithmetic errors appear in a network if its input is the output of some arithmetic device (adder, counter, multiplier, etc.).

We shall say that a system of LF *corrects* an error if the occurrence of the error in the argument vector leaves the value of the system unchanged. In other words, if a correctable error occurs at the input of any network (even a nonredundant network, of minimal complexity) realizing the system, the network will correct the error automatically, with no need for any redundancy. We reiterate that in general any nonredundant network will correct a certain class of errors. The delineation of this class is extremely important in many practical situations. The rest of this section is devoted to this problem—determination of the correcting capability of systems of LF.

5.1.2. Consider a system of completely specified LF

$$y^{(s)} = f^{(s)}(z^{(0)}, \ldots, z^{(m-1)}) \qquad (z^{(0)}, \ldots, z^{(m-1)} \in \{0, 1, \ldots, p-1\}; s = 0, 1, \ldots, k-1). \tag{5.1.1}$$

We first consider the correction of algebraic errors, omitting the adjective "algebraic" whenever there is no danger of confusion. We define an *error* for a system of LF as an ordered pair (z, z') of argument vectors. System (5.1.1) is said to *correct* a set of errors R if, for any $(z, z') \in R$,

$$f^{(s)}(z) = f^{(s)}(z') \qquad (s = 0, \ldots, k-1). \tag{5.1.2}$$

The *multiplicity* of an error (z, z') is the number $\|z \ominus z'\|$ (mod p), i.e., the number of nonzero components of the (componentwise) difference mod p between the vectors z and z'. This quantity is also known as the *Hamming distance* between z and z'. The number of l-fold errors that can be corrected by a system f is denoted by $\eta_f(l)$; the function $\eta_f(l)$ will be called the *correcting capability* (or *correcting power*) of the system f. Our problem is to construct the function $\eta_f(l)$ for a given system f.

We first construct a discrete function $y = f(z)$ representing system (5.1.1);

$$z = \sum_{s=0}^{m-1} z^{(s)} p^{m-1-s}, \qquad y = \sum_{s=0}^{k-1} f^{(s)}(z)\, p^{k-1-s}. \tag{5.1.3}$$

The characteristic functions of $y = f(z)$ are defined, as usual, by

$$f_i(z) = \begin{cases} 1 & \text{if } f(z) = i, \\ 0 & \text{otherwise} \end{cases} \qquad (i = 0, 1, \ldots, p^k - 1). \tag{5.1.4}$$

We now define the autocorrelation functions

$$B_i(\tau) = \sum_{z=0}^{p^m-1} f_i(z) f_i(z \ominus \tau) \qquad (\text{mod } p) \qquad (i = 0, 1, \ldots, p^k - 1) \tag{5.1.5}$$

and the total autocorrelation function

$$B_\Sigma(\tau) = \sum_{i=0}^{p^k-1} B_i(\tau). \tag{5.1.6}$$

The properties of autocorrelation functions $B_i(\tau)$ and their relationship to double Walsh–Chrestenson transforms were discussed in detail in Chapter I.

THEOREM 5.1.1.

$$\eta_f(l) = \sum_{\|\tau\| = l} B_\Sigma(\tau), \tag{5.1.7}$$

where $\|\tau\|$ denotes the number of nonzero components of the p-ary expansion of τ.

PROOF. It follows from (5.1.4), (5.1.5) and the definition of $\eta_f(l)$ that $\sum_{\|\tau\|=l} B_i(\tau)$ is the number of l-fold errors corrected by the system at points z for which $f(z) = i$. Hence, summing over the characteristic functions and using (5.1.6), we obtain (5.1.7). ∎

Theorem 5.1.1 provides a simple method for calculating the correcting capability of systems of LF in terms of their autocorrelation functions. The latter, in turn, may be evaluated using (5.1.5), but for large values of m it is often more convenient to express them in terms of double Chrestenson transforms. Recall that if $\chi^{(p)}(f_i)$ is the Chrestenson transform of f_i, then

$$B_i = p^m (\chi^{(p)})^{-1} (\chi^{(p)}(f_i) \cdot \overline{\chi^{(p)}(f_i)})$$

where $(\chi^{(p)})^{-1}$ and $\overline{\chi^{(p)}}$ are the inverse and the complex conjugate of the Chrestenson transformation $\chi^{(p)}$. The Chrestenson transform and its inverse may be evaluated by the fast Hadamard–Chrestenson transform algorithm, whose implementation involves $(p-1)mp^m$ elementary operations (see §§1.2, 1.3).

Computation of the correcting capability $\eta_f(l)$ for a given l may be simplified by the following device. Denote

$$\theta_l^{(p)}(z) = \sum_{\|\omega\| = l} \chi_\omega^{(p)}(z), \tag{5.1.8}$$

where $\chi_\omega^{(p)}(z)$ is the ω-th Chrestenson function (see § 1.3). $\{\theta_l^{(p)}(z)\}$ is a system of orthogonal functions, but it is not complete. Let $A_f(\omega)$ denote the total amplitude spectrum of the system of characteristic functions $\{f_i\}$, i.e., the value of the function

$$\sum_{i=0}^{p^k-1} \chi^{(p)}(f_i)\,\overline{\chi^{(p)}(f_i)}$$

at the point ω. In particular, if f is a system of Boolean functions, then $A_f(\omega)$ is the value of the function $\sum_{i=0}^{2k-1}(W(f_i))^2$ at ω, where $W(f_i)$ is the Walsh transform of f_i (since $W(f_i) = \chi^{(2)}(f_i) = \overline{\chi^{(2)}(f_i)}$).

COROLLARY 5.1.1.

$$\eta_f(l) = p^m \sum_{\omega=0}^{p^m-1} A_\omega(\omega)\,\theta_l^{(p)}(\omega). \tag{5.1.9}$$

PROOF. Using the definitions of $A_f(\omega)$ and $(\chi^{(p)})^{-1}$, we deduce from (5.1.5)–(5.1.8) that

$$\eta_f(l) = \sum_{\|\tau\|=l} B_\Sigma(\tau) = \sum_{\|\tau\|=l}\sum_{i=0}^{p^k-1} B_i(\tau) = \sum_{\|\tau\|=l} p^m \sum_{\omega=0}^{p^k-1} A_f(\omega)\,\chi_\tau^{(p)}(\omega)$$

$$= p^m \sum_{\omega=0}^{p^k-1} A_f(\omega) \sum_{\|\tau\|=l} \chi_\tau^{(p)}(\omega) = p^m \sum_{\omega=0}^{p^k-1} A_f(\omega)\,\theta_l^{(p)}(\omega). \quad ∎$$

Thus, according to Corollary 5.1.1, when calculating $\eta_f(l)$ there is no need to evaluate the Chrestenson spectrum of the function $A_f(\omega)$; it suffices to evaluate the l-th expansion coefficient of this function relative to the basis $\{\theta_l^{(p)}(z)\}$.

The computation of $B_\Sigma(\tau)$ by formula (5.1.5) for $p > 2$ is facilitated by the "evenness" property:

$$B_\Sigma(\tau) = B_\Sigma(\bar{\tau}), \qquad \tau \oplus \bar{\tau} = 0 \qquad (\text{mod}\, p). \tag{5.1.10}$$

This relation follows directly from (5.1.5), (5.1.6), and halves the necessary computations.

EXAMPLE 5.1.1. Consider the ternary function of two arguments ($p = 3, m = 2$, $k = 1$) defined by Table 5.1.1, which also gives the characteristic functions f_1, f_2 ($f_0 \equiv 0$), the autocorrelation functions B_1, B_2, and the total autocorrelation function B_Σ. Using Theorem 5.1.1, we see that $\eta_f(1) = 20$, $\eta_f(2) = 12$. Note that for $p = 3$, $m = 2$, $k = 1$ the total number of possible single errors is 36, and the number of double errors is also 36. Thus the above function corrects 5/9 of all single errors and a third of all double errors.

We now turn to a few properties of the function $\eta_f(l)$ characterizing the correcting capability of a system.

TABLE 5.1.1.

z, τ	$z^{(0)}$	$z^{(1)}$	$f(z)$	$f_1(z)$	$f_2(z)$	$B_1(\tau)$	$B_2(\tau)$	$B_\Sigma(\tau)$
0	0	0	1	1	0	5	4	9
1	0	1	2	0	1	2	1	3
2	0	2	1	1	0	2	1	3
3	1	0	1	1	0	4	3	7
4	1	1	2	0	1	2	1	3
5	1	2	2	0	1	2	1	3
6	2	0	1	1	0	4	3	7
7	2	1	2	0	1	2	1	3
8	2	2	1	1	0	2	1	3

First, the correcting capability is invariant under translations and permutations of the arguments, and also under translations of the function itself. In other words, for any $0 \leqq \tau < p^m, 0 \leqq q < p^k, \sigma$:

$$\eta_{f(z)}(l) = \eta_{f(z \oplus \tau)}(l) = \eta_{f(\sigma z)}(l) = \eta_{f(z) \oplus q}(l) \qquad (\text{mod } p), \tag{5.1.11}$$

where σz is the number obtained by applying the permutation σ to the components of z.

These invariance properties follow from the analogous invariance properties of the correlation functions.

One consequence of (5.1.11) is the *duality property* of the correcting capability of systems of Boolean functions ($p = 2$).

If $f^*(z)$ is the BF obtained from a BF $f(z)$ by interchanging disjunction and conjunction in any formula expressing $f(z)$ as a superposition of disjunction, conjunction, negation and addition modulo 2, then

$$\eta_{f(z)}(l) = \eta_{f^*(z)}(l)$$

for any l. This follows from (5.1.11) with $\tau = 2^m - 1$ and $q = 2^k - 1$ via De Morgan's laws (2.6.14), by reasoning similar to that proving the duality of spectral complexity (Corollary 2.7.1).

Our next task is to establish a bound on the number $\eta_f = \sum_{l=1}^{m} \eta_f(l)$ of errors of all multiplicities corrected by the system f.

THEOREM 5.1.2. *Let* $\{f_{m,k}^{(p)}\}$ *be the set of all systems of k p-valued logical functions of m arguments. Then*

$$\min_{f \in \{f_{m,k}^{(p)}\}} \eta_f = p^m(p^{m-k} - 1). \tag{5.1.12}$$

PROOF. By the definition of $B_i(\tau)$ (see (5.1.5)),

$$\sum_{\tau=1}^{p^m-1} B_i(\tau) = B_i(0)\,(B_i(0) - 1). \tag{5.1.13}$$

Hence, in view of Theorem 5.1.1, it follows that for any $f \in \{f_{m,k}^{(p)}\}$

$$\begin{aligned} \eta_f &= \sum_{l=1}^{m} \eta_f(l) = \sum_{l=1}^{m} \sum_{\|\tau\|=l} B_\Sigma(\tau) = \sum_{\tau=1}^{p^m-1} B_\Sigma(\tau) = \sum_{\tau=1}^{p^m-1} \sum_{i=0}^{p^k-1} B_i(\tau) \\ &= \sum_{i=0}^{p^k-1} B_i(0)\,(B_i(0) - 1). \end{aligned} \tag{5.1.14}$$

Minimizing the last expression in (5.1.14) over $B_i(0)$ ($i = 0, 1, \ldots, p^k - 1$) under the constraint $\sum_{i=0}^{p^k-1} B_i(0) = p^m$, we see that the minimum of η_f is achieved when

$$B_i(0) = p^{m-k} \qquad (i = 0, 1, \ldots, p^k - 1). \tag{5.1.15}$$

Inserting (5.1.15) into (5.1.14), we obtain (5.1.12). ▮

According to (5.1.14), the total correcting capability η_f of a system f depends only on the set of quantities $B_i(0) = \sum_{z=0}^{p^m-1} f_i(z)$ ($i = 0, 1, \ldots, p^k - 1$); in other words, if we call $\|f_i\| = \sum_{z=0}^{p^m-1} f_i(z)$ the *power* of f_i, then η_f depends only on the distribution of powers of the characteristic functions, and η_f is a minimum when the functions f_i ($i = 0, 1, \ldots, p^k - 1$) have equal powers.

5.1.3. We now specialize our results to the case of Boolean functions (BF). The following proposition is a corollary of Theorem 5.1.2.

COROLLARY 5.1.2. *Let $f(z)$ be a Boolean function of m arguments and power $\|f\| = \sum_z f(z)$. Then*

$$\eta_f = 2\|f\|^2 - 2^{m-1}\|f\| + 2^{2m} - 2^m. \tag{5.1.16}$$

PROOF. This follows from (5.1.4) and (5.1.14) with $p = 2$, $k = 1$, in view of the fact that the power distribution of the characteristic functions is in this case $\|f_1\| = \|f\|$, $\|f_0\| = 2^m - \|f\|$. ▮

According to (5.1.16), the total correcting capability of a BF depends only on its number of arguments and its power; with increasing power, η_f decreases for $\|f\| < 2^{m-1}$ and increases for $\|f\| > 2^{m-1}$, with a minimum at $f = 2^{m-1}$:

$$\min_{f \in \{f_{m,1}^{(2)}\}} \eta_f = 2^{2m-1} - 2^m.$$

On the one hand, for fixed m all BF of the same power correct the same number of errors. On the other, depending on the specific form of the function, the correcting capability of some functions is concentrated in the region of low-multiplicity errors, that of others in the region of high-multiplicity errors. It is thus important to analyze the correcting capability of various different classes of BF.

We first observe that for any BF f of m arguments ($p = 2, k = 1$)

$$f_1 = f, \qquad f_0 = 1 - f,$$

and if $B(\tau)$ is the autocorrelation function of f, then by (5.1.6)

$$B_\Sigma(\tau) = 2^m - 2\|f\| + 2B(\tau) \qquad (B(\tau) = B_1(\tau)). \tag{5.1.17}$$

Thanks to this formula, the autocorrelation functions of the 22 classes of BF listed in the Appendix may be used to analyze their correcting capabilities.

From the technical viewpoint, one is most interested in the three classes of BF defined by the following distributions of correcting capability: a) concentrated in the region of low-multiplicity errors, b) uniformly distributed over multiplicities, c) concentrated in the region of errors of maximal multiplicity. The reason for the prominence we give to these three classes is that devices realizing them possess high correcting capability with respect to three important classes of errors in the input signals: a) errors weakly correlated in time, the most probable errors being those of low multiplicity, b) errors of all multiplicities having equal probabilities, c) strongly correlated errors, the most probable errors being those of high multiplicity.

The distribution of correctable errors with respect to multiplicity may be represented by the relative frequency $\hat{\eta}_f(l)$ of l-fold errors corrected by the function f. The total number of l-fold errors for a BF of m arguments is $\binom{m}{l} \cdot 2^m$, and so

$$\hat{\eta}_f(l) = \frac{\eta_f(l)}{\binom{m}{l} \cdot 2^m}. \tag{5.1.18}$$

As an example of a class of BF with error-correcting capability concentrated mainly in the low-multiplicity region, we consider the class of majority functions.

A BF $f(z)$ of m arguments (where m is odd) is called a *majority function* if its value is 1 if and only if $\|z\| \geqq (m + 1)/2$.

For any majority function of m arguments,

$$\|f\| = \sum_{z=0}^{2^m - 1} f(z) = 2^{m-1},$$

and so, in view of (5.1.17), (5.1.7), (5.1.18),

$$B_\Sigma(\tau) = 2B_1(\tau) = 2B(\tau), \tag{5.1.19}$$

$$\hat{\eta}_f(l) = \binom{m}{l}^{-1} 2^{-m+1} \sum_{\|\tau\| = l} B(\tau). \tag{5.1.20}$$

A formula for the autocorrelation functions of majority functions is given in row 9 of the Appendix. The data in the Appendix, together with (5.1.20), imply the following formula for the relative frequency of errors corrected by majority functions:

$$\hat{\eta}_f(l) = \begin{cases} 2^{-m+2} \displaystyle\sum_{i=0}^{[l/2]} \binom{l}{i} \sum_{j=0}^{[m/2]} \binom{m-l}{m-j-i} & \text{if } l \text{ is odd,} \\ 2^{-m+1}\left(2 \displaystyle\sum_{i=0}^{l/2-1} \binom{l}{i} \sum_{j=0}^{[m/2]} \binom{m-l}{m-j-i} + \binom{l}{l/2} \sum_{j=0}^{[m/2]} \binom{m-l}{m-l/2-j} \right) & \text{if } l \text{ is even.} \end{cases} \tag{5.1.21}$$

(Here $\binom{p}{q} = 0$ if $p < q$, $\binom{p}{0} = 1$.)

Figure 5.1.1 illustrates the functions $\hat{\eta}_f(l)$ for majority functions of $m = 3, 5, 7$ arguments, according to (5.1.21).

An analysis of (5.1.21) shows that if f is a majority function, $\hat{\eta}_f(l)$ is a monotone decreasing function of l, and

$$\hat{\eta}_f(2s-1) - \hat{\eta}_f(2s) \qquad (s = 1, \ldots,]m/2[). \tag{5.1.22}$$

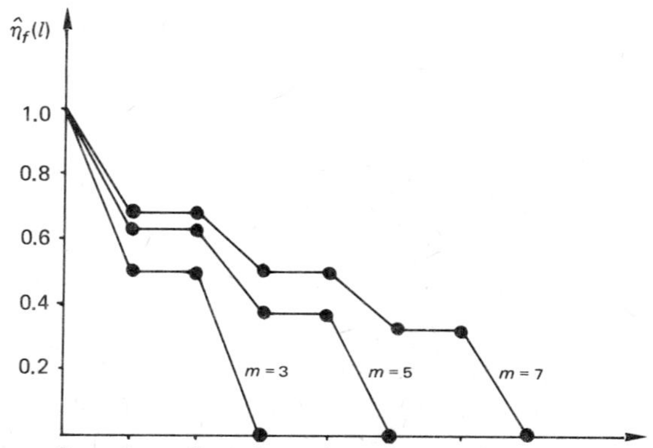

FIGURE 5.1.1.

Therefore, the correcting capability of a majority function is concentrated primarily in the region of low-multiplicity errors.

An example of a class of BF with correcting capability uniformly distributed over multiplicities is provided by the nonrepetitive quadratic forms in $m = 2k$ variables. We recall that a *nonrepetitive quadratic form* is a BF such that

$$f(z) = f(z^{(0)}, \dots, z^{(2k-1)}) = \bigoplus_{\substack{i,j=0 \\ i<j}}^{k-1} c_{ij} z^{(i)} z^{(j)} \pmod 2, \tag{5.1.23}$$

where each $z^{(i)}$ appears exactly once on the right. We know (see §2.1) that

$$\|f\| = 2^{2k-1} - 2^{k-1}, \tag{5.1.24}$$

and for any $\tau > 0$

$$B(\tau) = 2^{2k-2} - 2^{k-1}, \tag{5.1.25}$$

$$B_{\Sigma}(\tau) = 2^{2k-1}. \tag{5.1.26}$$

It follows from (5.1.7), (5.1.17), (5.1.18), (5.1.24) and (5.1.25) that if f is a nonrepetitive quadratic form, then, for $l > 0$,

$$\hat{\eta}_f(l) = \tfrac{1}{2} \tag{5.1.27}$$

Finally, as an example of a class of BF which correct mainly errors of maximal multiplicity, we consider the elementary symmetric BF of $m = 2k$ arguments with operating number k.

A BF $f(z)$ is an *elementary symmetric function with operating number* k if $f(z) = 1$ if and only if $\|z\| = k$. We then have

$$\|f\| = \binom{2k}{k}. \tag{5.1.28}$$

The autocorrelation function for an elementary symmetric BF with operating number k is (see Appendix, row 8)

$$B(\tau) = \begin{cases} \dbinom{2k - \|\tau\|}{k - \|\tau\|/2} \dbinom{\|\tau\|}{\|\tau\|/2} & \text{if } \|\tau\| \text{ is even,} \\ 0 & \text{if } \|\tau\| \text{ is odd.} \end{cases} \tag{5.1.29}$$

It follows from (5.1.7), (5.1.17), (5.1.18), (5.1.28) and (5.1.29) that if f is an elementary symmetric function then

$$\hat{\eta}_f(l) = \begin{cases} \binom{2k}{l}^{-1} 2^{-2k}\left(2^{2k} - 2\binom{2k}{k} + 2\binom{2k-l}{k-l/2}\binom{l}{l/2}\binom{2k}{l}\right) & \text{if } l \text{ is even,} \\ \binom{2k}{l}^{-1} 2^{-2k}\left(2^{2k} - 2\binom{2k}{k}\right) & \text{if } l \text{ is odd.} \end{cases} \quad (5.1.30)$$

Figure 5.1.2 illustrates the functions $\hat{\eta}_f(l)$ for elementary symmetric BF of $m = 2k$ arguments with operating numbers $k = 2, 3, 4$.

An analysis of (5.1.30) shows that for any k the function $\hat{\eta}_f(l)$ goes through a maximum at $l = 2k$ and has local maxima at $l = 2i$ ($i = 1, 2, \ldots, k - 1$). Another conclusion from (5.1.30) is that $\hat{\eta}_f(l)$ is symmetric about $l = k$, i.e., if $l < k$, then

$$\hat{\eta}_f(l) = \hat{\eta}_f(l - k). \quad (5.1.31)$$

It follows from (5.1.30) that the asymptotic distribution of correctable errors is

$$\lim_{k\to\infty} \hat{\eta}_f(l) = 0 \qquad (l = 1, \ldots, 2k - 1). \quad (5.1.32)$$

Since

$$\hat{\eta}_f(2k) = 1 \quad (5.1.33)$$

for any k, it follows from (5.1.32) that the correcting capability of an elementary symmetric BF of $2k$ arguments with operating number k is concentrated in the region of maximal multiplicity.

To summarize: devices realizing majority functions are reliable with respect to weakly temporally correlated errors, since the most probable errors in such cases

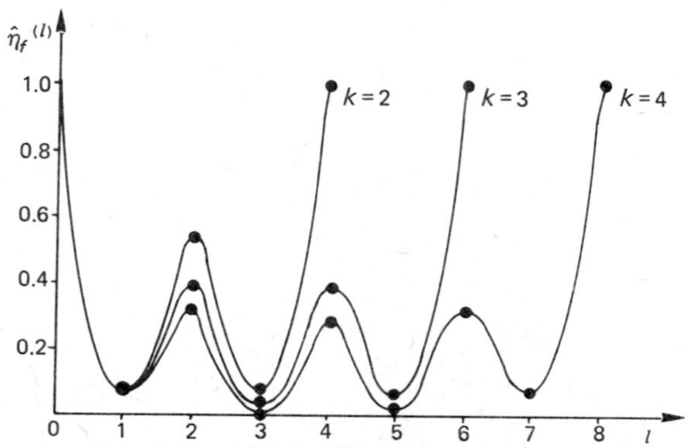

FIGURE 5.1.2.

are of low multiplicity; devices realizing nonrepetitive quadratic forms display good correcting capability when errors of all multiplicities are equally probable; and devices realizing elementary symmetric BF are most reliable with respect to strongly correlated errors, since then errors of maximal multiplicity are most probable.

There is an interesting relationship between correcting capability and complexity for completely specified BF. Recall (see §2.3) that the complexity of a BF may be measured by the number of unordered pairs of matched constituents of unity or zero: the complexity criterion $\tilde{\eta}(f)$ for a BF $f(z)$ is the number of pairs $\{z_1, z_2\}$ such that $f(z_1) = f(z_2)$ and $\|z_1 \oplus z_2\| = 1 \pmod 2$ (see (2.3.10)). As shown in §2,3, for almost all BF $f(z)$ of m arguments ($m \to \infty$) the number of two-input logic elements necessary for a minimal network realizing $f(z)$ decreases with increasing $\tilde{\eta}(f)$. A comparison of the definitions of $\tilde{\eta}(f)$ and of the correcting capability $\eta_f(l)$ reveals the simple relationship

$$\tilde{\eta}(f) = \eta_f(1). \tag{5.1.34}$$

Thus, the higher the complexity criterion $\tilde{\eta}(f)$ for the function (i.e., the simpler its realization), the greater the number of single errors the function is capable of correcting, and vice versa.

5.1.4. We now consider the correction of arithmetic errors by systems of LF, hence also by memoryless logic networks. Given a system (5.1.1), we define an *arithmetic error*, as before, to be any pair (z, z') of argument vectors; an error (z, z') is said to be *corrected* if $f^{(i)}(z) = f^{(i)}(z')$ $(i = 0, 1, \ldots, k-1)$.

We shall consider two types of arithmetic errors: symmetric and nonsymmetric. The multiplicity of a *symmetric error* (z, z'), denoted by $\|z \ominus z'\|^{(s)}$, is defined as the minimal number of terms in the representation of $z \ominus z' \pmod{p^m}$ or $z' \ominus z \pmod{p^m}$ as a linear combination of powers of p with exponents at most $m-1$ and coefficients in $\{\pm 1, \pm 2, \ldots, \pm(p-1)\}$. For example, if $p = 2$, $m = 4$, and $z \ominus z' = 7 \pmod{2^4}$, then $\|7\|^{(s)} = 2$, since $2^3 - 2^0 = 7$ (but, if $m = 3$ and $z \ominus z' = 7 \pmod{2^3}$, then $\|7\|^{(s)} = 1$, since $7 = -2^0 \pmod{2^3}$!).

The multiplicity of a *nonsymmetric error* (z, z') is defined as the number of nonzero terms in the p-ary expansion of the number $z \ominus z' \pmod{p^m}$. For example, if $p = 2$, then $\|7\|^{(n)} = 3$ for any m. (Note that $\|\tau\|^{(n)} = \|\tau\|$ for any $\|\tau\|$; we nevertheless find it convenient to use the notation $\|\tau\|^{(n)}$ instead of $\|\tau\|$ throughout this subsection.)

It is easy to see that for any error (z, z') and any p, m,

$$\begin{aligned} \|z \ominus z'\|^{(s)} &\leq \|z \ominus z'\|^{(n)}, \\ \|z \ominus z'\|^{(s)} &= \|z' \ominus z\|^{(s)}. \end{aligned} \pmod{p^m} \tag{5.1.35}$$

In general, however,

$$\|z \ominus z'\|^{(n)} \neq \|z' \ominus z\|^{(n)} \qquad (\mathrm{mod}\ p^m).$$

An examination of the definition of multiplicity for symmetric and nonsymmetric errors reveals the following difference between the two types. In terms of a physical realization of the system of LF, symmetric errors will both increase and decrease the input to the device (mod p^m), whereas nonsymmetric errors will only increase (or only decrease) the input. These two cases correspond to situations in which the input to the logic network is the output of a p-ary m-digit network implementing an arithmetical operation, according as the additive noise produced by the network is two-way or one-way.

Any given system of LF corrects a certain class of algebraic errors (see §5.1.2) and certain classes of symmetric and nonsymmetric arithmetic errors; the latter are generally not the same. Our immediate task is therefore to find ways of constructing the classes of arithmetic errors corrected by a system.

We denote the number of l-fold symmetric (nonsymmetric) arithmetic errors corrected by a system f by $\lambda_f^{(s)}(l)$ $\left(\lambda_f^{(n)}(l)\right)$. To calculate the correcting capabilities $\lambda_f^{(s)}(l)$ and $\lambda_f^{(n)}(l)$ we shall use a method similar to that used in §5.1.2 to calculate the correcting capability $\eta_f(l)$ for algebraic errors.

Let $\{f_i\}$ be the system of characteristic functions for f (see (5.1.3), (5.1.4)). Now set

$$B_i(\tau) = \sum_{z=0}^{p^m-1} f_i(z)\, f_i(z \ominus \tau) \qquad (\mathrm{mod}\ p^m) \qquad (i = 0, \ldots, p^k - 1). \tag{5.1.36}$$

Reasoning as in the proof of Theorem 5.1.1, one can show that if $B_\Sigma(\tau) = \sum_{i=0}^{p^k-1} B_i(\tau)$, then

$$\begin{aligned} \lambda_f^{(s)}(l) &= \sum_{\|\tau\|^{(s)}=l} B_\Sigma(\tau), \\ \lambda_f^{(n)}(l) &= \sum_{\|\tau\|^{(n)}=l} B_\Sigma(\tau). \end{aligned} \tag{5.1.37}$$

Based on formulas (5.1.36), (5.1.37), we have a simple computational procedure for the arithmetic-error correcting capabilities of systems of LF.

EXAMPLE 5.1.2. Table 5.1.2 defines a BF f of 4 arguments ($m = 4, p = 2, k = 1$). Here $f = f_1$, $1 - f = f_0$. Also shown in the table are the autocorrelation function $B(\tau) = B_1(\tau)$ of f, the total autocorrelation function

$$B_\Sigma(\tau) = 2^m - 2\sum_z f(z) + 2B(\tau) = 4 + 2B(\tau)$$

(see (5.1.17)), and the values of $\|\tau\|^{(s)}$ and $\|\tau\|^{(n)}$. For this example, formula (5.1.37) gives: $\lambda_f^{(s)}(1) = 40$, $\lambda_f^{(s)}(2) = 80$, $\lambda_f^{(s)}(3) = \lambda_f^{(s)}(4) = 0$, $\lambda_f^{(n)}(1) = 22$, $\lambda_f^{(n)}(2) = 60$, $\lambda_f^{(n)}(3) = 60$, $\lambda_f^{(n)}(3) = 32$, $\lambda_f^{(n)}(4) = 6$.

TABLE 5.1.2.

z, τ	$\|\tau\|^{(s)}$	$\|\tau\|^{(n)}$	$z^{(0)}$	$z^{(1)}$	$z^{(2)}$	$z^{(3)}$	$f(z)$	$B(\tau)$	$B_\Sigma(\tau)$
0	0	0	0	0	0	0	1	6	16
1	1	1	0	0	0	1	0	1	6
2	1	1	0	0	1	0	0	0	4
3	2	2	0	0	1	1	1	5	14
4	1	1	0	1	0	0	0	2	8
5	2	2	0	1	0	1	0	0	4
6	2	2	0	1	1	0	1	4	12
7	2	3	0	1	1	1	0	3	10
8	1	1	1	0	0	0	0	0	4
9	2	2	1	0	0	1	1	3	10
10	2	2	1	0	1	0	0	4	12
11	2	3	1	0	1	1	0	0	4
12	1	2	1	1	0	0	1	2	8
13	2	3	1	1	0	1	0	5	14
14	1	3	1	1	1	0	0	0	4
15	1	4	1	1	1	1	1	1	6

The following theorem indicates the relationship between the numbers of correctable symmetric and nonsymmetric errors, for the case of most practical importance —single errors.

THEOREM 5.1.3. *For any system f of p-valued logical functions of m arguments,*

$$\lambda_f^{(s)}(1) = \begin{cases} 2\lambda_f^{(n)}(1) - B_\Sigma(p^m/2) & \textit{if} \quad p \textit{ is even}, \\ 2\lambda_f^{(n)}(1) & \textit{if} \quad p \textit{ is odd}. \end{cases} \tag{5.1.38}$$

PROOF. For any τ, $\|\tau\|^{(s)} = \|p^m \ominus \tau\|^{(s)} \pmod{p^m}$. Hence, if $\|\tau\|^{(n)} = 1$, then $\|\tau\|^{(s)} = 1$ and $\|p^m \ominus \tau\|^{(s)} = 1 \pmod{p^m}$. Moreover, for any τ and the functions $B_\Sigma(\tau)$ as defined here, we have the analog of (5.1.10):

$$B_\Sigma(\tau) = B_\Sigma(p^m \ominus \tau) \qquad \pmod{p^m}. \tag{5.1.39}$$

If p is odd, then $\tau \neq p^m \ominus \tau \pmod{p^m}$ for any τ; hence, by (5.1.37) and (5.1.39),

$$\lambda_f^{(s)}(1) = 2\lambda_f^{(n)}(1).$$

If p is even, then, since $\tau = p^m \ominus \tau \pmod{p^m}$, we have $\|p^m/2\|^{(n)} = \|p^m/2\|^{(s)} = 1$ and, by (5.1.37), (5.1.39),

$$\lambda_f^{(s)}(1) = 2\lambda_f^{(n)}(1) - B_\Sigma(p^m/2),$$

proving (5.1.38). ∎

We note that the total correcting capabilities of any system of LF with respect to symmetric and nonsymmetric errors are the same:

$$\sum_{l=1}^{m} \lambda_f^{(s)}(l) = \sum_{l=1}^{m} \lambda_f^{(n)}(l) = \lambda_f. \tag{5.1.40}$$

This follows directly from (5.1.37). The quantity λ_f satisfies estimates analogous to those established for algebraic errors in Theorem 5.1.2 and Corollary 5.1.2.

The correcting capability of a system of LF with respect to arithmetic errors is not invariant under permutation of the arguments of the LF, but it remains invariant under translation of the arguments of the function itself: for any $0 \leqq \tau < p^m$, $0 \leqq q < p$,

$$\left\{\begin{array}{l} \lambda_{f(z)}^{(n)}(l) = \lambda_{f(z\oplus\tau)}^{(n)}(l) \\ \lambda_{f(z)}^{(s)}(l) = \lambda_{f(z\oplus\tau)}^{(s)}(l) \end{array}\right. \pmod{p^m}, \tag{5.1.41}$$

and

$$\left\{\begin{array}{l} \lambda_{f(z)}^{(n)}(l) = \lambda_{f(z\oplus_q)}^{(n)}(l) \\ \lambda_{f(z)}^{(s)}(l) = \lambda_{f(z)\oplus_q}^{(s)}(l) \end{array}\right. \pmod{p}. \tag{5.1.42}$$

The autocorrelation functions $B_i(\tau) \bmod p^m$ may be calculated by using either (5.1.36) or the double Chrestenson transforms $\chi^{(p^m)}$, which degenerate in this case into discrete Fourier transforms, with the interval of definition of the functions divided into p^m equal subintervals. One can then apply the fast Fourier transform algorithm [47, 140], using an analog of Corollary 5.1.1. If formula (5.1.36) is employed, one can use the symmetry relation (5.1.39) for $B_i(\tau)$, thus halving the computational labor involved, and in the case of symmetric errors, where $\|\tau\|^{(s)} = \|p^m \ominus \tau\|^{(s)}$, there is no need at all to evaluate $B_i(\tau)$ and $B_\Sigma(\tau)$ for $\tau > [p^m/2]$: using Theorem 5.1.3 we can express (5.1.37) as

$$\lambda_f^{(s)}(l) = \begin{cases} 2\displaystyle\sum_{\substack{\|\tau\|^{(s)}=1 \\ 0 \leqq \tau \leqq p^m/2}} B_\Sigma(\tau) - B_\Sigma(p^m/2) & \text{if } l = 1 \text{ and } p \text{ is even}, \\ 2\displaystyle\sum_{\substack{\|\tau\|^{(s)}=l \\ 0 \leqq \tau \leqq [p^m/2]}} B_\Sigma(\tau) & \text{otherwise}. \end{cases} \tag{5.1.43}$$

5.1.5. We now analyze the correcting capability of systems of LF on the assumption that the probabilities of individual errors are given. Here the correcting capability will be measured by the expectation of the number of corrected errors.

We confine attention to algebraic and symmetric arithmetic errors. Each error (z, z') is assigned a certain probability; errors (z, z') and (z', z) are assumed to have the same probability.

Given a system of p-valued logical functions of m arguments, we shall represent an algebraic error by $z \ominus z' \pmod p$ and an arithmetic error by $z \ominus z' \pmod{p^m}$. For each error $z \ominus z' \pmod p$ or $z \ominus z' \pmod{p^m}$, we assume given a probability $p(z \ominus z') \pmod p$ or $q(z \ominus z') \pmod{p^m}$, such that

$$\begin{aligned} p(z \ominus z') &= p(z' \ominus z) \qquad \pmod p, \\ q(z \ominus z') &= q(z' \ominus z) \qquad \pmod{p^m}. \end{aligned} \tag{5.1.44}$$

Thus, for any $\tau \in \{1, 2, \ldots, p^m - 1\}$ we can define functions $p(\tau)$ and $q(\tau)$ giving the probabilities of algebraic and arithmetic errors (z, z') such that $\tau = z \ominus z' \pmod p$ and $\tau = z \ominus z' \pmod{p^m}$, respectively.

We note once more that the probabilities of two errors (z, z') and (z_1, z_1') are assumed to be equal if $z \ominus z' = z_1 \ominus z_1' \pmod p$ (algebraic errors) or $z \ominus z' = z_1 \ominus z_1'$ $\pmod{p^m}$ (arithmetic errors).

Formulas (5.1.7) and (5.1.37) may be generalized to this case. Let $\mathrm{M}\{\eta_f(l)\}$ and $\mathrm{M}\{\lambda_f^{(s)}(l)\}$ denote the expectations of the numbers of l-fold algebraic and symmetric arithmetic errors, respectively, that can be corrected by a system f. Then

$$\mathrm{M}\{\eta_f(l)\} = \sum_{\|\tau\| = l} p(\tau)\, B_\Sigma(\tau), \tag{5.1.45}$$

where $B_\Sigma(\tau)$ is the total autocorrelation function modulo p (see (5.1.5), (5.1.6)), and

$$\mathrm{M}\{\lambda_f^{(s)}(l)\} = \sum_{\|\tau\|^{(s)} = l} q(\tau)\, B_\Sigma(\tau) \tag{5.1.46}$$

where $B_\Sigma(\tau)$ is the total autocorrelation function modulo p^m.

Thus, the expected numbers of correctable errors may also be analyzed in terms of the appropriate autocorrelation functions.

REMARK. The approach to analysis of the correcting capability of systems of LF for l-fold errors, as described in this section, carries over in a natural way to arbitrarily specified sets of errors.

5.2. SPECTRAL METHODS FOR SYNTHESIS OF RELIABLE (FAULT-TOLERANT) DIGITAL DEVICES FOR INFORMATION TRANSMISSION AND PROCESSING

5.2.1. The last section was devoted to spectral methods for analysis of the correcting capability of systems of logical functions. We are now going to use the results for the synthesis of reliable systems for the transmission and logical processing of information.

Throughout this section we shall use only binary signals, though the results may of course be generalized to p-ary signals ($p > 2$).

The structure of the most general information transmission and processing system is shown in Figure 5.2.1. We shall assume that the message generator produces information in the form of binary vectors $(z^{(0)}, \ldots, z^{(m-1)})$ of length m. This information may be distorted because of errors in the communication channel or in the message generator itself; it then proceeds to the input of the receiver. We shall consider parallel transmission, i.e., all the components $z^{(i)}$ $(i = 0, 1, \ldots, m - 1)$ reach the receiver input simultaneously, and the receiver itself is a switching network with no memory elements.

The reader should note, however, that all the synthesis methods to be examined below may be extended to cases in which the information is transmitted in serial code and the receiver includes memory elements (see §5.3).

We shall assume that the errors appearing at the receiver input are algebraic, that is to say, an l-fold error will be an error distorting exactly l components of the input signal.

If the receiver includes no memory elements, its operation may be described by a system of Boolean functions (**BF**):

$$y^{(s)} = f^{(s)}(z^{(0)}, \ldots, z^{(m-1)}) \qquad (s = 0, \ldots, k - 1). \tag{5.2.1}$$

Note that the information processor realizing system (5.2.1) could have been combined with the message generator, only the results $\{y^{(s)}\}$ $(s = 0, 1, \ldots, k - 1)$ being

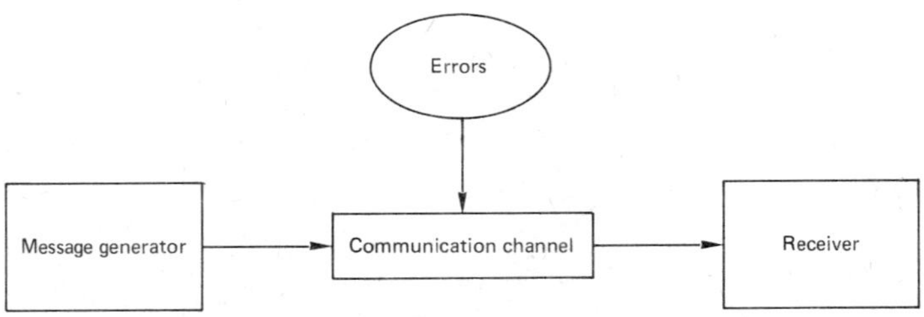

FIGURE 5.2.1.

transmitted along the communication channel. In many important cases, however, this is impossible (for example, when the message generator is installed on board an aircraft or spacecraft and its complexity is therefore restricted). In addition, transmission of the result $\{y^{(s)}\}$ $(s = 0, 1, \ldots, k-1)$ may prove to be no more reliable than that of the message $\{z^{(i)}\}$ $(i = 0, 1, \ldots, m-1)$ (for example, when $m \leqq k$).

The reliability of the system illustrated in Figure 5.2.1 may be increased by using the sophisticated methods of the theory of error-correcting codes [14, 116]. The information $(z^{(0)}, \ldots, z^{(m-1)})$ is then transmitted in a redundant code (i.e., additional arguments $z^{(m)}, \ldots, z^{(m+r-1)}$ appear) and quite complex encoders and decoders are needed.

Our system (Figure 5.2.1) differs from those familiar from classical coding theory in that the information received is processed in the receiver, as evidenced by the form of system (5.2.1).

We saw in §5.1 that the system of BF describing the operation of the receiver has a certain correcting capability. Therefore, instead of transmitting information in redundant code and introducing encoders and decoders, we shall represent the information by a nonredundant code, trying at the same time to make full use of the correcting capability of the system of BF describing the operation of the receiver.

We shall assume that the errors in the communication channel are algebraic (see §5.1); for the sake of generality we allow the probabilities of distortion of the individual transmitted symbols $z^{(i)}$ $(i = 0, 1, \ldots, m-1)$ or of their combinations to be different. (This will be the situation, for example, if the information is transmitted in parallel code along m channels with different statistical properties.) We consider the case in which the system (5.2.1) describing the receiver is completely specified. Our problem is to minimize the expected number of uncorrected errors of the type in question at the output of the entire system of Figure 5.2.1.

In order to increase the reliability of the entire transmission and processing system, we shall linearize the system of BF describing the receiver. As usual, we represent system (5.2.1) by a discrete function $f(z)$ $(z = 0, 1, \ldots, 2^m - 1)$ and then express $f(z)$ as

$$f(z) = f_\sigma(\sigma \otimes z), \tag{5.2.2}$$

where σ is some nonsingular $m \times m$ matrix and $\otimes$ is the symbol for matrix multiplication over $GF(2)$.

We now replace our original system of Figure 5.2.1 by that represented schematically in Figure 5.2.2.

Here we have introduced an additional linear block implementing multiplication by the matrix σ (the complexity of this block is comparatively low, involving a number of two-input mod 2 adders of the order of $m^2/\log_2 m$; see §3.2). The modified receiver now realizes the function f_σ.

Thus, based on the superposition (5.2.2), we have a decomposition of the function

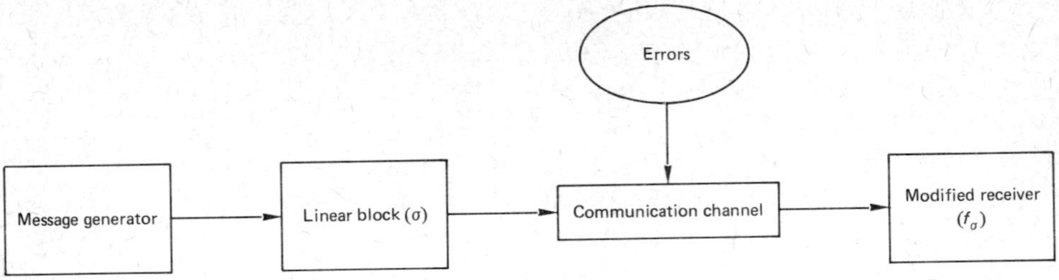

FIGURE 5.2.2.

f into blocks σ and f_σ, the task of correcting errors in the communication channel being assigned to f_σ.

Our problem is now to determine a matrix σ minimizing the expected number of channel errors uncorrected by f_σ. The solution of this problem for various classes of errors is the main topic of this section.

5.2.2. *Correction of single errors.* We start with a system of completely specified BF:

$$y^{(s)} = f^{(s)}(z^{(0)}, \ldots, z^{(m-1)}) \qquad (s = 0, \ldots, k-1),$$

and let p_i $(i = 0, 1, \ldots, m-1)$ denote the probability of distortion of the argument $z^{(i)}$. Let Ξ be the class of all nonsingular $m \times m$ matrices over $GF(2)$ (i.e., with elements in $\{0, 1\}$), and define f_σ by

$$f_\sigma(\sigma \otimes z) = f(z) \qquad (\text{mod } 2),$$

where $\sigma \in \Xi$ and $f(z)$ is the discrete function representing the system.

Let $\mathrm{M}\{\eta_{f_\sigma}(l)\}$ denote the expected number of l-fold errors corrected by f_σ (for the relevant definitions, see §5.1). We may now state our problem thus: find a matrix $\sigma_1 \in \Xi$ such that

$$\mathrm{M}\{\eta_{f_{\sigma_1}}(1)\} = \max_{\sigma\in\Xi} \mathrm{M}\{\eta_{f_\sigma}(1)\}, \tag{5.2.3}$$

where f_σ is defined by (5.2.2).

As usual, let $\{f_t(z)\}$ $(t = 0, 1, \ldots, 2^k - 1)$ denote the system of characteristic functions of $f(z)$ and $B_\Sigma(\tau)$ the total autocorrelation function of $\{f_t(z)\}$:

$$B_\Sigma(\tau) = \sum_{t=0}^{2^k-1} B_t(\tau) = \sum_{t=0}^{2^k-1} \sum_{z=0}^{2^m-1} f_t(z)\, f_t(z \oplus \tau) \qquad (\text{mod } 2). \tag{5.2.4}$$

Given any system $(\tau_0, \ldots, \tau_{m-1})$ of $m-1$ binary vectors of length m, let T denote the matrix with columns $\tau_0, \ldots, \tau_{m-1}$.

THEOREM 5.2.1. *Let*

$$\max_{T \in \Xi} \sum_{i=0}^{m-1} p_i B_\Sigma(\tau_i) = \sum_{i=0}^{m-1} p_i B_\Sigma(\tilde{\tau}_i). \tag{5.2.5}$$

Then

$$\tilde{T} \otimes \sigma_1 = E_m \qquad (\text{mod } 2), \tag{5.2.6}$$

where E_m is the $m \times m$ identity matrix.

The proof follows immediately from the observation that

$$\sum_{i=0}^{m-1} p_i B_\Sigma(2^i) = \mathrm{M}\{\eta_f(1)\}, \tag{5.2.7}$$

after which Theorem 1.5.4 may be used.

Note that if $p_i = 1$ $(i = 0, 1, \ldots, m-1)$ and $l = 1$ formula (5.2.7) follows from Theorem 5.1.1.

It follows from Theorem 5.2.1 that in order to find σ_1 we need only determine m vectors $\tilde{\tau}_0, \ldots, \tilde{\tau}_{m-1}$, linearly independent over $GF(2)$, which maximize $\sum_{i=0}^{m-1} p_i B_\Sigma(\tau_i)$. These vectors may be determined by the following algorithm.

Arrange the probabilities p_i $(i = 0, 1, \ldots, m-1)$ in decreasing order:

$$p_{i_0} \geqq p_{i_1} \geqq \cdots \geqq p_{i_{m-1}} \qquad (i_s \neq i_r \quad \text{if} \quad s \neq r). \tag{5.2.8}$$

Now put

$$\max_{\tau \neq 0} B_\Sigma(\tau) = B_\Sigma(\tilde{\tau}_{i_0}). \tag{5.2.9}$$

Assuming that $\tilde{\tau}_{i_0}, \ldots, \tilde{\tau}_{i_s}$ $(s < m-1)$ have already been found, let $\tilde{\tau}_{i_{s+1}}$ be a vector satisfying the condition

$$\max_{\tau \notin L_s} B_\Sigma(\tau) = B_\Sigma(\tilde{\tau}_{i_{s+1}}), \tag{5.2.10}$$

where L_s is the set of all linear combinations of $\tilde{\tau}_{i_0}, \ldots, \tilde{\tau}_{i_s}$ over $GF(2)$.

Denote the algorithm just described by A. We note that the output $\tilde{\tau}_0, \tilde{\tau}_1, \ldots, \tilde{\tau}_{m-1}$ of algorithm A applied to $B_\Sigma(\tau)$ is independent of the probabilities p_i $(i=0,1,\ldots,m-1)$; it depends only on their ordering (5.2.8). If $p_i = 1$ $(i = 0, 1, \ldots, m-1)$, algorithm A is simply the linearization algorithm for a system of BF with respect to $\eta(f)$ (see §2.4). This provides yet another indication of the relationship between the concepts of complexity and error-correcting capability (see 5.1.3).

EXAMPLE 5.2.1. Table 5.2.1 defines two BF $f^{(0)}$, $f^{(1)}$ of three arguments, giving also the discrete function $f(z)$, characteristic functions $f_i(z)$, autocorrelation functions $B_i(\tau)$ and total autocorrelation function $B_\Sigma(\tau)$.

TABLE 5.2.1.

z, τ	$z^{(0)}$	$z^{(1)}$	$z^{(2)}$	$f^{(0)}$	$f^{(1)}$	f	f_0	f_1	f_2	f_3	B_0	B_1	B_2	B_3	B_Σ	$2A_{01}$	$2A_{02}$	$2A_{03}$	$2A_{12}$	$2A_{13}$	$2A_{23}$	A_Σ
0	0	0	0	0	0	0	1	0	0	0	1	2	2	3	6	0	0	0	0	0	0	0
1	0	0	1	1	1	3	0	0	0	1	0	0	0	0	0	0	0	2	2	2	2	12
2	0	1	0	0	1	1	0	1	0	0	0	0	0	2	2	2	0	0	2	0	2	8
3	0	1	1	1	0	2	0	0	1	0	0	0	0	0	0	0	2	0	0	4	2	8
4	1	0	0	1	1	3	0	0	0	1	0	0	2	0	2	0	0	2	0	4	0	8
5	1	0	1	0	1	1	0	1	0	0	0	0	0	2	2	2	0	0	2	0	2	8
6	1	1	0	1	1	3	0	0	0	1	0	0	0	0	0	0	0	2	2	2	2	12
7	1	1	1	1	0	2	0	0	1	0	0	2	0	2	4	0	2	0	0	0	2	4

Let the distortion probabilities be $p_0 = 0.7$, $p_1 = 0.8$, $p_2 = 0.9$. Then, by algorithm A:

$$\tilde{\tau}_2 = (1, 1, 1), \quad \tilde{\tau}_1 = (1, 0, 0), \quad \tilde{\tau}_0 = (0, 1, 0),$$

$$\tilde{T} = \begin{pmatrix} 0 & 1 & 1 \\ 1 & 0 & 1 \\ 0 & 0 & 1 \end{pmatrix},$$

and by (5.2.6),

$$\sigma_1 = \tilde{T}^{-1} = \begin{pmatrix} 0 & 1 & 1 \\ 1 & 0 & 1 \\ 0 & 0 & 1 \end{pmatrix} \pmod 2.$$

Using (5.2.7), we obtain the following expected values for the number of correctable errors:

$$\mathrm{M}\{\eta_f(1)\} = 0.7 \cdot 2 + 0.8 \cdot 2 + 0.9 \cdot 0 = 3;$$

$$\mathrm{M}\{\eta_{f_{\sigma_1}}(1)\} = 0.9 \cdot 4 + 0.8 \cdot 2 + 0.7 \cdot 2 = 6.6.$$

In order to determine an optimal linear transformation σ_1 maximizing the expected number of correctable single errors and to estimate the expected values themselves, one can use the data collected in the Appendix; in some cases these data also yield analytical expressions for σ_1, $\mathrm{M}\{\eta_f(1)\}$ and $\mathrm{M}\{\eta_{f_{\sigma_1}}(1)\}$.

5.2.3. *Correction of burst errors.* A *burst of width* b in a signal $(z^{(0)}, \ldots, z^{(m-1)})$ is defined as the simultaneous distortion of b arguments $z^{(i)}, z^{(i+1)}, \ldots, z^{(i+b-1)}$ if $i \leq m - 1 - b$, or the arguments $z^{(i)}, z^{(i+1)}, \ldots, z^{(m-1)}$ if $i > m - 1 - b$. Thus, if

(z_1, z_2) is a burst of width b, then

$$z_1 \oplus z_2 \in \{(\overbrace{1, 1, \ldots, 1}^{b}, 0, 0, \ldots, 0), (0, \overbrace{1, 1, \ldots, 1}^{b}, 0, 0, \ldots, 0), \ldots,$$
$$(0, 0, \ldots, 0, \overbrace{1, 1, \ldots, 1}^{b}), \ldots, (0, 0, \ldots, 0, 1, 1), (0, 0, \ldots, 0, 1)\} \pmod 2.$$

For example, if $b = 2, m = 4$, then

$$z_1 \oplus z_2 \in \{(1, 1, 0, 0), (0, 1, 1, 0), (0, 0, 1, 1), (0, 0, 0, 1)\} \pmod 2.$$

As we have defined them, burst errors are "solid," in the sense that the distortion of an argument $z^{(i)}$ invariably entails distortion of the arguments $z^{(i+1)}, \ldots, z^{(i+b-1)}$ if $i \leqq m-1-b$, or of $z^{(i+1)}, \ldots, z^{(m-1)}$ if $i > m-1-b$. The concept of burst is a generalization of single error: a burst of width $b = 1$ is simply a single error.

Now let $p_i(b)$ be the probability of a burst of width b for which the first distorted argument is $z^{(i)}$. We shall treat the parameters b and $p_i(b)$ $(i = 0, 1, \ldots, m-1)$ as given quantities, together with the system (5.2.1) of Boolean functions. If $f(z)$ is the function representing system (5.2.1) and $f_\sigma(z)$ is defined as in (5.2.2), we denote the expected numbers of corrected bursts of width b by $\mathrm{M}\{\eta_f^{\mathrm{B}}(b)\}$, $\mathrm{M}\{\eta_{f_\sigma}^{\mathrm{B}}(b)\}$. Let σ_b^{B} denote an optimal linear transformation of arguments, maximizing $\mathrm{M}\{\eta_{f_\sigma}^{\mathrm{B}}(b)\}$.

Our problem may thus be stated as follows: given $f(z)$, b and $p_i(b)$ $(i = 0, 1, \ldots, m-1)$, find $m \times m$ matrices σ_b^{B} such that

$$\max_{\sigma \in \Xi} \mathrm{M}\{\eta_{f\sigma}^{\mathrm{B}}(b)\} = \mathrm{M}\{\eta_{f\sigma_b^{\mathrm{B}}}^{\mathrm{B}}(b)\}. \tag{5.2.11}$$

Note that $\mathrm{M}\{\eta_{f_\sigma}(1)\} = \mathrm{M}\{\eta_{f_\sigma}^{\mathrm{B}}(1)\}, \sigma_1^{\mathrm{B}} = \sigma_1$.

Let $E_m^{(b)}$ denote the $m \times m$ matrix whose columns are

$$(\overbrace{11 \ldots 1}^{b}0 \ldots 0), (0\overbrace{11 \ldots 1}^{b}0 \ldots 0), \ldots, (0 \ldots 0\overbrace{11 \ldots 1}^{b}), (0 \ldots 0\overbrace{11 \ldots 1}^{b}), \ldots, (00 \ldots 01).$$

For example, if $m = 3, b = 2$, then

$$E_3^{(2)} = \begin{pmatrix} 1 & 0 & 0 \\ 1 & 1 & 0 \\ 0 & 1 & 1 \end{pmatrix}.$$

For any b and m, $E_m^{(b)}$ is nonsingular over $GF(2)$, and $E_m^{(1)} = E_m$.

THEOREM 5.2.2. *Let $B_\Sigma(\tau)$ be the total autocorrelation function of system* (5.2.1) (*see* (5.2.4)), *and $\tilde{\tau}_0, \tilde{\tau}_1, \ldots, \tilde{\tau}_{m-1}$ linearly independent vectors of length m over GF*(2) *maximizing the function $\sum_{i=0}^{m-1} p_i(b)\, B_\Sigma(\tau_i)$. Let $\tilde{T}$ be the matrix with columns $\tilde{\tau}_0, \ldots, \tilde{\tau}_{m-1}$, Then*

$$\tilde{T} \otimes \sigma_b^{\mathrm{B}} = E_m^{(b)} \qquad (\mathrm{mod}\ 2). \tag{5.2.12}$$

The proof follows from the observation that

$$\sum_{i=0}^{m-1} p_i(b)\, B_\Sigma(e_{i,m}^{(b)}) = \mathrm{M}\{\eta_f^{\mathrm{B}}(b)\} \tag{5.2.13}$$

where $e_{i,m}^{(b)}$ is the number whose binary expansion is the i-th column of $E_m^{(b)}$, after which one uses Theorem 1.5.4.

Theorem 5.2.1 is the special case of Theorem 5.2.2 with $b = 1$.

Thus, in order to determine σ_b^{B} we need only apply algorithm A to $B_\Sigma(\tau)$ and then use (5.2.12).

As an example, consider the system of BF defined in Table 5.2.1 ($m = 3$). Set $b = 2$, $p_0(2) = 0.9$, $p_1(2) = 0.8$, $p_2(2) = 0.7$. (Thus the probability of a simultaneous distortion of $z^{(0)}$ and $z^{(1)}$ is 0.9; for $z^{(1)}$ and $z^{(2)}$ the probability is 0.8; and for $z^{(2)}$ it is 0.7.) B_Σ is given in Table 5.2.1. Applying algorithm A, we have

$$\tilde{T} = \begin{pmatrix} 1 & 0 & 1 \\ 1 & 1 & 0 \\ 1 & 0 & 0 \end{pmatrix}.$$

Since

$$E_3^{(2)} = \begin{pmatrix} 1 & 0 & 0 \\ 1 & 1 & 0 \\ 0 & 1 & 1 \end{pmatrix},$$

it follows from (5.2.12) that

$$\sigma_2^{\mathrm{B}} = \begin{pmatrix} 1 & 0 & 1 \\ 1 & 1 & 0 \\ 1 & 0 & 0 \end{pmatrix}^{-1} \otimes \begin{pmatrix} 1 & 0 & 0 \\ 1 & 1 & 0 \\ 0 & 1 & 1 \end{pmatrix} = \begin{pmatrix} 0 & 0 & 1 \\ 1 & 1 & 1 \\ 1 & 0 & 1 \end{pmatrix} \qquad (\mathrm{mod}\ 2).$$

Thus, using (5.2.13), we obtain

$$\mathrm{M}\{\eta_f^{\mathrm{B}}(2)\} = 0\cdot 0.9 + 0\cdot 0.8 + 0\cdot 0.7 = 0;$$

$$\mathrm{M}\{\eta_{f\sigma_2^{\mathrm{B}}}^{\mathrm{B}}(2)\} = 4\cdot 0.9 + 2\cdot 0.8 + 2\cdot 0.7 = 6.6.$$

5.2.4. *Correction of errors with different costs.* We now wish to characterize errors not only by their probabilities but also by their costs. In general, different uncorrected errors will incur different costs. The cost of an error may be measured, for example, by its significance in a given system relative to other, subsequent systems: the cost of an error leading to distortion of all components of the output may usually be assumed to be larger than that of an error distorting only some of the components. Thus, all the results of this subsection will be generalizations of those of §§5.2.2 and 5.2.3. We proceed to the formal discussion.

We begin once again with a system of BF

$$y^{(s)} = f^{(s)}(z^{(0)}, \dots, z^{(m-1)}) \qquad (s = 0, \dots, k-1)$$

and let $f(z)$ be a discrete function representing the system.

Let $C(\alpha, \beta)$ be some given real-valued function, $\alpha, \beta \in \{0, 1, \dots, 2^k - 1\}$.

The *cost* of an error (z_1, z_2) $(z_1, z_2 \in \{0, 1, \dots, 2^m - 1\})$ is defined to be the quantity $C(f(z_1), f(z_2))$. The function $C(\alpha, \beta)$ is assumed to have the properties

$$C(\alpha, \alpha) = 0, \tag{5.2.14}$$

$$C(\alpha, \beta) = C(\beta, \alpha). \tag{5.2.15}$$

Formula (5.2.14) means that corrected errors cost nothing, and (5.2.15) that errors (z_1, z_2) and (z_2, z_1) have equal costs.

We shall consider burst errors (and also single errors, which are special cases of bursts; see §5.2.3). As before, every error is assigned a probability $p_i(b)$. Let $\mathrm{M}\{C_f^{\mathrm{B}}(b)\}$ denote the expected cost of uncorrected errors. Our problem is now to find a nonsingular matrix $\tilde{\sigma}_b^{\mathrm{B}}$ minimizing $\mathrm{M}\{C_{f_\sigma}^{\mathrm{B}}(b)\}$, where f_σ is defined by (5.2.2):

$$\min_{\sigma\in\Xi} \mathrm{M}\{C_{f_\sigma}^{\mathrm{B}}(b)\} = \mathrm{M}\{C_{f\tilde{\sigma}_b^{\mathrm{B}}}^{\mathrm{B}}(b)\}. \tag{5.2.16}$$

Our previous problem (5.2.11) is a special case of (5.2.16); set

$$C(\alpha, \beta) = \begin{cases} 1 & \text{if } \alpha \neq \beta, \\ 0 & \text{if } \alpha = \beta. \end{cases}$$

Instead of the autocorrelation functions, our solution of problem (5.2.16) will use the cross-correlation functions of the original system. Let $A_\Sigma(\tau)$ denote the *weighted total cross-correlation function* of the characteristic functions $\{f_t(z)\}$ $(t = 0, 1, \ldots, 2^k - 1)$;

$$A_\Sigma(\tau) = \sum_{t_1 \neq t_2} C(t_1, t_2)\, A_{t_1, t_2}(\tau)$$

$$= \sum_{t_1 \neq t_2} C(t_1, t_2) \sum_{z=0}^{2^m - 1} f_{t_1}(z)\, f_{t_2}(z \oplus \tau) \qquad (\mathrm{mod}\ 2). \tag{5.2.17}$$

THEOREM 5.2.3. *Let*

$$\min_{T \in \Xi} \sum_{i=0}^{m-1} p_i(b)\, A_\Sigma(\tau_i) = \sum_{i=0}^{m-1} p_i(b)\, A_\Sigma(\tilde{\tau}_i). \tag{5.2.18}$$

Then

$$\tilde{T} \otimes \tilde{\sigma}_b^{\mathrm{B}} = E_m^{(b)} \qquad (\mathrm{mod}\ 2). \tag{5.2.19}$$

where τ_i and $\tilde{\tau}_i$ are the columns of the matrices T and $\tilde{T}$, respectively.

The proof is immediate: we note that

$$\sum_{i=0}^{m-1} p_i(b)\, A_\Sigma(e_{i,m}^{(b)}) = \mathrm{M}\,\{C_f^{\mathrm{B}}(b)\}, \tag{5.2.20}$$

where $e_{i,m}^{(b)}$ is the i-th column of $E_m^{(b)}$, and then use Theorem 1.5.4.

This theorem is a further generalization of Theorems 5.2.1 and 5.2.2: for any $f(z)$ and any τ, if the cost function is

$$C(\alpha, \beta) = \begin{cases} 1 & \text{if} \quad \alpha \neq \beta, \\ 0 & \text{if} \quad \alpha = 0, \end{cases}$$

then

$$A_\Sigma(\tau) + B_\Sigma(\tau) = 2^m. \tag{5.2.21}$$

To determine $\tilde{T}$ from condition (5.2.18), one can use a modified version of the basic algorithm A, the only change being to replace maximization of $B_\Sigma(\tau)$ in (5.2.9) and (5.2.10) by minimization of $A_\Sigma(\tau)$.

As an example, we determine the optimal linear transformation $\tilde{\sigma}_1$ for correction of single errors ($b = 1$) in Table 5.2.1, with the cost function defined by $C(\alpha, \beta) = \|\alpha \oplus \beta\|$. The function $A_{t_1 t_2}(\tau)$ and $A_\Sigma(\tau)$ are also shown in Table 5.2.1 (since $A_{t_1 t_2}(\tau) = A_{t_2 t_1}(\tau)$, the table specifies the values of $2A_{t_1 t_2}(\tau)$ for $t_1 < t_2$). Suppose that the distortion probabilities are $p_0(1) = 0.9$, $p_0(1) = 0.8$, $p_2(1) = 0.7$. Then the modified algorithm A gives

$$\tilde{T} = \begin{pmatrix} 1 & 0 & 1 \\ 1 & 1 & 0 \\ 1 & 0 & 0 \end{pmatrix}, \qquad \tilde{\sigma}_1^{\mathrm{B}} = \tilde{\sigma}_1 = \begin{pmatrix} 1 & 0 & 1 \\ 1 & 1 & 0 \\ 1 & 0 & 0 \end{pmatrix}^{-1} = \begin{pmatrix} 0 & 0 & 1 \\ 0 & 1 & 1 \\ 1 & 0 & 1 \end{pmatrix} \quad (\mathrm{mod}\ 2)$$

and formula (5.2.20) yields the following expected single-error costs for the systems f and $f_{\tilde{\sigma}_1}$:

$$\mathrm{M}\{C_f(1) = 12 \cdot 0.9 + 8 \cdot 0.8 + 8 \cdot 0.7 = 22.8;$$
$$\mathrm{M}\{C_{f_{\tilde{\sigma}_1}}(1)\} = 4 \cdot 0.9 + 8 \cdot 0.8 + 8 \cdot 0.7 = 15.6.$$

To minimize the expected error costs, then, one simply constructs the weighted total cross-correlation function A_Σ and applies the modified algorithm A'. For large values of k, construction of the function A_Σ by (5.2.17) may become quite cumbersome. We therefore devote some attention to ways and means of simplifying the construction.

We first observe that since $C(t_1, t_2) = C(t_2, t_1)$ and $A_{t_1 t_2}(\tau) = A_{t_2 t_1}(\tau)$, it follows from (5.2.17) that

$$A_\Sigma(\tau) = 2 \sum_{t_1 < t_2} C(t_1, t_2)\, A_{t_1 t_2}(\tau) = 2 \sum_{t_1 < t_2} C(t_1, t_2) \sum_{z=0}^{2^m - 1} f_{t_1}(z)\, f_{t_2}(z \oplus \tau) \qquad (\mathrm{mod}\ 2). \tag{5.2.22}$$

Thus the functions $A_{t_1 t_2}(\tau)$ may be derived from the characteristic functions $f_t(z)$ in terms of double Walsh transforms, using the fast Hadamard–Walsh transform algorithm.

Next, the cost of an error in a system $y^{(s)} = f^{(s)}(z^{(0)}, \ldots, z^{(m-1)})$ $(s = 0, 1, \ldots, k - 1)$ is usually determined by the number of functions in the system that change value on appearance of the error; in other words, if $y(z)$ is the value vector of the system, the cost of an error (z_1, z_2) is some function (usually monotone increasing) of $\|y(z_1) \oplus y(z_2)\|$. Then $C(\alpha, \beta) = C(\|\alpha \oplus \beta\|)$ and A_Σ may be expressed as

$$A_\Sigma(\tau) = \sum_{z=0}^{2^m - 1} C(\|y(z) \oplus y(z \oplus \tau)\|) \qquad (\mathrm{mod}\ 2). \tag{5.2.23}$$

This formula may be more convenient for large values of k.

5.2.5. *Correction of multiple errors.* In the previous subsections we studied the correction of single errors and bursts with different probabilities and error costs. We now consider the correction of errors of arbitrary multiplicity.

Recall that the multiplicity of an error (z_1, z_2) is the number $\|z_1 \oplus z_2\|$. Suppose that, given a system of BF (5.2.1), we wish to effect simultaneous correction of all errors of multiplicity not exceeding some number $l \geqq 1$. The previous results (Theorems 5.2.1, 5.2.3) may be generalized to the case of arbitrary l, but the computational complexity of the algorithms is greatly increased.

As before, we shall assume that each error is assigned a probability and a cost. By the probability of an error (z_1, z_2) we mean the probability of simultaneous distortion of all arguments $z^{(i_1)}, \ldots, z^{(i_r)}$ for which $(z_1 \oplus z_2)^{(i_1)} = \cdots = (z_1 \oplus z_2)^{(i_r)} = 1$. Thus, any two errors (z_1, z_2), (z_3, z_4) such that $z_1 \oplus z_2 = z_3 \oplus z_4$ have equal probabilities, so that we can treat the error probability as a function $p(z)$ defined for $z \in \{1, 2, \ldots, 2^m - 1\}$. If we are interested only in errors of multiplicity $\leqq l$, then $p(z) = 0$ if $\|z\| > l$.

The cost of an error (z_1, z_2) will again be a quantity $C(f(z_1), f(z_2))$, where $C(\alpha, \beta)$ is a function satisfying (5.2.14), (5.2.15).

Let $\mathrm{M}\{C_{f_\sigma}(l)\}$ be the expected cost of uncorrected errors of multiplicity $\leqq l$ for f_σ. Our problem is thus to determine a matrix $\tilde{\sigma}_l$ such that

$$\min_{\sigma \in \Xi} \mathrm{M}\{C_{f_\sigma}(l)\} = \mathrm{M}\{C_{f_{\tilde{\sigma}_l}}(l)\}. \tag{5.2.24}$$

Let $\tau_0, \tau_1, \ldots, \tau_{m-1}$ be an m-tuple of vectors, and T the matrix with columns $\tau_0, \tau_1, \ldots, \tau_{m-1}$. Let $L_l(\tau_0, \ldots, \tau_{m-1})$ denote the set of all linear combinations $d_0\tau_0 \oplus \cdots \oplus d_{m-1}\tau_{m-1}$ (mod 2), where $d_0, \ldots, d_{m-1} \in \{0, 1\}$ and the number $\|d\|$ of nonzero components of the vector $(d_0, \ldots, d_{m-1})$ satisfies the inequality $0 < \|d\| \leqq l$. As before, we denote the weighted total cross-correlation function of the characteristic functions by $A_\Sigma(\tau)$ (see (5.2.22)).

THEOREM 5.2.4. *Let*

$$\min_{T \in \Xi} \sum_{q \in L_l(\tau_0, \ldots, \tau_{m-1})} p(q)\, A_\Sigma(q) = \sum_{q \in L_l(\tilde{\tau}_0, \ldots, \tilde{\tau}_{m-1})} p(q)\, A_\Sigma(q). \tag{5.2.25}$$

Then

$$\tilde{T} \otimes \tilde{\sigma_l} = E_m \qquad (\mathrm{mod}\ 2). \tag{5.2.26}$$

The proof is based on the relation

$$\sum_{q \in L_l(e_0, \ldots, e_{m-1})} p(q)\, A_\Sigma(q) = \mathrm{M}\{C_f(l)\} \tag{5.2.27}$$

where $e_i = (\underbrace{0, \ldots, 0}_{i}, 1, 0, \ldots, 0)$.

Setting $l = 1$ in Theorem 5.2.4 and using (5.2.21), we get Theorem 5.2.1, and also the case $b = 1$ of Theorem 5.2.3.

As an example, consider the correction of double errors by the system of Table

5.2.1. We let $C(\alpha, \beta) = \|\alpha \oplus \beta\|$ and

$$p(q) = \begin{cases} 0.9 & \text{if } \|q\| = 1, \\ 0.8 & \text{if } \|q\| = 2, \\ 0 & \text{otherwise.} \end{cases}$$

The function A_Σ is given in Table 5.2.1. By Table 5.2.1 and formula (5.2.25), we can set

$$\tilde{T} = \begin{pmatrix} 1 & 0 & 1 \\ 1 & 1 & 0 \\ 1 & 0 & 0 \end{pmatrix}.$$

Then

$$\tilde{\sigma}_2 = \begin{pmatrix} 0 & 0 & 1 \\ 0 & 1 & 1 \\ 1 & 0 & 1 \end{pmatrix}$$

and by (5.2.27)

$$\mathrm{M}\{C_f(2)\} = \sum_{q \in L_2(e_0, e_1, e_2)} p(q)\, A_\Sigma(q) = 0.9(12 + 8 + 8) + 0.8(8 + 8 + 12) = 47.6$$

and

$$\mathrm{M}\{C_{f_\sigma}(2)\} = \sum_{q \in L_2((111),(010),(100))} p(q)\, A_\Sigma(q) = 0.9(4+8+8) + 0.8(8+8+12) = 40.4.$$

To summarize: in order to find the matrix $\tilde{\sigma}_l$ of an optimal linear transformation, minimizing the expected cost of uncorrected errors of multiplicity $\leqq l$, it suffices to construct $A_\Sigma(\tau)$ and to find $\tilde{T}$ from condition (5.2.25). Note, however, that in the general case ($l > 1$) the determination of $\tilde{T}$ is a fairly complicated problem, which can no longer be tackled through use of the basic algorithm A.

5.3. SPECTRAL METHODS FOR ANALYSIS OF THE CORRECTING CAPABILITY OF DIGITAL DEVICES WITH MEMORY

5.3.1. As in Chapter IV, our mathematical model of a digital device with memory is the finite automaton (FA). For the reader's convenience, we recall that a finite automaton M is a sequence of six objects:

$$M = \{X, A, Y, a_0, \varphi(x, a), g(a)\},$$

where X is the set of input signals, A the set of (internal) states, Y the set of output signals, a_0 the initial state ($a_0 \in A$), $\varphi(x, a)$ the next-state function ($X \times A \xrightarrow{\varphi} A$), and $g(a)$ the output function ($A \xrightarrow{g} Y$). We may assume without loss of generality that $Y = A$ and $g(a) = a$. Methods for specification and realization of FA were considered in §4.1.

Let $q = x_0 x_1 \dots x_{m-1}$ be an input word (i.e., string of input signals) of length m ($x_i = X, i = 0, 1, \dots, m - 1$), and set

$$\varphi(q, a) = \varphi(x_{m-1}, \dots, \varphi(x_2, \varphi(x_1, \varphi(x_0, a))) \dots), \tag{5.3.1}$$

$$\varphi(q, a_0) = \varphi(q). \tag{5.3.2}$$

For example,

$$\varphi(x_0 x_1 x_2, a) = \varphi(x_2, \varphi(x_1, \varphi(x_0, a))).$$

We shall say that two states $a_i, a_j \in A$ are *equivalent* if, for any input word p,

$$\varphi(p, a_i) = \varphi(p, a_j).$$

Henceforth we shall confine attention to *minimal automata*, that is to say, automata all of whose states are pairwise inequivalent. (A standard theorem of automata theory states that for any automaton one can construct a minimal automaton implementing the same mapping of the set of input words into the set of output words; see, e.g., [70, 92].)

An *error* in a FA is defined as an arbitrary ordered pair (q, q') of input words $(q \neq q')$. We shall say that the FA *corrects* an error (q, q') if $\varphi(q) = \varphi(q')$. It will be shown below that every automaton corrects a certain (generally nonempty) set of errors.

5.3.2. Let Q be some set of errors for an automaton M; suppose that each error $(q, q') \in Q$ is assigned a probability $p(q, q')$.

Our problem is to calculate the expectation $\mathrm{M}\{\eta(Q)\}$ of the number of errors in Q corrected by M.

We first construct a system of characteristic functions $\varphi_i(q)$, defined on all input words:

$$\varphi_i(q) = \begin{cases} 1 & \text{if} \quad \varphi(q) = a_i, \\ 0 & \text{otherwise} \end{cases} \tag{5.3.3}$$

$$(i = 0, 1, \ldots, n_a - 1),$$

where n_a is the number of states.

We now define the total "two-dimensional" autocorrelation function of the characteristic functions:

$$B_\Sigma(q, q') = \sum_{i=0}^{n_a - 1} \varphi_i(q)\,\varphi_i(q'). \tag{5.3.4}$$

The following theorem is a direct consequence of the above definitions.

THEOREM 5.3.1.

$$\mathrm{M}\{\eta(Q)\} = \sum_{(q,q')\in Q} p(q, q')\,B_\Sigma(q, q'). \tag{5.3.5}$$

This theorem provides a formula for the expected number of errors in any set Q corrected by a given automaton. Its practical use, however, is limited by the fact that the computation of $\mathrm{M}\{\eta(Q)\}$ is quite tedious even for relatively small values of n_a. The difficulties stem both from the need to calculate the characteristic functions $\varphi_i(q)$ $(i = 0, 1, \ldots, n_a - 1)$ by (5.3.1)–(5.3.3), and from the cumbersome definition of the two-dimensional autocorrelation function by formula (5.3.4).

5.3.3. We first consider how to simplify calculation of the characteristic functions $\varphi_i(q)$. For fixed m, let X^m denote the set of all input words of length m.

We shall assume that the original automaton M is defined by an $n_a \times n_a$ matrix (M_{ij}), where $M_{ij} = x_{s0} \cup x_{s_1} \cup \cdots \cup x_{s_\mu}$ $(x_{s_k} \in X,\ k = 1, 2, \ldots, \mu)$ if and only if

$$\varphi(x_{s0}, a_i) = \varphi(x_{s_1}, a_i) = \cdots = \varphi(x_{s_\mu}, a_i) = a_j.$$

The symbol $\cup$ stands for set-theoretical union of sets; we are using the same notation for a singleton (= set with one element) and for the (single) element of the set. If there is no $x_s \in X$ such that $\varphi(x_s, a_i) = a_j$, we put $M_{ij} = \varnothing$ (where $\varnothing$ is the symbol for the empty set).

Given two $n_a \times n_a$ matrices (M_{ij}) and (M'_{ij}), we define their *logical product* (M''_{ij}) as the matrix with

$$M''_{ij} = \bigcup_{s=0}^{n_a - 1} M_{is} M'_{sj} \tag{5.3.6}$$

where, if $M_{is} = x_{\alpha_0} \cup x_{\alpha_1} \cup \cdots \cup x_{\alpha_\lambda}$ and $M'_{sj} = x_{\beta_0} \cup x_{\beta_1} \cup \cdots \cup x_{\beta_\gamma}$, then

$$M_{is}M'_{sj} = \bigcup_{g=0}^{\lambda} \bigcup_{q=0}^{\gamma} x_{\alpha_g} x_{\beta_q} \tag{5.3.7}$$

(where $\varnothing x = x\varnothing = \varnothing$ for any $x \in X$; $x_{\alpha_g} x_{\beta_q}$ is a word of length 2 over X). Thus, logical multiplication is analogous to ordinary multiplication, except that summation is replaced by set-theoretical union and multiplication by concatenation of words. If $(M''_{ij}) = (M_{ij})(M'_{ij})$, the elements of (M''_{ij}) are sets of input words of length 2. Similarly, the elements of a logical product of m matrices are sets of input words of length m.

Let (M_{ij}) be a matrix defining an automaton, and (M^m_{ij}) its m-th logical power.

LEMMA 5.3.1. *Let* $\{\varphi_i(q)\}$ $(i = 0, 1, \ldots, n_a - 1;\ q \in X^m)$ *be the system of characteristic functions of the automaton defined by* (M_{ij}). *Then*

$$\varphi_i(q) = \begin{cases} 1 & \text{if } q \in M^m_{0i}, \\ 0 & \text{otherwise} \end{cases} \qquad (i = 0, 1, \ldots, n_a - 1). \tag{5.3.8}$$

The proof follows from the definition of (M_{ij}) and formulas (5.3.6), (5.3.7), by induction on m.

Thus, in order to construct the characteristic functions we need only determine the zeroth row of the m-th logical power of the matrix (M_{ij}).

EXAMPLE 5.3.1. Let M be the automaton with the state diagram of Figure 5.3.1. Here $A = \{a_0, a_1, a_2, a_3, a_4, a_5\}$ $(n_a = 6)$ is the set of states, $X = \{0, 1\}$ the set of input signals, a_0 the initial state; an arrow labeled x_s $(x_s \in X)$ joins a_i to a_j $(a_i, a_j \in A)$ if and only if $\varphi(x_s, a_i) = a_j$.

Construct the characteristic functions $\varphi_i(q)$, $q \in X^m$ $(i = 0, 1, \ldots, 5)$ of M for arbitrary m. we have

$$(M_{ij}) = \begin{pmatrix} \varnothing & 1 & 0 & \varnothing & \varnothing & \varnothing \\ \varnothing & \varnothing & 1 & 0 & \varnothing & \varnothing \\ \varnothing & \varnothing & \varnothing & 1 & 0 & \varnothing \\ \varnothing & \varnothing & \varnothing & \varnothing & 1 & 0 \\ 0 & \varnothing & \varnothing & \varnothing & \varnothing & 1 \\ 1 & 0 & \varnothing & \varnothing & \varnothing & \varnothing \end{pmatrix}.$$

This matrix satisfies the condition

$$M_{ij} = \begin{cases} 1 & \text{if } i = j \ominus 1 \pmod 6, \\ 0 & \text{if } i = j \ominus 2 \pmod 6, \\ \varnothing & \text{otherwise.} \end{cases} \tag{5.3.9}$$

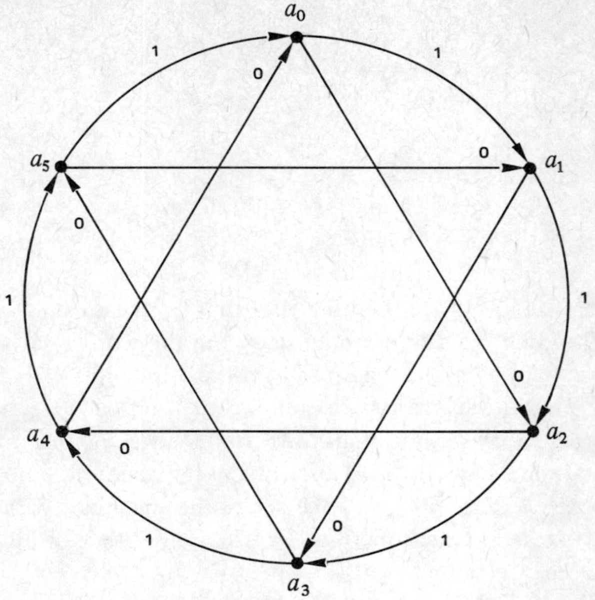

FIGURE 5.3.1.

We now determine the set M^m_{0i} for arbitrary m and i. Let S^α_m denote the set of binary sequences of length m containing α units.

We claim that for any m

$$M^m_{0i} = \bigcup_{\alpha = 2m \ominus i} S^\alpha_m \quad (\text{mod } 6) \qquad (i = 0, 1, \dots, 5). \tag{5.3.10}$$

We prove this by induction on m. Set

$$M^{m-1}_{0t} = \bigcup_{\alpha = 2m \ominus t \ominus 2} S^\alpha_{m-1} \quad (\text{mod } 6) \qquad (t = 0, 1, \dots, 5). \tag{5.3.11}$$

Note that

$$S^{\alpha-1}_{m-1}1 \cup S^\alpha_{m-1}0 = S^\alpha_m, \tag{5.3.12}$$

where (see (5.3.7)) $S^{\alpha-1}_{m-1}1$ and $S^\alpha_{m-1}0$ are the sets obtained by concatenating 1 and 0, respectively, to the right of vectors of $S^{\alpha-1}_{m-1}$ and S^α_{m-1}. It now follows from (5.3.6), (5.3.7), (5.3.9), (5.3.11), (5.3.12) that

$$M_{0i}^m = \bigcup_{t=0}^{5} M_{0t}^{m-1} M_{ti} = \bigcup_{t=0}^{5} \left(\bigcup_{\alpha = 2m \ominus t \ominus 2} S_{m-1}^{\alpha} \right) M_{ti}$$

$$= \left(\bigcup_{\alpha = 2m \ominus i \ominus 2} S_{m-1}^{\alpha} \right) 1 \cup \left(\bigcup_{\alpha = 2m \ominus i} S_{m-1}^{\alpha} \right) 0 = \bigcup_{\alpha = 2m \ominus i} S_{m-1}^{\alpha - 1} 1 \cup \bigcup_{\alpha = 2m \ominus i} S_{m-1}^{\alpha} 0$$

$$= \bigcup_{\alpha = 2m \ominus i} (S_{m-1}^{\alpha-1} 1 \cup S_{m-1}^{\alpha} 0) = \bigcup_{\alpha = 2m \ominus i} S_m^{\alpha} \qquad (\text{mod } 6).$$

It thus follows from (5.3.10) and Lemma 5.3.1 that for the automaton of our example $\varphi_i(q) = 1$ if and only if the number of units $\|q\|$ in the word q satisfies the condition $\|q\| = 2m \ominus i \pmod 6$ (m is the length of q); otherwise $\varphi_i(q) = 0$.

Lemma 5.3.1 provides a method calculating the number $\eta_{M,m}$ of errors in words of length m corrected by an automaton M and a lower bound on this number for the set of all automata with given n_a and n_x. Let $|R|$ denote the number of elements of an arbitrary set R, and $M(n_a, n_x)$ the set of all automata with n_a states and n_x input signals whose next-state functions $\varphi(x, a)$ are defined for all $x \in X$, $a \in A$.

COROLLARY 5.3.1. *For any automaton $M \in M(n_a, n_x)$, defined by a matrix (M_{ij}),*

$$\eta_{M,m} = \sum_{i=0}^{n_a - 1} |M_{0i}^m| (|M_{0i}^m| - 1) \tag{5.3.13}$$

and

$$\min_{M \in M(n_a, n_x)} \eta_{M,m} = n_x^m \left(\frac{n_x^m}{n_a} - 1 \right). \tag{5.3.14}$$

PROOF. Formula (5.3.13) follows from Theorem 5.3.1 and Lemma 5.3.1 in view of the relation

$$\sum_{q \in X^m} \varphi_i(q) = |M_{0i}^m|.$$

To prove (5.3.14), we minimize (5.3.13) with respect to the variables $|M_{0i}^m|$ ($i = 0, 1, \ldots, n_a - 1$), subject to the constraint

$$\sum_{i=0}^{n_a - 1} |M_{0i}^m| = n_x^m. \quad \blacksquare \tag{5.3.15}$$

To illustrate, we calculate the function $\eta_{M,m}$ for all m, for the automaton M of Example 5.3.1 (see Figure 5.3.1).

By (5.3.10),

$$|M_{0i}^m| = \left| \bigcup_{\alpha = 2m \ominus i} S_m^{\alpha} \right| \qquad (\text{mod } 6). \tag{5.3.16}$$

Since $S_m^{\alpha_1} \cap S_m^{\alpha_2} = \varnothing$* if $\alpha_1 \neq \alpha_2$ by the definition of S_m^α, and $|S_m^\alpha| = \binom{m}{\alpha}$, it follows from (2.3.11) that

$$|M_{0i}^m| = \sum_{\alpha = 2m \ominus i} \binom{m}{\alpha} \quad (\text{mod } 6) \tag{5.3.17}$$

(where $C_m^\alpha = 0$ if $\alpha > m$), and finally, by (5.3.13),

$$\eta_{M,m} = \sum_{i=0}^{5} \left(\sum_{\alpha = 2m \ominus i} \binom{m}{\alpha} \right) \left(\sum_{\alpha = 2m \ominus i} \binom{m}{\alpha} - 1 \right) \quad (\text{mod } 6). \tag{5.3.18}$$

5.3.4. We now try to simplify calculation of the "two-dimensional" autocorrelation function $B_\Sigma(q, q')$ defined by (5.3.4). To achieve this end, we impose certain restrictions on the set Q of errors.

First, as before, we fix the length m of the input words. We shall also assume that the set of input words of length m is a commutative group G_m, whose structure is such that the class Q of errors may be described in terms of the group operation $\circ$. (It will be shown below that this can indeed be done for most cases of practical importance.) In this new situation, the "two-dimensional" autocorrelation function $B_\Sigma(q, q')$ may be replaced by the usual "one-dimensional" function $B_\Sigma(\tau)$, and the correcting capability of automata may be analyzed in exactly the same way as for systems of LF.

Let G_m be the group of input words, $\circ$ the group operation and e the identity element of G_m. We assume that the set of errors Q is described by a subset Q_m of G_m with the following properties: If $(q, q') \in Q$, then $q \circ q' \in Q_m$; in addition we assume the existence of a function $p(q, q') = p(q', q) = p(\tau)$, where $\tau = q \circ q'$ (i.e., we are assuming that $p(q, q') = p(q_1, q_1')$ if $q \circ q' = q_1 \circ q_1'$). $p(q, q')$ is the probability of the error $(q, q') \in Q$.

Let $M\{\eta_m(Q)\}$ be the expected number of errors of the set Q in input words of length m that can be corrected by the automaton.

COROLLARY 5.3.2.

$$M\{\eta_m(Q)\} = \sum_{\tau \in Q_m} p(\tau) B_\Sigma(\tau), \tag{5.3.19}$$

where

$$B_\Sigma(\tau) = \sum_{i=0}^{n_a - 1} B_i(\tau) = \sum_{i=0}^{n_a - 1} \sum_{q \in G_m} \varphi_i(q) \varphi_i(q \circ \tau^{-1}) \tag{5.3.20}$$

and $\tau \circ \tau^{-1} = e$.

* $A \cap B$ denotes the (set-theoretical) intersection of A and B.

Corollary 5.3.2 is the analog of formula (5.1.45) for systems of LF, and permits calculation of $M\{\eta_m(Q)\}$ by a generalization of the fast Hadamard–Chrestenson transform, since

$$B_i(\tau) = \sum_{q \in G_m} \varphi_i(q)\, \varphi_i(q \circ \tau^{-1}) \tag{5.3.21}$$

is the autocorrelation function of $\varphi_i(q)$ on the group G_m (see §1.6). Generalized fast Hadamard–Chrestenson transform algorithms for arbitrary finite commutative groups may be found, e.g., in [12, 20].

We shall now consider four important classes of groups G_m of input words of given length m. These classes correspond to serial or parallel input of data to the device, with algebraic or arithmetic errors (see §5.1) in the input data.

1) *Serial input, algebraic errors.* The input set X is $\{0, 1, \dots, n_x - 1\}$. G_m is the group of all vectors $(x_0, \dots, x_{m-1})$, $x_i \in \{0, 1, \dots, n_x - 1\}$, the group operation is componentwise addition modulo n_x. The multiplicity of an error (q, q') is the number of nonzero components of $q \oplus q' \pmod{n_x}$,

2) *Serial input, arithmetic errors.* The input set X is as before. G_m is the set of all numbers $\sum_{i=0}^{m-1} x_i (n_x)^i$, $x_i \in \{0, 1, \dots, n_x - 1\}$; the group operation is addition modulo n_x^m.

This situation corresponds to the FA receiving input from an m-digit n_x-ary arithmetic device. For a symmetric error (q, q') the multiplicity is defined as the minimal number of terms in the representation of $q \ominus q'$ or $q' \ominus q \pmod{n_x^m}$ as a linear combination of powers of n_x with exponents $\leqq m - 1$ and coefficients in $\{\pm 1, \pm 2, \dots, \pm(n_x - 1)\}$. The multiplicity of a nonsymmetric error is the number of terms in the n_x-ary representation of $q \ominus q' \pmod{n_x^m}$ (see also §5.1.4).

3) *Parallel input, algebraic errors.* The input set X is the set of all r-ary vectors of length t, where $1 < r < n_x$, $n_x \leqq r^t$, i.e., $X = \{0, 1, \dots, r - 1\}^t$. G_m is the set of all r-ary $t \times m$ matrices (q_{ij}), $q_{ij} \in \{0, 1, \dots, r - 1\}$; the group operation is componentwise addition of matrices modulo r. The multiplicity of an error (q, q') is the number of nonzero components in the matrix $q \oplus q' \pmod r$.

4) *Parallel input, arithmetic errors.* As before, $X = \{0, 1, \dots, r - 1\}^t$. G_m is the set of all vectors

$$\left(\sum_{i=0}^{t-1} q_{0i} r^i, \sum_{i=0}^{t-1} q_{1i} r^i, \dots, \sum_{i=0}^{t-1} q_{m-1,i} r^i \right)$$

where $q_{ij} \in \{0, 1, \dots, r - 1\}$; the group operation is componentwise addition modulo r^t.

This situation corresponds to parallel input of information to the FA from a t-digit r-ary arithmetic device. The multiplicities of symmetric and nonsymmetric errors are defined exactly as for serial input (case (2) above), with t in place of m and r in place of n_x.

The methods for analysis of correcting capability described by (5.3.19)–(5.3.21) may be applied to each of these four groups of input words.

Note that if the input word is fed to an automaton from a communication channel (in which algebraic errors may occur), the expected number of corrected errors in the channel may be increased by using linearization methods similar to those described in §5.2. The linearization procedure in such cases should be applied to the system $\{\varphi_i(q)\}$ $(i = 0, 1, \ldots, n_a - 1)$ of characteristic functions defined on the groups G_m of (2) and (4) above. These systems may be constructed with the aid of Lemma 5.3.1.

To illustrate the analysis of error-correcting capability of automata, let us calculate $\mathrm{M}\{\eta_5(l)\}$ for the automaton of Example 5.3.1, with input words of length $m = 5$. We consider the case of serial input and algebraic errors (case 1). Since $X = \{0, 1\}$, the group $G_m = G_5$ is the set of binary vectors of length 5 under componentwise addition modulo 2. We showed in §5.3.3 that in this case $\varphi_i(q) = 1$ if and only if $\|q\| = 2m \ominus i = 4 \ominus i \pmod 6$ $(i = 0, 1, \ldots, 5)$.

Recall that a Boolean function $\varphi(q)$ is called an elementary symmetric function with operating number α if $\varphi(q) = 1$ if and only if $\|q\| = \alpha$.

Thus, each characteristic function $\varphi_i(q)$ $(q = 0, 1, \ldots, 5)$ of this automaton is an elementary symmetric function, with operating number $4 \ominus i \pmod 6$ $(i = 0, 1, \ldots, 5)$. To determine the autocorrelation functions

$$B_i(\tau) = \sum_{q \in G_5} \varphi_i(q)\, \varphi_i(q \oplus \tau) \qquad \pmod 2 \tag{5.3.22}$$

we use the data in row 8 of the Appendix. This gives

$$B_i(\tau) = \begin{cases} \dbinom{5 - \|\tau\|}{(4 \ominus i) - \|\tau\|/2} \dbinom{\|\tau\|}{\|\tau\|/2} & \text{if } \|\tau\| \text{ is even,} \\ 0 & \text{if } \|\tau\| \text{ is odd.} \end{cases} \qquad \pmod 6 \tag{5.3.23}$$

(As usual, we put $\dbinom{t}{q} = 0$ if $t < q$ or $q < 0$, and $\|\tau\|$ is the number of units in the binary vector τ.)

Let us assume that the probability of an l-fold error in the input words is p_l, i.e., $p(\tau) = p_l$ for $\|\tau\| = l$. Then, by Corollary 5.3.2,

$$\mathrm{M}\{\eta_5(l)\} = \sum_{\|\tau\| = l} p(\tau)\, B_\Sigma(\tau) = \sum_{\|\tau\| = l} p(\tau) \sum_{i=0}^{5} B_i(\tau). \tag{5.3.24}$$

Inserting (5.3.23) into (5.3.24), we finally obtain the result

$$\mathrm{M}\{\eta_5(l)\} = 0 \tag{5.3.25}$$

if l is odd, and

$$\begin{aligned}\mathrm{M}\{\eta_5(l)\} &= \sum_{\|\tau\|=l} p(\tau) \sum_{i=0}^{5} \binom{5-\|\tau\|}{(4\ominus i)-\|\tau\|/2}\binom{\|\tau\|}{\|\tau\|/2} \\ &= p_l \binom{5}{l}\binom{l}{l/2} \sum_{i=0}^{5} \binom{5-l}{(4\ominus i)-l/2} \\ &= p_l \binom{5}{l}\binom{l}{l/2} 2^{5-l} \pmod 6 \end{aligned} \tag{5.3.26}$$

is l is even.

5.3.5. We now estimate the correcting capability of some classes of automata which are important from the engineering standpoint. This will be done for the case of serial input and algebraic errors (case 1 of §5.3.4). Thus the multiplicity of an error (q, q') is the number of unequal components (letters) in the words q and q'.

We first consider the class of linear automata over an arbitrary finite field $GF(p)$ (where p is a prime). We put $X = A\{0, 1, \ldots, p-1\}^t$—the set of all p-ary vectors of length t, and

$$\varphi(x_s, a_j) = T_1 \otimes x_s \oplus T_2 \otimes a_j \pmod p \tag{5.3.27}$$

where T_1 and T_2 are nonsingular $t \times t$ matrices over $GF(p)$ and $\otimes$ is the symbol for matrix multiplication over $GF(p)$. Thus the number of input signals is the same as the number of states: $n_x = n_a = n = p^t$.

Let $\eta_{M,m}(l)$ denote the number of l-fold errors in input words of length m corrected by the automaton M.

THEOREM 5.3.2. *For any linear automaton M with n input signals and n states,*

$$\eta_{M,m}(l) = ((n-1)^l + (-1)^l(n-1))\, n^{m-1}\binom{m}{l}. \tag{5.3.28}$$

PROOF. Let $q = x_0 x_1 \ldots x_{m-1}$ $(x_i \in X)$. Then by (5.3.2), (5.3.27),

$$\begin{aligned}\varphi(q) = T_2^m \otimes a_0 \oplus T_2^{m-1} \otimes T_1 \otimes x_0 \oplus T_2^{m-2} \otimes T_1 \otimes x_1 \oplus \cdots \\ \oplus T_2^1 \otimes T_1 \otimes x_{m-2} \oplus T_2^0 \otimes T_1 \otimes x_{m-1} \pmod p.\end{aligned} \tag{5.3.29}$$

Let $q' = z_0 z_1 \dots z_{m-1}$ ($z_i \in X$). Then the error (q, q') is corrected if and only if $\varphi(q) = \varphi(q')$ i.e.,

$$T_2^{m-1} \otimes x_0 \oplus T_2^{m-2} \otimes x_1 \oplus \cdots \oplus T_2^1 \otimes x_{m-2} \oplus T_2^0 \otimes x_{m-1}$$
$$= T_2^{m-1} \otimes z_0 \oplus T_2^{m-2} \otimes z_1 \oplus \cdots \oplus T_2^1 \otimes z_{m-2} \oplus T_2^0 \otimes z_{m-1} \pmod p. \qquad (5.3.30)$$

On the assumption that (q, q') is an l-fold error, there exist l distinct numbers $i_0, i_1, \dots, i_{l-1}$ ($0 \leqq i_s \leqq m-1$, $s = 0, 1, \dots, l-1$) such that $z_{i_s} \neq x_{i_s}$ ($s = 0, 1, \dots, l-1$) and $z_j = x_j$ if there is no s such that $j = i_s$. Then, by (5.3.30),

$$\bigoplus_{s=0}^{l-1} (T_2^{m-i_s-1} \otimes x_{i_s}) = \bigoplus_{s=0}^{l-1} (T_2^{m-i_s-1} \otimes z_{i_s}) \pmod p. \qquad (5.3.31)$$

We shall now calculate the number $N(l)$ of solutions of equation (5.3.31) for fixed $x_{i_0}, \dots, x_{i_{l-1}}$ such that $z_{i_s} \neq x_{i_s}$ ($s = 0, 1, \dots, l-1$). Fix $z_{i_0}, \dots, z_{i_{l-2}}$ arbitrarily so as to satisfy the conditions

$$\begin{cases} \bigoplus_{s=0}^{l-2} (T_2^{m-i_s-1} \otimes x_{i_s}) \neq \bigoplus_{s=0}^{l-2} (T_2^{m-i_s-1} \otimes z_{i_s}) \pmod p, \\ z_{i_s} \neq x_{i_s} \qquad (s = 0, 1, \dots, l-2). \end{cases} \qquad (5.3.32)$$

The number of ways in which this can be done is $(n-1)^{l-1} - N(l-1)$, where $N(0) = 1$. For each choice of $z_{i_0}, \dots, z_{i_{l-2}}$ satisfying (5.3.32), we can determine $z_{i_{l-1}}$ such that $z_{i_{l-1}} \neq x_{i_{l-1}}$ from (5.3.31), and conversely; thus

$$N(l) = (n-1)^{l-1} - N(l-1). \qquad (5.3.33)$$

Solving the difference equation (5.3.33), we get

$$N(l) = n^{-1}((n-1)^l + (-1)^l (n-1)). \qquad (5.3.34)$$

Since the number of all possible choices of a word $x_0 x_1 \dots x_{m-1}$ and numbers $i_0, \dots, i_{l-1}$ as stipulated above is $n^m \binom{m}{l}$, it finally follows that

$$\eta_{M,m}(l) = n^m \binom{m}{l} N(l) = ((n-1)^l + (-1)^l (n-1))\, n^{m-1} \binom{m}{l}. \qquad ∎ \qquad (5.3.35)$$

COROLLARY 5.3.3. *Linear automata over $GF(p)$ do not correct single errors.*

PROOF. For any n and m, setting $l = 1$ in (5.3.28), we have

$$\eta_{M,m}(1) = 0. \qquad ∎ \qquad (5.3.36)$$

COROLLARY 5.3.4. *For a linear automaton, the relative abundance $\hat{\eta}_{M,m}(l)$ of corrected l-fold errors in input words of length m is independent of m:*

$$\hat{\eta}_{M,m}(l) = n^{-1}(n-1)^{-l}((n-1)^l + (-1)^l(n-1)). \tag{5.3.37}$$

PROOF. Formula (5.3.37) follows from (5.3.28), in view of the fact that the total number of l-fold errors in input words of length m for an automaton with n input signals is $n^m \binom{m}{l}(n-1)^l$. ▮

Figure 5.3.2 illustrates $\hat{\eta}_{M,m}(l)$ as a function of l for linear automata with $n = 2, 3, 4, 5, 6$.

Letting $n \to \infty$ in (5.3.37), we see that for any $l > 1$

$$\lim_{n\to\infty} \hat{\eta}_{M,m}(l) = n^{-1}. \tag{5.3.38}$$

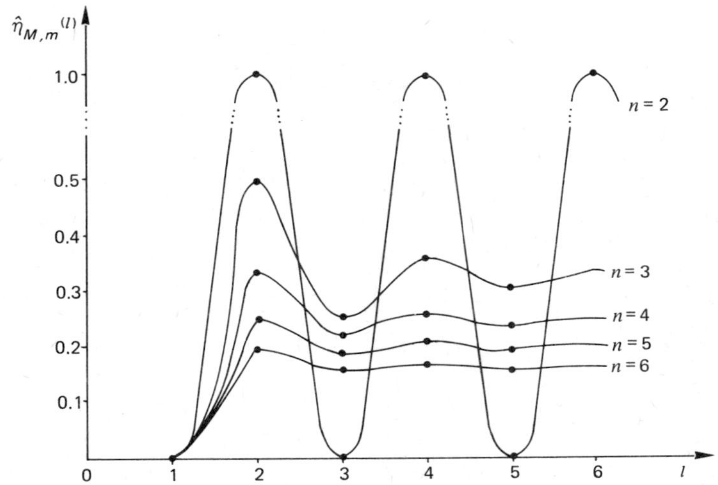

FIGURE 5.3.2.

In other words, if the number n of input signals and states is sufficiently large, the correcting capability of a linear automaton for $l > 1$ is uniformly distributed with respect to multiplicities.

The results of this subsection may be generalized to other types of automaton.

A *group automaton* is an automaton M, whose sets of inputs and internal states are identified ($X = A$), such that

$$\varphi(x_s, a_j) = x_s \circ a_j \circ b, \tag{5.3.39}$$

where $\circ$ is some commutative group operation on X and b is independent of x_s and a_j.

For example, the linear automata considered above are group automata. Similarly, automata representing adders or multipliers may be interpreted as group automata, for in these cases, respectively,

$$\varphi(x_s, a_j) = x_s \oplus a_j \oplus b \qquad (\text{mod } R), \tag{5.3.40}$$

$$\varphi(x_s, a_j) = x_s \otimes a_j \otimes b \qquad (\text{mod } R) \tag{5.3.41}$$

($x_s, a_j, b \in \{0, 1, \ldots, R-1\}$ where R is a prime), with b playing the role of the initial state.

Theorem 5.3.2 is also valid for arbitrary group automata (the proof is analogous to that presented above). The same is true of Corollaries 5.3.3, 5.3.4, the asymptotic relation (5.3.38) and the curves $\hat{\eta}_{M,m}(l)$ of Figure 5.3.2.

The correcting capabilities $\eta_{M,m}(l)$ of linear and group automata may thus be determined (see the proof of Theorem 5.3.2) without actually constructing the characteristic functions $\varphi_i(q)$ (see (5.3.3)) and the autocorrelation functions $B_i(\tau)$ (see (5.3.21)) ($i = 0, 1, \ldots, n_a - 1$).

We now consider yet another important class of automata for which analytical calculation of the correcting capability is more conveniently done using characteristic functions $\varphi_i(q)$ and autocorrelation functions $B_i(\tau)$.

Given an automaton M with input set $X = \{0, 1\}$, we let $N_1(q)$ and $N_0(q)$ denote the numbers of units and zeros, respectively, in an input word q; $\varphi(q)$ will denote the the state to which the input word q brings M from the initial state.

An automaton M with $X = \{0, 1\}$, $A = \{0, 1, \ldots, n_a - 1\}$ (where n_a is a prime), and initial state 0 will be called a *counting automaton* if

$$\varphi(q) \equiv C_1 N_1(q) + C_2 N_0(q) \qquad (\text{mod } n_a), \tag{5.3.42}$$

where C_1, C_2 ($C_1 \neq C_2$) are arbitrary (possibly even negative) integers (the notation $\alpha \equiv \beta \pmod{\gamma}$ has the usual number-theoretical meaning of congruence modulo γ —the numbers α and β have the same remainders upon division by γ).

For example, binary pulse counters ($C_1 = 1$, $C_2 = 0$) and reversible (up-down) counters ($C_1 = 1$, $C_2 = -1$) are counting automata. The automaton of Example 5.3.1 (see Figure 5.3.1) is also a counting automaton (with $C_1 = 1$, $C_2 = 2$).

Let us consider the correcting capability $\eta_{M,m}(l)$ of a counting automaton, assuming serial input and algebraic errors.

The group G_m of input words of length m is the set of all binary vectors of length m under componentwise addition modulo 2.

THEOREM 5.3.3. *If M is a counting automaton, then for any C_1, C_2 and n_a, $m < n_a$,*

$$\eta_{M,m}(l) = \begin{cases} 2^{m-l} \binom{l}{l/2} \binom{m}{l} & \text{if } l \text{ is even,} \\ 0 & \text{if } l \text{ is odd;} \end{cases} \tag{5.3.43}$$

$$\hat{\eta}_{M,m}(l) = \begin{cases} 2^{-l}\binom{l}{l/2} & \text{if } l \text{ is even,} \\ 0 & \text{if } l \text{ is odd.} \end{cases} \tag{5.3.44}$$

PROOF. Let q be an input word of length m. Then for any $i \in \{0, 1, \ldots, n_a - 1\}$, we have a system

$$\begin{aligned} N_i(q) + N_0(q) &= m, \\ C_1 N_1(q) + C_2 N_0(q) &\equiv i \pmod{n_a}. \end{aligned} \tag{5.3.45}$$

Consequently, for fixed m and any $N_1(q) \in \{0, 1, \ldots, m\}$, there exists a unique $i \in \{0, 1, \ldots, n_a - 1\}$ satisfying (5.3.45); moreover, these i's are different for different $N_1(q)$ $(m < n_a)$. We may therefore write $N_1(q) = n(i, m)$. Then, in view of (5.3.1)–(5.3.3) and (5.3.42), it follows that if $i \in \{0, 1, \ldots, n_a - 1\}$ then $\varphi_i(q)$ is an elementary symmetric Boolean function of m arguments with operating number $n(i, m)$. Since

$$\begin{aligned} \eta_{M,m}(l) &= \sum_{\|\tau\| = l} B_\Sigma(\tau) = \sum_{\|\tau\| = l} \sum_{i=0}^{n_a - 1} B_i(\tau) \\ &= \sum_{\|\tau\| = l} \sum_{i=0}^{n_a - 1} \sum_{q=0}^{2^m - 1} \varphi_i(q)\, \varphi_i(q \oplus \tau) \pmod{2}, \end{aligned} \tag{5.3.46}$$

where $\|\tau\| = N_1(\tau)$ is the number of units in the binary vector τ of length m, we may use the data of row 8 in the Appendix to evaluate $B_i(\tau)$, so that for odd l

$$\eta_{M,m}(l) = 0 \tag{5.3.47}$$

and for even l

$$\begin{aligned} \eta_{M,m}(l) &= \sum_{i=0}^{n_a - 1} \sum_{\|\tau\| = l} \binom{m - \|\tau\|}{n(i, m) - \|\tau\|/2} \binom{\|\tau\|}{\|\tau\|/2} \\ &= \sum_{i=0}^{n_a - 1} \binom{m - l}{n(i, m) - l/2} \binom{l}{l/2} \binom{m}{l} = \binom{l}{l/2} \binom{m}{l} \sum_{i=0}^{n_a - 1} \binom{m - l}{n(i, m) - l/2}. \end{aligned} \tag{5.3.48}$$

Now, it follows from (5.3.45) that for fixed m the function $n(i, m)$ runs through all values in $\{0, 1, \ldots, m\}$ as i runs over $\{0, 1, \ldots, n_a - 1\}$. Thus, for $m < n_a$, we deduce from (5.3.48) that for even l

$$\hat{\eta}_{M,m}(l) = \binom{l}{l/2} \binom{m}{l} \sum_{i=0}^{m-l} \binom{m - l}{i} = 2^{m-l} \binom{l}{l/2} \binom{m}{l}. \tag{5.3.49}$$

Using the definition of $\hat{\eta}_{M,m}(l)$ and formulas (5.3.47), (5.3.49), we obtain

$$\hat{\eta}_{M,m}(l) = \frac{\eta_{M,m}(l)}{2^m \binom{m}{l}} = \begin{cases} 2^{-l}\binom{l}{l/2} & \text{if } l \text{ is even,} \\ 0 & \text{if } l \text{ is odd.} \end{cases} \tag{5.3.50}$$

The assertion of our theorem now follows from (5.3.47), (5.3.49) and (5.3.50). ∎

Note that by Theorem 5.3.3 the relative abundance $\hat{\eta}_{M,m}(l)$ of errors corrected by a counting automaton is independent of the parameters C_1, C_2, of the number of states n_a and of the length m of input sequences (provided $m < n_a$).

Figure 5.3.3 illustrates the behavior of the function $\hat{\eta}_{M,m}(l)$ for a counting automaton with $m < n_a$.

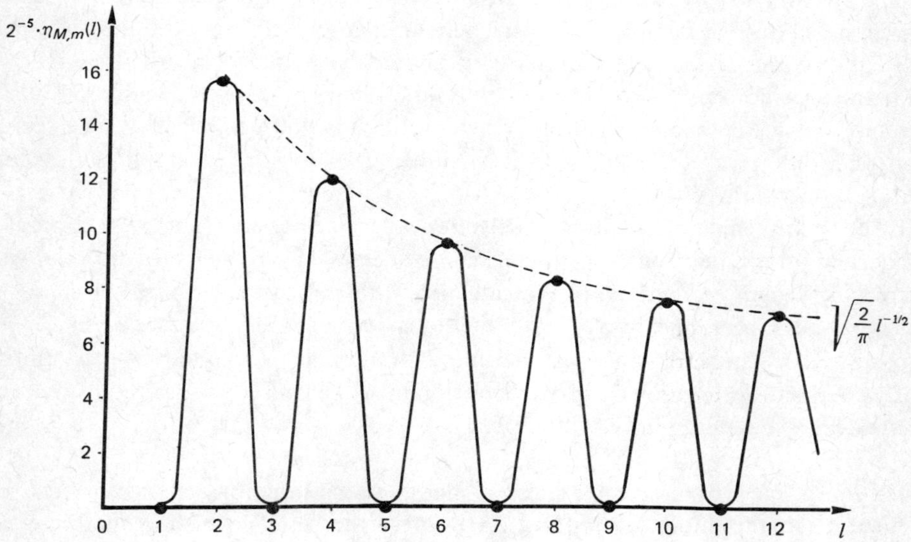

FIGURE 5.3.3.

The asymptotic behavior of $\hat{\eta}_{M,m}(l)$ for even l is readily determined from (5.3.44), using Stirling's formula:

$$\lim_{s\to\infty} \hat{\eta}_{M,m}(2s) = \sqrt{\frac{2}{\pi}}\, l^{-1/2}. \tag{5.3.51}$$

Thus, while the correcting capabilities of linear and group automata ($l > 1$) are approximately uniformly distributed over multiplicities (see (5.3.37), (5.3.38) and Figure 5.3.2), the correcting capability of a counting automaton is concentrated mainly in the region of errors of low even multiplicity (see (5.3.44), (5.3.51) and Figure 5.3.3).

5.4. SYNTHESIS OF FAULT-TOLERANT AUTOMATA BY SPECTRAL METHODS

5.4.1. In general, the requirement that a digital device (DD) be capable of correcting (or detecting) errors in a prescribed class implies the need for redundancy. Nevertheless, by incorporating error-correcting means at the very earliest stages in design of the device one can usually lower the necessary degree of redundancy [78, 79, 81]. In this section we shall confine the discussion for the most part to the case of error correction, but everything carries over without modification to error detection. We shall assume throughout that the excitation functions of the automaton A_0 to be synthesized are realized by expansion in orthogonal series (see §4.1) relative to the Walsh or Haar basis; A_0 itself need not be completely specified.

The FA will be provided with error-correcting ability through the use of error-correcting codes for state assignment. There are two principal conditions to be met:

a) The parameters of the selected code should be matched in a flexible manner with the specific features of the next-state function. (For example, this condition is not fulfilled in the familiar method of state redundancy; indeed, in that case correction of l-fold errors in any FA with n_a states, $2^{k-1} < n_a \leqq 2^k$, employs the same $((2l + 1)k, k)$-code [116].)

b) Easy implementation of code selection and state assignment by a computer. Underlying this condition is the increase in efficiency of error-correcting codes with increasing length of code words, hence also with increasing number of states n_a.

The present section is devoted to methods for introducing redundancy in FA synthesized by spectral methods, assuming the above two conditions to hold. A fairly complete description of error-correction methods in FA synthesized by classical methods may be bound in [78, 79, 81, 121].

5.4.2. In an FA synthesized by spectral methods, the functions of memory are implemented in the adder computing the sum of the series; we shall therefore use arithmetic $(AN + B)$-codes [116] for error correction. A binary m-digit $(AN + B)$-*code* is a set of numbers $N_1, N_2, \ldots, N_\theta$ such that

$$\left.\begin{array}{ll} N_i \equiv B & \pmod{A} \\ 0 \leqq N_i < 2^m & \end{array}\right\} \quad (i = 1, 2, \ldots, \theta). \tag{5.4.1}$$

The binary expansions of the numbers $N_1, N_2, \ldots, N_\theta$ will represent the codes of the states of the FA for error-correcting purposes.

We define the *distance* $\rho(N_p, N_q)$ between numbers N_p, N_q as the minimal number of terms in the representation of $|N_p - N_q|$ as a linear combination of powers of 2 with exponents $\leqq m - 1$ and coefficients ± 1 (i.e., in this section we consider only symmetric arithmetic errors; see §5.1.4). If the result of an error is to distort the number N_p, giving N_q, we shall say that the multiplicity of the error is $\rho(N_p, N_q)$. The *code distance* $\rho(V)$ of an $(AN + B)$-code V is defined as

$$\rho(V) = \min_{N_p, N_q \in V} \rho(N_p, N_q). \tag{5.4.2}$$

It can be shown that a code V detects (corrects) an l-fold error if and only if $\rho(V) \geqq l + 1$ ($\rho(V) \geqq 2l + 1$). Methods for construction of optimal $(AN + B)$-codes are considered in [116].

Note that the $(AN + B)$-codes as defined here perform error correction for symmetric errors in binary computing channels. Analogous $(AN + B)$-codes may be constructed for nonsymmetric errors and bursts, and also for error-correction in nonbinary channels (the latter should be used for DD synthesized from multistable elements) [36]. All the error-correction methods described below carry over directly to any class of $(AN + B)$-codes.

5.4.3. We are now going to examine error correction within the network realizing a given FA, that is to say, correction of errors not in the input signals of the automaton (as before) but in the internal states (memory) of the automaton.

The definitions of errors and multiplicity in this case are entirely analogous to the definitions for input signals. As for the input signals themselves, we shall assume here that they are free of errors (if the input to the FA in question is the output of a preceding FA, error correction is effected in the latter; we may also assume that the classical methods of error-correcting codes [14, 116] have been used to correct errors in the input signals).

Let A_0 be an automaton defined by a state table (state diagram) (see §4.1) and $\lambda = \{\lambda_0, \lambda_1, \ldots, \lambda_{n_\lambda - 1}\}$ some *partition* of its input set X: $\lambda_i \subseteqq X$, $\lambda_i \cap \lambda_s = \varnothing$ $(i \neq s)$, $\bigcup_{i=0}^{n_\lambda - 1} \lambda_i = X$. We shall call the subsets λ_s $(s = 0, 1, \ldots, n_\lambda - 1)$ the *blocks of the partition* λ. We shall construct a redundant FA A_λ, equivalent to A_0, in such a way that A_λ will be in any of its states only upon application of input signals from the same block of the partition λ. To construct A_λ, we "split" each state a_i of A_0 into $n_\lambda(a_i)$ equivalent states of A_λ, where $n_\lambda(a_i)$ is the number of blocks of λ that contain input signals passing A_0 into state a_i. To any state $\tilde{a}_i$ of these $n_\lambda(a_i)$ states of A_λ corresponds the same signal as for a_i in A_0; moreover, if $\varphi_0(x_s, a_i) = a_q$ in A_0, we put $\varphi_\lambda(x_s, \tilde{a}_i) = \tilde{a}_q$ in A_λ, where $\tilde{a}_i$ and $\tilde{a}_q$ are states of A_λ obtained by "splitting" states a_i and a_q of A_0: φ_0 and φ_λ are the next-state functions of A_0 and A_λ, respectively.

For example, consider Figure 5.4.1. Let $\lambda = \{\lambda_0, \lambda_1\}$, where $\lambda_0 = \{x_0, x_1\}$, $\lambda_1 = \{x_2\}$, $X = \{x_0, x_1, x_2\}$; then the state a_4 is replaced by $n_\lambda(a_4) = 2$ states in A_λ.

We say that a state of A_λ is λ_s*-reachable* if the input signals of block λ_s pass A_λ into this state. To correct errors in A_λ, it is sufficient to assign to all the λ_s-reachable states of A_λ different elements of a suitable $(AN + s)$-code $(s = 0, 1, \ldots, n_\lambda - 1)$

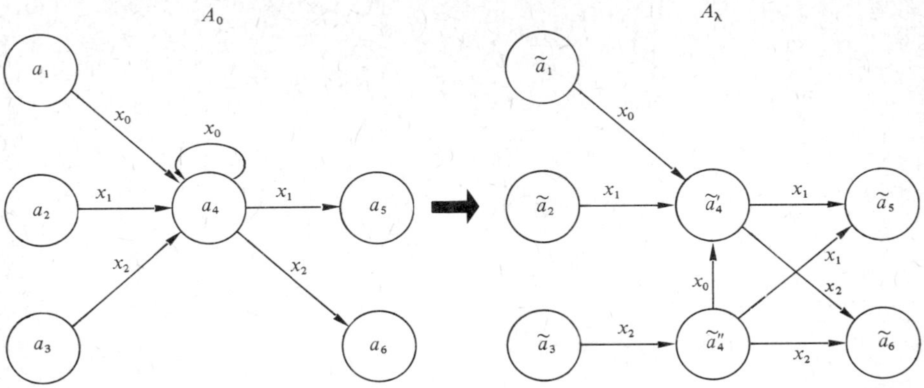

FIGURE 5.4.1.

for which the modulus A ensures the desired correcting capability. The required redundancy for the modulus A then depends on the specific properties of the next-state function φ_0 of the FA A_0, and for any multiplicity of corrected errors there exists a class of automata for which correction of errors of this multiplicity requires no redundant memory elements (see Example 5.4.1). A formal description of this error-correction method in the language of automata theory, including necessary and sufficient conditions for the existence of a DD with the appropriate correcting capability implementing the given input-output mapping and having the stipulated number of memory elements, may be set up in the same way as for implementation of the combinational part by the usual methods of Boolean algebra [78, 81].

The algorithms computing an optimal partition λ, selecting a number A generating the appropriate code, and constructing decoders which directly effect the error-correction procedure, are analogous to those considered in detail in [78, 81, 85].

EXAMPLE 5.4.1. Consider the FA A_0 defined by Table 5.4.1. Suppose that A_0 is realized by a binary network, and it is required to detect single errors. One possible optimal partition for A_0 is $\lambda = \{\lambda_0, \lambda_1, \lambda_2\}$, where $\lambda_0 = \{x_0, x_3\}$, $\lambda_1 = \{x_1\}$, $\lambda_2 = \{x_3\}$ $(n_\lambda = 3)$.

The automaton A_λ coincides with A_0. To detect errors, we put $A = 3$; this yields the state assignment of Table 5.4.2. Recall that $K^a(q) = r$ $(K^x(q) = r)$ if the code assigned to the state a_q (input signal x_q) is the binary expansion of the number r. The values of the function $\Phi(z)$ representing the excitation functions when the input assignment is $K^x(q) = q$ and the order of arguments in the excitation functions is defined by the permutation $T = (0, 1, 2, 3, 4)$ (see §4.1) are given in Table 5.4.3. Also shown in the table are the expansion coefficients $c(t)$ of $\Phi(z)$ relative to the Haar basis $\{H_l^{(q)}(z)\}$, where $t = 2^p + q - 1$. It is evident from Table 5.4.3 that computation of the ex-

TABLE 5.4.1.

a	x			
	x_0	x_1	x_2	x_3
a_0	a_0	a_3	a_2	a_0
a_1	a_5	a_3	a_7	a_1
a_2	a_5	a_3	a_7	a_1
a_3	a_0	a_3	a_2	a_5
a_4	a_1	a_6	a_7	a_0
a_5	a_5	a_4	a_2	a_5
a_6	a_0	a_4	a_7	a_5
a_7	a_1	a_3	a_7	a_0

citation functions of the FA of Table 5.4.1 with single-error detection entails storage of 13 nonvanishing coefficients, of which only 6 are distinct.

5.4.4. The block-diagram of a DD whose excitation functions are realized by expansion in orthogonal series involves a basis-function generator, storage block for expansion coefficients, and an adder computing the sum of the series.

The use of arithmetic codes makes it possible to correct (or detect) errors in the adder, in the storage block, and sometimes also in the basis-function generator (see §§6.1, 6.2).

If the memory of the FA is combined with the adder, errors may also be corrected

TABLE 5.4.2.

q	0	1	2	3	4	5	6	7
$K^a(q)$	0	3	2	1	4	6	7	5

TABLE 5.4.3.

z, t	$\Phi(z)$	$4c(t)$	z, t	$\Phi(z)$	$4c(t)$	z, t	$\Phi(z)$	$4c(t)$	z, t	$\Phi(z)$	$4c(t)$
0	0	12	8	1	−12	16	2	0	24	0	0
1	0	−1	9	1	0	17	2	0	25	6	0
2	6	1	10	1	0	18	5	0	26	3	0
3	6	1	11	1	0	19	5	12	27	3	−6
4	3	0	12	7	−6	20	5	0	28	0	−12
5	3	−6	13	1	6	21	5	0	29	0	0
6	6	0	24	4	0	22	2	12	30	6	0
7	0	0	25	4	−12	23	5	0	31	6	0

or detected by methods not based on the use of arithmetic codes. Indeed, the binary codes of the expansion coefficients and the numbers stored in the adder during the summation procedure have integral and fractional parts. At the end of the computations, the fractional part of the number stored in the adder should be zero. This provides a fairly simple and efficient technique for detecting errors in the appropriate adder bits or for correcting them, by rounding off the sum to the nearest integer. Thus, arithmetic codes should be used only to correct errors in the digits of the integral part of the coefficient codes and in the corresponding bits of the adder. In addition, one can detect errors in these digits by using overflow signals in the adder and the value of the sign digit once the computation of the excitation functions has been completed.

A final remark: in the realization of automata by spectral methods combined with the use of $(AN + s)$-codes, one can effectively organize simultaneous detection and/or correction of errors in any system of automata containing arithmetic units.

Chapter Six

TECHNICAL REALIZATION OF SPECTRAL METHODS OF SYNTHESIS

This last chapter is concerned with question of technical realization and bounds on the complexity of the spectral methods of the previous chapters for the synthesis of combinational networks and networks with memory based on bistable elements. The complexity bounds established in this chapter are approximate; their main object is to allow comparison of the various spectral methods of synthesis. We shall also propose block diagrams demonstrating an approach to solution of the major technical problems that arise when spectral methods are employed. As a rule, the basis system will be the system of Haar functions.

6.1. SERIAL REALIZATION OF SPECTRAL METHODS OF SYNTHESIS

6.1.1. We use the term "serial realization" of spectral methods for the design technique in which the value of the function describing the operation of a network is obtained by serial (i.e., sequential in time) summation of the appropriate series.

Once a state assignment has been selected, the realization of a network with memory is equivalent to realization of a suitable combinational network (see §4.1); for this reason, the discussion of this chapter will be limited to the technical realization of systems of Boolean functions (BF), i.e., (m, k)-terminal logic networks.

A system of k BF of m arguments in fixed order is represented by a single step function $\Phi(z)$, which is then expanded as an orthogonal series (see §1.1). However, a system of k BF of m arguments may also be treated as 2^l $(0 \leqq l < m)$ systems of $m - l$ arguments, obtained by fixing l arguments of the original system in all possible ways. Alternatively, we can speak of s $(0 < s \leqq k)$ systems of m arguments, where the i-th system contains p_i functions $(i = 1, 2, \ldots, s)$, $\sum_{i=1}^{s} p_i = k$; and yet another interpretation is to take some combination of these two cases. Whatever the alternative chosen, each of the systems may be realized independently by any desired method. It will become clear from the sequel that, depending on the specific alternative, there are two possibilities: either (1) the total number of nonvanishing expansion coefficients of the representative step functions is increased, or (2) the total number of bits in the adders computing the values of the step functions is increased. We shall therefore confine ourselves throughout the sequel to the case of a system of k BF of m arguments represented by a single step function $\Phi(z)$, defined as in formulas (1.1.2)–(1.1.4).

6.1.2. Our first topic will be serial summation of the expansion of $\Phi(z)$ in Haar series.

Figure 6.1.1 illustrates a suitable block-diagram, which operates as follows. The binary code of the argument z is recorded in the shift register together with a "1" in the first position on the right; this register receives clock pulses (CP) which shift the code recorded therein. At consecutive instants, the portion of the shift register controlling the decoder will hold the code sequences $(1, 0, \ldots, 0)$, $(z^{(0)}, 1, 0, \ldots, 0)$, $(z^{(1)}, z^{(0)}, 1, 0, \ldots, 0), \ldots, (z^{(m-2)}, z^{(m-3)}, \ldots, z^{(0)}, 1)$, where $z^{(s)}$ is the s-th digit of the binary code z. Each of these sequences activates exactly one output of the decoder, depending on the code of z; if the decoder outputs are numbered from 1 through $2^m - 1$, the output activated at the l-th time is one of those numbers in the set $\{2^{l-1}, 2^{l-1} + 1, \ldots, 2^l - 1\}$ $(l = 1, 2, \ldots, m)$.

The decoder controls the operation of the encoder—a permanent storage unit (PSU) holding the binary codes of the expansion coefficients of $\Phi(z)$ in Haar series.

When the δ-th input $(\delta = 1, 2, \ldots, 2^m - 1)$ of the encoder is activated, the coefficient $c_l^{(q)}$ is fed from the encoder output to the input of an adder-accumulator Σ,

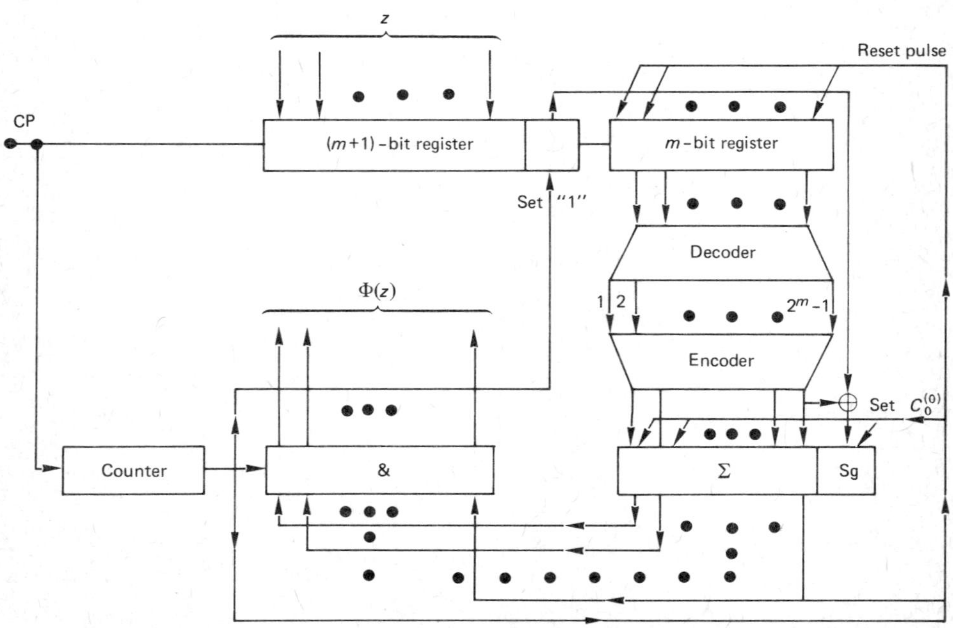

FIGURE 6.1.1.

where $l \in \{1, \ldots, m\}$ and $q \in \{1, \ldots, 2^l\}$ are uniquely determined by the condition:

$$2^l + q - 1 = \delta. \tag{6.1.1}$$

The sign of the coefficient is fed through a mod 2 adder into the sign bit "Sg" of the adder Σ.

To compute $\Phi(z)$ for any fixed z, one must generally sum $m + 1$ coefficients $c_l^{(q)}$ (see Theorem 1.2.10), and so the value of $\Phi(z)$ is produced after m elementary addition operations. (Before summation begins, the adder Σ holds the code of the coefficient $c_0^{(0)}$, which is thus the initial state of Σ.) Upon each elementary addition, the number held in the register is shifted in accordance with the clock pulses.

We now consider the realization of the encoder in greater detail.

If one of the coefficients $c_l^{(q)}$ is zero, there is no need to store it in the encoder or to realize the correspond decoder output. Thus the complexity of both encoder and decoder depends essentially on the number of nonvanishing coefficients.

We have already stated that the encoder is a PSU, while the decoder computes the address for the PSU. The PSU itself may be either a single-coordinate or a many-coordinate device [60, 93]. Economical and asymptotically optimal methods for the design of decoder-encoder networks, with the PSU organized on the single-coordinate principle, from logical sum $\vee$ (disjunction) elements (e.g., as a diode network) are described in [86]. Magnetic core implementation of single-coordinate and many-coordinate PSU is considered in [60, 65, 93]. In the case of a two-coordinate PSU, e.g., ferrite matrix, the decoder is split into two decoders, one for each coordinate (x-decoder and y-decoder) with $[m/2]$ and $]m/2[$ arguments, respectively (see Figure 6.1.2). It will be shown below (§6.1.3) that this approach sharply reduces the complexity of the decoder.

Each cell of the PSU holds one of the coefficients $c_l^{(q)}$. The length of the binary code of this coefficient is at most $m + k + 1$, and so, if the PSU is a ferrite-core device, each memory cell requires at most $m + k + 1$ cores. Moreover, if the s-th digit of the binary code of $c_l^{(q)}$ is 0, no special core is required to store it. (This, however, leads to a nonuniform load on the decoder lines and an unfavorable signal-to-noise ratio at the encoder output, thus possibly necessitating the introduction of additional "compensating" cores; we shall not dwell on such questions here.)

At each instant of time, the code of only one coefficient should appear at the decoder output; therefore, the output windings of each core (say the s-th core) in the cell of each coefficient is connected in series with the output windings of the s-th cores in the cells of the other coefficients. For example, if $c_0^{(0)} = 011111101$ and $c_0^{(1)} = 110000110$, the two cells storing $c_0^{(0)}$ and $c_0^{(1)}$ should be connected as shown in Figure 6.1.3. The core $\Phi_s^{(q)}$ ($s = 1, 2, \ldots, 9$; $q = 0, 1$) stores the s-th digit of the binary code of $c_0^{(q)}$ (the rewrite windings, which copy the initial state into the cores after reading, are not shown in Figure 6.1.3).

Returning to Figure 6.1.1, we now examine the formation of the sign digit "Sg"

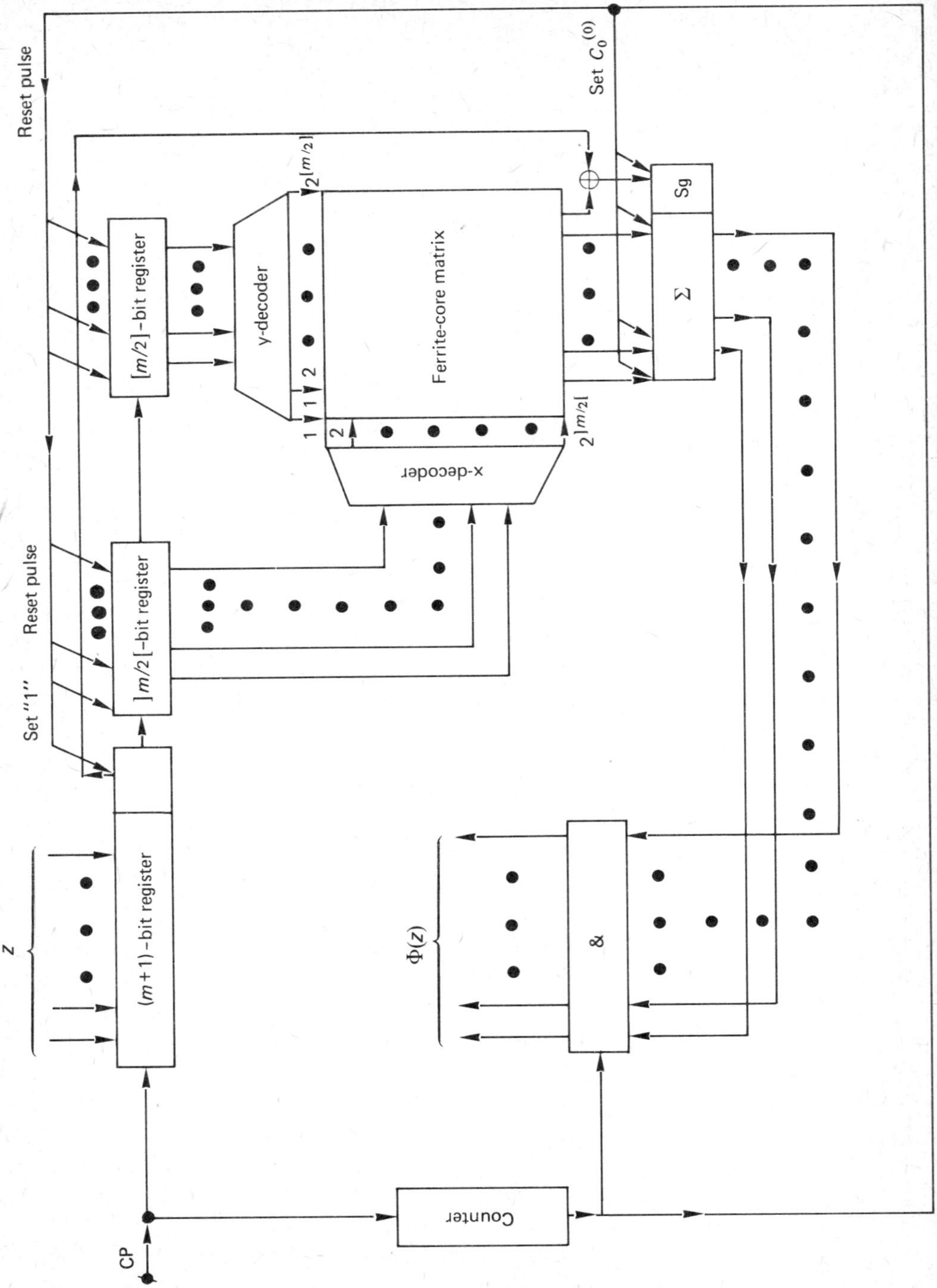
Reset pulse
Set $C_0^{(0)}$
[m/2]-bit register
y-decoder
Ferrite-core matrix
Sg
Σ
x-decoder
]m/2[-bit register
Set "1"
Reset pulse
z
(m+1)-bit register
Φ(z)
&
CP
Counter

FIGURE 6.1.2.

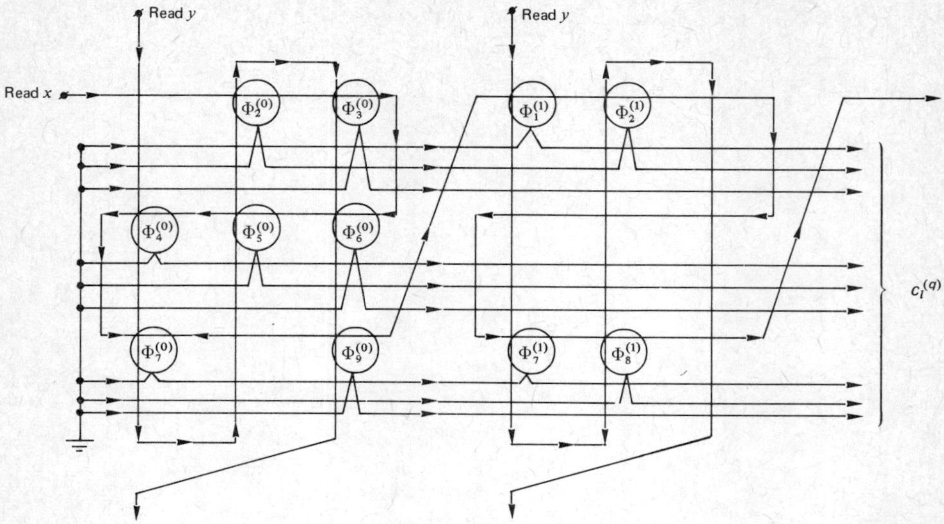

FIGURE 6.1.3.

of the adder Σ, assuming, as usual, that "1" in the sign digit implies a negative value of the number stored in Σ, and "0" a positive value. Since (see §1.2)

$$H_l^{(q)}(z) = R_{l+1}(z) = (-1)^{z^{(l)}} \tag{6.1.2}$$

for $z \in [(q-1)\,2^{m-l}, q \cdot 2^{m-l})$, it follows in view of the previously shown relation that the value of $H_l^{(q)}(z)$ is equal to the l-th digit $z^{(l)}$ of the argument code, and multiplication of the sign of the coefficient by the sign of $H^{(q)}(z)$ is thus accomplished by the mod 2 adder, whose output is fed to the sign digit of adder Σ.

The computation of $\Phi(z)$ ends after m elementary additions, and a mod m counter (see Figure 6.1.1) sends a reading signal to a network of "&" gates, a "reset" signal to the decoder control register, a "set" signal to the register bit controlling the sign in adder Σ, and a signal restoring the code $c_0^{(0)}$ in Σ.

EXAMPLE 6.1.1. Let us design a network implementing the incompletely specified automaton of Example 4.3.1, with serial summation of series.

The optimal completion of the state matrix is the matrix A_5'' of §4.3.3, for which $n_a = 8$, $n_x = 8$; the state assignment is shown in Table 4.3.3, the order of arguments in the excitation functions is $(x^{(0)}, a^{(0)}, a^{(1)}, x^{(1)}, x^{(2)}, a^{(2)})$, and bounds on the number of nonvanishing coefficients are given in Table 4.3.4.

Using A_5'' and the state assignment of Table 4.3.1, we construct a step function $\Phi(z)$ representing the system of excitation functions for the selected order of arguments. This function and the nonvanishing coefficients $c_l^{(q)}$ of its expansion are given in Tables 6.1.1 and 6.1.2, respectively. A network realizing this automaton as described above is illustrated in Figure 6.1.4. The timing of the input signal coincides with the timing of signals at the counter output.

TABLE 6.1.1.

z	$\Phi(z)$	z	$\Phi(z)$
0	0	32	5
1	0	33	5
2	0	34	5
3	0	35	5
4	2	36	1
5	2	37	1
6	2	38	7
7	2	39	7
8	0	40	2
9	0	41	2
10	0	42	2
11	0	43	2
12	0	44	1
13	0	45	5
14	0	46	7
15	0	47	7
16	4	48	0
17	4	49	0
18	6	50	0
19	6	51	5
20	6	52	2
21	6	53	2
22	6	54	2
23	6	55	2
24	3	56	4
25	3	57	4
26	3	58	4
27	3	59	4
28	7	60	1
29	7	61	1
30	7	62	1
31	7	63	1

6.1.3. We now estimate the complexity of a serial realization of a combinational network by spectral methods.

The block-diagram of Figure 6.1.1 contains the following components:

a) $(2m + 1)$-position shift register;

b) decoder;

c) encoder (PSU);

TABLE 6.1.2.

$\delta = 2^l + q - 1$	(l, q)	$64c_l^{(q)}$	Binary code of $c_l^{(q)}$								
			sign	2^1	2^0	2^{-1}	2^{-2}	2^{-3}	2^{-4}	2^{-5}	2^{-6}
0	(0, 0)	189	0	1	0	1	1	1	1	0	1
1	(0, 1)	− 5	1	0	0	0	0	0	1	0	1
2	(1, 1)	−152	1	1	0	0	1	1	0	0	0
3	(1, 2)	62	0	0	0	1	1	1	1	1	0
4	(2, 1)	32	0	0	0	1	0	0	0	0	0
5	(2, 2)	16	0	0	0	0	1	0	0	0	0
6	(2. 3)	32	0	0	0	1	0	0	0	0	0
7	(2, 4)	− 28	1	0	0	0	1	1	1	0	0
8	(3, 1)	− 64	1	0	1	0	0	0	0	0	0
10	(3, 3)	− 32	1	0	1	1	0	0	0	0	0
11	(3, 4)	−128	1	1	0	0	0	0	0	0	0
12	(3, 5)	− 32	1	0	0	1	0	0	0	0	0
13	(3, 6)	96	0	0	1	1	0	0	0	0	0
14	(3, 7)	− 24	1	0	0	0	1	1	0	0	0
15	(3, 8)	64	0	0	1	0	0	0	0	0	0
20	(4, 5)	− 64	1	0	1	0	0	0	0	0	0
25	(4, 10)	−192	1	1	1	0	0	0	0	0	0
27	(4, 12)	−128	1	1	0	0	0	0	0	0	0
28	(4, 13)	− 80	1	0	1	0	1	0	0	0	0
54	(5, 23)	−128	1	1	0	0	0	0	0	0	0

d) adder;
e) mod m counter;
f) gate block ("&" networks).

We now estimate the complexity of each block.

a) The complexity of the shift register L_{SR} is

$$L_{\mathrm{SR}} = (2m + 1)\, L_{\mathrm{SR}}^{(1)}, \tag{6.1.3}$$

where $L_{\mathrm{SR}}^{(1)}$ is the complexity of one stage of the shift register.

b) If the encoder is a two-coordinate PSU (such as a ferrite matrix), the complexity of the decoder is:

$$L_{\mathrm{Dc}} \leqq L_{\mathrm{Dc}}\left(\left[\frac{m}{2}\right], 2^{[m/2]}\right) + L_{\mathrm{Dc}}\left(\left]\frac{m}{2}\right[, 2^{]m/2[}\right) \leqq (2^{[m/2]+1} + 2^{]m/2[+1} - 8)\, L_{\&}, \tag{6.1.4}$$

where $L_{\mathrm{Dc}}(p, q)$ is the complexity of a decoder with p inputs and q outputs, and $L_{\&}$ is that of a two-input coincidence network (logical product "&").

To prove (6.1.4) we note that in a two-coordinate PSU the decoder splits into

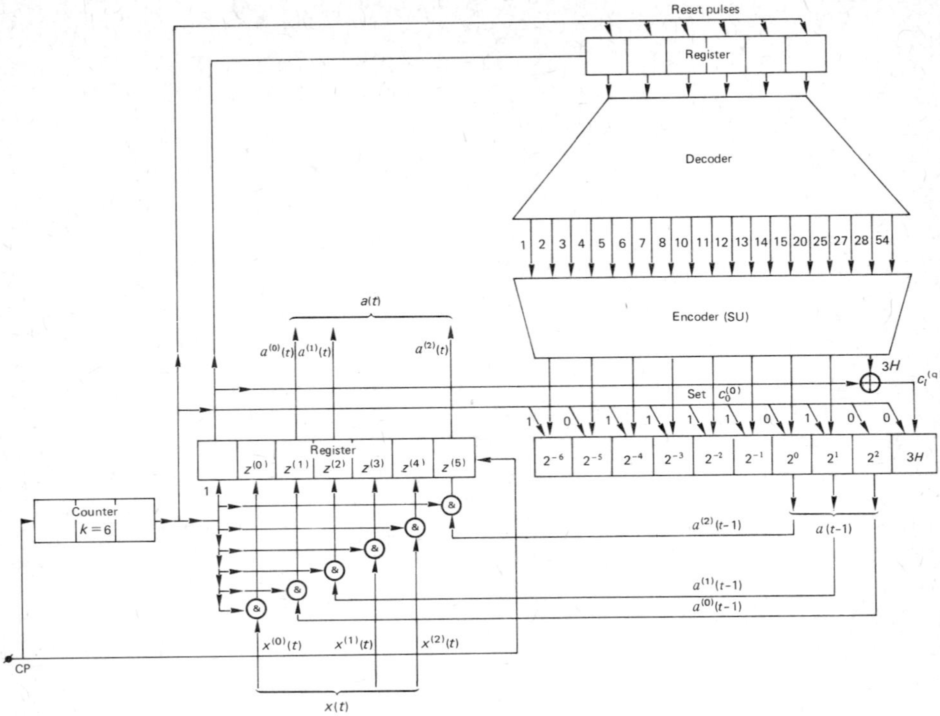

FIGURE 6.1.4.

two address decoders, with $[m/2]$ and $]m/2[$ inputs, respectively, and $2^{[m/2]}$ and $2^{]m/2[}$ outputs. Let us suppose that each of these decoders is of the so-called tree type: a tree decoder with λ inputs $z^{(0)}, \ldots, z^{(\lambda-1)}$ and 2^λ outputs is a network obtained by serial connection of λ separate networks ("stages"), the l-th stage receiving the output of the $(l-1)$-th and the input variable $z^{(l-1)}$. The zeroth stage is a constant "1" signal, i.e., a clock pulse. In a tree decoder, each stage, from the zeroth to the $(l-1)$-th inclusive, is a full decoder for the variables $z^{(0)}, \ldots, z^{(l-1)}$. It is readily shown by induction on λ that the complexity $L_{\mathrm{Dc}}(\lambda, 2^\lambda)$ of a tree decoder with λ inputs and 2^λ outputs is

$$L_{\mathrm{Dc}}(\lambda, 2^\lambda) = (2^{\lambda+1} - 4)\, L_{\&}.$$

The estimate (6.1.4) now follows easily.

If the encoder is a single-coordinate PSU, the decoder has m inputs and L outputs, where L is the number of nonzero coefficients of the code being used; reasoning as in the case of a two-coordinate PSU, one shows that its complexity is bounded by $2L \cdot L_{\&}$.

c) The encoder is a PSU which stores the nonvanishing coefficients $c_l^{(q)}$. The complexity of the encoder is given approximately by

$$L_{\mathrm{En}} \cong \frac{1}{2} \sum_{l=1}^{m} L_{m-l}(k + l)\, L_{\text{“}1\text{”}} \tag{6.1.5}$$

where L_{m-l} is the number of nonvanishing coefficients $c_{m-l}^{(q)}$ $(q = 1, 2, \ldots, 2^{m-l})$ and $L_{\text{“}1\text{”}}$ the complexity of the element necessary to store one bit of information in the PSU. Indeed, if $\Phi(z)$ is the step function representing a system of k BF of m arguments, then $\max_z \Phi(z) < 2^k$ and so $|c_{m-l}^{(q)}| = 2^{-l} \tilde{c}_{m-l}^{(q)}$, where $]\log_2 \tilde{c}_{m-l}^{(q)}[\leqq k < l$. In addition, it is assumed in (6.1.5) that the numbers of units and zeros in the binary expansion of each nonvanishing coefficient $c_{m-l}^{(q)}$ are equal.

REMARK. Both in (6.1.5) and in further complexity estimates for encoders, if $l = m$ then $L_0 \leqq 1$, since the coefficient $c_0^{(0)}$ is not produced in the encoder but fed directly to the adder as an initial state.

d) In serial summation of the terms of the series, one uses an adder-accumulator containing at most $m + k + 1$ stages. The complexity $L_{\mathrm{a}\Sigma}$ of the adder-accumulator is

$$L_{\mathrm{a}\Sigma} \leqq (m + k + 1)\, L_{\mathrm{a}\Sigma}^{(1)} \tag{6.1.6}$$

where $L_{\mathrm{a}\Sigma}^{(1)}$ is the complexity of one stage of the adder-accumulator.

e) The complexity of a mod m counter is

$$L_{\mathrm{Co}} =]\log_2 m[\; L_{\mathrm{Co}}^{(1)}, \tag{6.1.7}$$

where $L_{\mathrm{Co}}^{(1)}$ is the complexity of one stage of the counter.

f) The complexity $L_{\mathrm{B\&}}$ of the gate block is

$$L_{\mathrm{B\&}} = m L_{\&}. \tag{6.1.8}$$

Finally, the complexity L_{Sq} of the serial variant is obtained by setting $L_{\mathrm{RS}}^{(1)} = L_{\mathrm{a}\Sigma}^{(1)} = L_{\mathrm{Co}}^{(1)} = L_{\mathrm{ff}}$, where L_{ff} is the complexity of one flip-flop, and $L_{\text{“}1\text{”}} = L_{\&}$ for the case of a single-coordinate storage unit. Noting that $L = \sum_{l=1}^{m} L_{m-l}$, we have

$$L_{\mathrm{Sq}} = L_{\mathrm{SR}} + L_{\mathrm{Dc}} + L_{\mathrm{En}} + L_{\mathrm{a}\Sigma} + L_{\mathrm{Co}} + L_{\mathrm{B\&}} \leqq (3m + k +]\log_2 m[+ 2)\, L_{\mathrm{ff}}$$
$$+ \left((\tfrac{1}{2}k + 2)\, L + m + \frac{1}{2} \sum_{l=1}^{m} L_{m-l} \cdot l \right) L_{\&}. \tag{6.1.9}$$

For the case of a two-coordinate storage unit, we have

$$L_{\text{Sq}} \leqq (3m + k +]\log_2 m[+ 2)\, L_{\text{ff}}$$

$$+ \left(\frac{1}{2} L \cdot k + \frac{1}{2} \sum_{l=1}^{m} L_{m-l} \cdot l + m + 2^{[m/2]+1} + 2^{]m/2[+1} - 8 \right) L_{\&}. \quad (6.1.10)$$

A comparison of the estimates (6.1.9) and (6.1.10) shows that the use of single-coordinate storage is advisable when $L < 2^{m/2}$; in other cases a many-coordinate system should be used.

For the example in §4.3, we have $m = 6$, $k =]\log_2 n_a[= 3$, $L_{\text{Sq}} = 26L_{\text{ff}} + 96L_{\&}$ when a single-coordinate PSU of $\vee$-networks is used, and $L_{\text{Sq}} = 26L_{\text{ff}} + 76L_{\&}$ for a two-coordinate PSU. The upper bounds provided by formulas (6.1.9) and (6.1.10) in these cases are, by contrast:

$$L_{\text{Sq}} \leqq 26L_{\text{ff}} + 103L_{\&} \quad \text{and} \quad L_{\text{Sq}} \leqq 26L_{\text{ff}} + 89L_{\&}.$$

6.1.4. To end this section, we outline a method whereby the complexity L_{Sq} of the above block-diagrams for sequential realization of spectral methods may be minimized. Since by definition the step function $\Phi(z)$ being implemented takes only integral values, it will suffice to compute another step function $\Psi(z)$ such that

$$\max_z |\Phi(z) - \Psi(z)| < \tfrac{1}{2} \quad (6.1.11)$$

and subsequently round off $\Psi(z)$ to the nearest integer (this approach may require a fairly simple automatic rounding-off network, but the realization of $\Psi(z)$ may turn out to be substantially simpler than that of $\Phi(z)$). We consider two ways of doing this.

The first construction of $\Psi(z)$ involves deleting some of the expansion coefficients of $\Phi(z)$. Note that if $c_{m-1}^{(q)} \neq 0$, then $|c_{m-1}^{(q)}| \geqq 1/2$ and so coefficients with subscript $m - 1$ cannot be deleted.

Now let P_{m-l} $(l = 2, 3, \ldots, m)$ be some subset of coefficients $c_{m-l}^{(q)}$ for $\Phi(z)$ and P the union of all the sets P_{m-l}. Then the coefficients in P may be deleted in computing $\Phi(z)$ if and only if

$$\max_z \left| \sum_{c_{m-l}^{(q)} \in P} c_{m-l}^{(q)} H_{m-l}^{(q)}(z) \right| < \frac{1}{2}. \quad (6.1.12)$$

If m is large, however, the use of this condition to determine a maximal set P of deletable coefficients is not very practical. We therefore replace (6.1.12) by a somewhat weaker condition which is much easier to verify. Set

$$\hat{c}_{m-l} = \max_{c_{m-l}^{(q)} \in P_{m-1}} \left| c_{m-l}^{(q)} \right|. \tag{6.1.13}$$

Then the coefficients of P may be deleted if

$$\sum_{l=2}^{m} \hat{c}_{m-l} < \frac{1}{2}. \tag{6.1.14}$$

Going back to the example of this section (see Table 6.1.2), we see that condition (6.1.12) allows us to delete the coefficients $c_0^{(1)}, c_2^{(2)}, c_3^{(7)}$. The complexity of the encoder is thereby reduced by about 13 %, and the complexity of the decoder is also reduced.

The second construction of $\Psi(z)$ involves deleting the least significant digits of the codes of $c_{m-l}^{(q)}$ for $\Phi(z)$. This reduces the length of the coefficient codes and thereby the necessary capacity of the adder.

Let $c_{m-l}^{(q)}(s)$ denote the value of the digit of weight 2^{-s} in the binary expansion of $c_{m-l}^{(q)}$ ($s \leqq m$, $c_{m-l}^{(q)}(s) \in \{0, 1\}$). For example, $c_{m-l}^{(q)}(m)$ is the value of the least significant digit in the binary expansion of $c_{m-l}^{(q)}$. Then the digits of the coefficients $c_{m-l}^{(q)}$ of weights $-m, -m+1, \ldots, -m+r$ may be deleted from the expansion of $\Phi(z)$ if

$$\sum_{s=m-r}^{m} \sum_{l,q} c_{m-l}^{(q)}(s)\, 2^{-s} < \frac{1}{2}. \tag{6.1.15}$$

This condition generates a simple procedure yielding the maximal number r of deletable code digits of coefficients $c_{m-l}^{(q)}$. For the example of this section, one can delete the digits of weights $2^{-6}, 2^{-5}, 2^{-4}$ ($r = 3$). The complexity of the adder is reduced by 30 % and that of the encoder is also reduced.

6.2. PARALLEL REALIZATION OF SPECTRAL METHODS OF SYNTHESIS

6.2.1. The use of time-parallel summation to compute the series expansion of a step function $\Phi(z)$ representing a system of BF in m arguments yields an m-fold gain in speed. This advantage is achieved, however, at the cost of a concomitant increase in complexity, due mainly to the fact that the adder-accumulator must be replaced by a parallel adder for m terms. In this section we shall discuss the technical realization of parallel summation and establish complexity estimates.

6.2.2. A block-diagram for parallel summation of series is shown in Figure 6.2.1. Its operation is as follows. The binary code $(z^{(0)}, z^{(1)}, \ldots, z^{(m-1)})$ of the argument z is recorded in m memory elements. For any fixed $z = z^*$, there are exactly $m + 1$

Haar functions $H_l^{(q)}(z)$ such that $H_l^{(q)}(z^*) \neq 0$ (in fact, of all functions with the same nonzero subscript, exactly one does not vanish); hence only the coefficients of these functions have to be accessed from the encoder. This is ensured by introducing a tree-type decoder (see §6.1.3). At the output of the l-th stage of the decoder ($l = 0, 1, \ldots, m-1$), one line is activated, corresponding to the coefficients $c_l^{(q)}$ such that $H_l^{(q)}(z) \neq 0$ for the z in question. Thus at most m codes of coefficients $c_l^{(q)}$ (for different l) are produced simultaneously at the encoder output (the coefficient $c_0^{(0)}$ is fed directly to the adder Σ). These $m+1$ coefficients are summed by the parallel adder Σ.

The decoder in this network has $m-1$ stages (and correspondingly $m-1$ inputs $z^{(0)}, \ldots, z^{(m-2)}$), since the output of the decoder corresponding to $c_0^{(0)}$ is taken directly from its input (see Figure 6.2.1).

Each of the m coefficients produced at the encoder output is multiplied by the values of the appropriate basis function, by means of mod 2 adders; the process is precisely the same as in the serial variant.

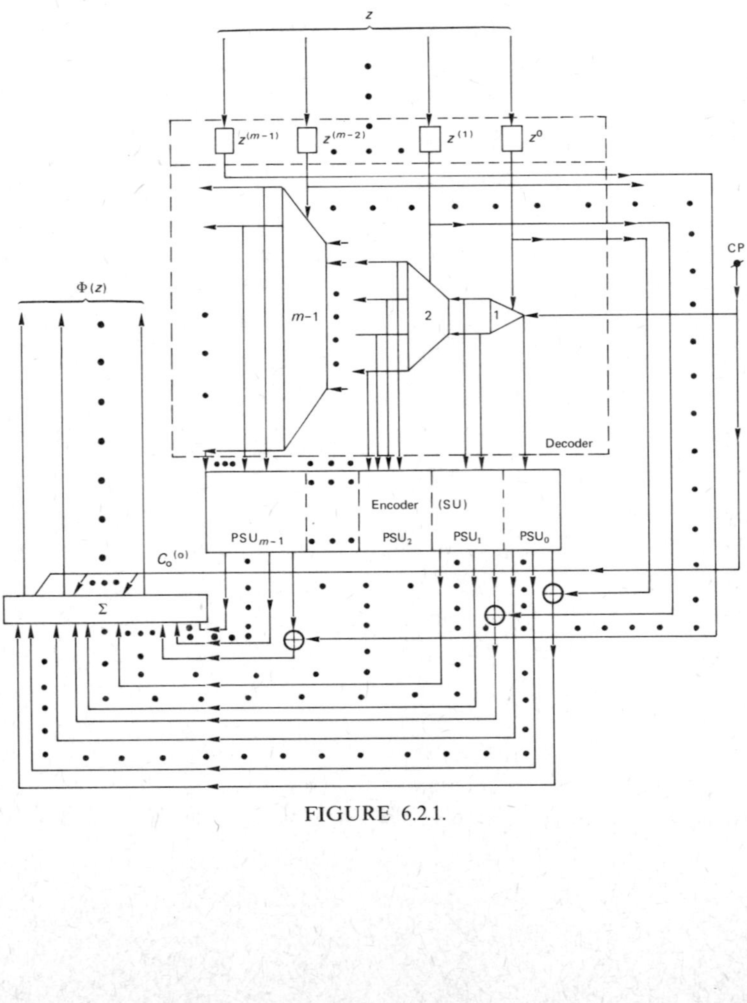

FIGURE 6.2.1.

The encoder is implemented by the same methods as in the serial variant. If the encoder is a ferrite-core memory, one must remember that at any given instant only one coefficient with given subscript l is to be activated at the encoder output; thus, for each s, the output winding of the s-th core, corresponding to the s-th digit of the code of $c_l^{(q)}$, must be connected in series to the output windings of the s-th cores of the other coefficients with subscript l.

Thus, for parallel summation, the encoder consists of m independent single-coordinate PSU, where the l-th PSU ($l = 0, 1, \ldots, m - 1$) holds the codes of the coefficients $c_l^{(q)}$ ($q = 1, 2, \ldots, 2^l$) and the l-th stage of the tree decoder acts as address decoder for the l-th PSU.

Figure 6.2.2 is a block-diagram of a parallel-type realization of the automaton considered in §6.1.2.

6.2.3. We now estimate the complexity L_{Pr} of a parallel realization of an (m, k)-terminal network by spectral methods. The block-diagram of Figure 6.2.1 contains

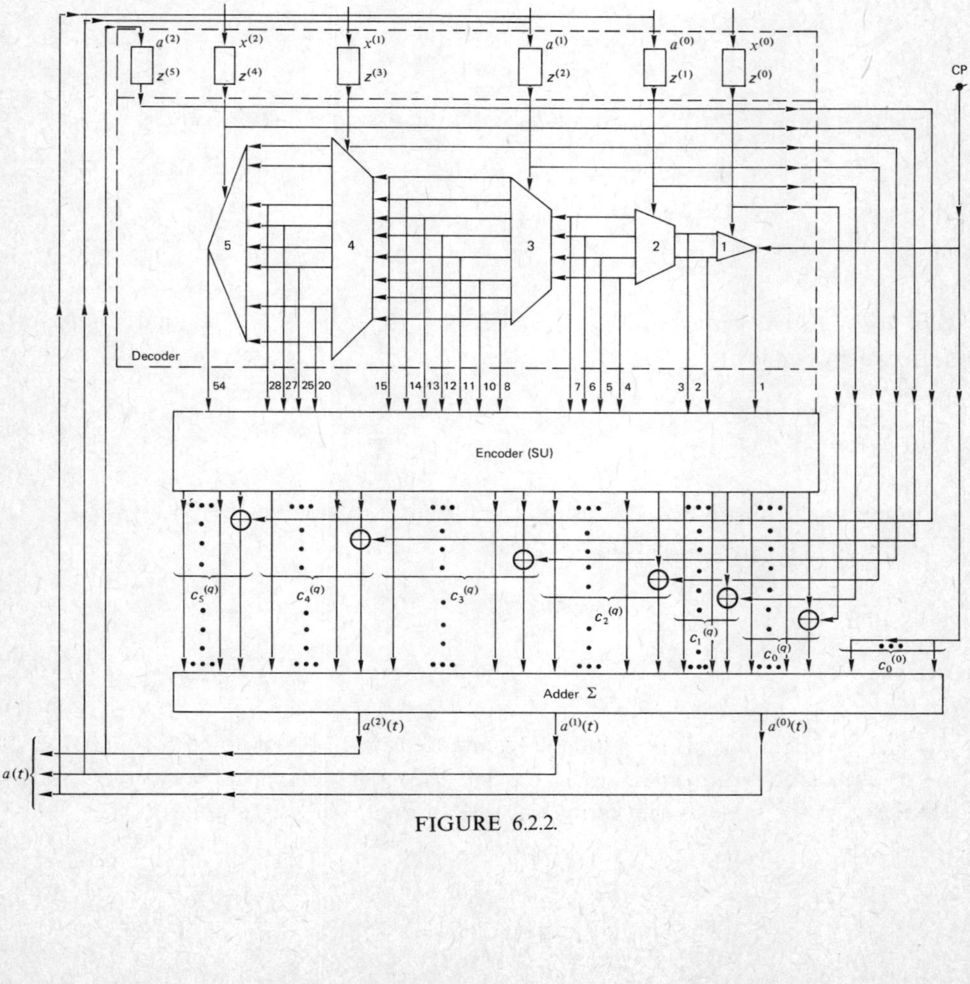

FIGURE 6.2.2.

the following components:

a) Storage block for argument code $(z^{(0)}, \ldots, z^{(m-1)})$;

b) decoder;

c) encoder (PSU);

d) sign-production block (mod 2 adders);

e) adder.

We proceed to evaluate the complexity of each block.

a) Storage block for the argument code:

$$L_{ma} = m \cdot L_{ff}. \tag{6.2.1}$$

(Both here and in §6.3 we retain the notation of §6.1.)

b) The decoder of Figure 6.2.1 has $m - 1$ arguments and is of tree type, the $(m - l)$-th stage having L_{m-l} outputs. Suppose that the decoder is designed so that the first $m - 2$ stages (for the first $m - 2$ arguments) form a full tree decoder, while the $(m - 1)$-th stage includes L_{m-1} two-input &-elements whose outputs correspond to the L_{m-1} nonvanishing coefficients $c_{m-1}^{(q)}$; then the complexity of the decoder satisfies the inequality (see §6.1.3)

$$L_{Dc} \leqq (2^{m-1} + L_{m-1} - 4)\, L_{\&}. \tag{6.2.2}$$

Similarly, if the full tree decoder consists of the first $m - l$ stages, then

$$L_{Dc} \leqq (2^{m-l+1} + L_{m-l+1} + 2L_{m-l+2} + 3L_{m-l+3} + \cdots + (l-1)\, L_{m-1} - 4)\, L_{\&}$$

$$= \left(2^{m-l+1} - 4 + \sum_{s=1}^{l-1} sL_{m-l+s}\right) L_{\&}. \tag{6.2.3}$$

Since l may be chosen arbitrarily from the set $\{2, 3, \ldots, m - 1\}$, the final bound for the decoder complexity is

$$L_{Dc} \leqq \min_{l\in\{2,3,\ldots,m-1\}} \left(2^{m-l+1} - 4 + \sum_{s=1}^{l-1} s \cdot L_{m-l+s}\right) L_{\&}. \tag{6.2.4}$$

For practical purposes, it is usually sufficient to examine the cases $l = 2, 3, 4$, thus obtaining the following rough estimate;

$$L_{Dc} \leqq \min\, (2^{m-1} + L_{m-1} - 4;\ 2^{m-2} + L_{m-2} + 2L_{m-1} - 4;$$

$$2^{m-3} + L_{m-3} + 2L_{m-2} + 3L_{m-1} - 4)\, L_{\&}. \tag{6.2.5}$$

c) The encoder consists of m independent encoders, of which the l-th ($l = 0, 1, \ldots, m - 1$) is a single-coordinate PSU storing the codes of $c_l^{(q)}$ ($q = 1, 2, \ldots, 2^l$). Its complexity is the same as that of the encoder in Figure 6.1.1 (see (6.1.5)).

d) The sign block has complexity

$$L_{\mathrm{Sg}} = m \cdot L_{\oplus}, \tag{6.2.6}$$

where $L_{\oplus}$ is the complexity of a two-input mod 2 adder.

e) In parallel realization, summation is performed by a combinational parallel adder for m terms, each of which contains at most $m + k + 1$ digits. Thus the complexity of the adder satisfies the inequality

$$L_{\mathrm{c}\Sigma} \leqq (m - 1)(m + k + 1)\, L_{\mathrm{c}\Sigma}^{(1)} \tag{6.2.7}$$

where $L_{\mathrm{c}\Sigma}^{(1)}$ is the complexity of one stage of a combinational adder for two terms (full adder) [70].

Finally, the complexity L_{Pr} of the entire network, assuming $L_{\text{“1”}} = L_{\oplus} = L_{\&}$, $L_{\mathrm{c}\Sigma}^{(1)} = 9L_{\&}$ [70] and using the estimate $L_{\mathrm{Dc}} \leqq (2^{m-1} + L_{m-1} - 4)\, L_{\&}$, is

$$\begin{aligned} L_{\mathrm{Pr}} &= L_{ma} + L_{\mathrm{Dc}} + L_{\mathrm{En}} + L_{\mathrm{Sg}} + L_{\mathrm{c}\Sigma} \\ &\leqq mL_{\mathrm{ff}} + \Bigg(2^{m-1} + 9m^2 + m(9k + 1) + L_{m-1} + \frac{1}{2} \sum_{l=1}^{m} l \cdot L_{m-l} \\ &\qquad + (\tfrac{1}{2}L - 9)\, k - 13 \Bigg) L_{\&}. \end{aligned} \tag{6.2.8}$$

For the example of §4.3, $m = 6$, $k =]\log_2 n_a[= 3$ and $L_{\mathrm{Pr}} = 6L_{\mathrm{ff}} + 519L_{\&}$ (the bound following from (6.2.8) in this case would be $L_{\mathrm{Pr}} \leqq 6L_{\mathrm{ff}} + 544L_{\&}$).

A comparison of (6.1.9), (6.1.10) and (6.2.8) shows that the m-fold gain in speed over the serial variant is achieved at the cost of an increase in the upper bound on complexity; when the serial variant uses single-coordinate storage, the increase is

$$\begin{aligned} \Delta L = {} & (2^{m-1} + 9m^2 + 9k(m - 1) - 2L + L_{m-1} - 13)\, L_{\&} \\ & - (2m + k +]\log_2 m[+ 2)\, L_{\mathrm{ff}}, \end{aligned} \tag{6.2.9}$$

and for two-coordinate storage

$$\begin{aligned} \Delta L = {} & (2^{m-1} + 9m^2 + 9k(m - 1) + L_{m-1} - 2^{]m/2[+1} - 2^{[m/2]+1} - 5)\, L_{\&} \\ & - (2m + k +]\log_2 m[+ 2)\, L_{\mathrm{ff}}. \end{aligned} \tag{6.2.10}$$

Turning once again to the example of §4.3, we see that the increase in complexity over the serial realization is $\Delta L = 423L_{\&} - 20L_{\mathrm{ff}}$ for single-coordinate storage, $\Delta L = 443L_{\&} - 20L_{\mathrm{ff}}$ for two-coordinate storage.

In most practical situations, the complexity of the parallel realization is increased by a factor of approximately m in comparison with the serial variant.

REMARK. A parallel realization of spectral methods of synthesis may be minimized by either of the two methods described in §6.1.4 (approximation of $\Phi(z)$ by a step function $\Psi(z)$ such that $\max |\Phi(z) - \Psi(z)| < \frac{1}{2}$).

6.3. TECHNICAL REALIZATION OF SPECTRAL METHODS USING EXPANSIONS OVER FINITE FIELDS

6.3.1. The techniques of §§6.1, 6.2 will now be modified, yielding a reduction in the number of stages of the adder Σ and in the lengths of the codes of the coefficients $c_l^{(q)}$ stored in the PSU. The modification will be based on Haar expansions over finite fields (see §1.6).

As we know, the Haar expansion of a step function $\Phi(z)$ representing a system of BF of m arguments has the form (see §§4.1.3, 4.2.2)

$$\Phi(z) = c_0^{(0)} + \sum_{l=0}^{m-1} \sum_{q=1}^{2^l} c_l^{(q)} H_l^{(q)}(z), \tag{6.3.1}$$

where

$$c_l^{(q)} = 2^{-m+l} \left(\sum_{s=2^{m-l-1}+1}^{2^{m-l}} \Phi(2^{m-l}q - s) - \sum_{s=1}^{2^{m-l-1}} \Phi(2^{m-l}q - s) \right). \tag{6.3.2}$$

Let us assume now that all operations on $\Phi(z)$ in (6.3.1) and (6.3.2) are performed modulo some prime number p. We shall refer to the result as a Haar expansion over the finite field $GF(p)$ (the field of residues modulo p). We know (see §1.6) that if

$$p > \max_z \Phi(z) \tag{6.3.3}$$

the Haar functions constitute a complete orthogonal system over $GF(p)$.

For any system of k BF,

$$\max_z \Phi(z) \leqq 2^k - 1,$$

and so condition (6.3.3) becomes simply

$$p > 2^k. \tag{6.3.4}$$

There is always at least one prime in the interval $(2^k, 2^{k+1})$, and so it follows from (6.3.4) that

$$]\log_2 p[= k + 1. \tag{6.3.5}$$

In general, there is a whole set of primes $\Pi_k = \{p_1, p_2, \dots, p_\theta\}$ such that $2^k < p_i < 2^{k+1}$ $(i = 1, 2, \dots, \theta)$; we are thus faced with the problem of how to select an optimal prime

$p_{opt} \in \Pi_k$ for a given $\Phi(z)$. If $\bar{L}_\Phi(p_i)$ $(p_i \in \Pi_k)$ denotes the number of coefficients $c_l^{(q)}$ for $\Phi(z)$ such that $c_l^{(q)} \equiv 0 \pmod{p_i}$, a natural condition for p_{opt} is

$$\bar{L}_\Phi(p_{opt}) = \max_{p \in \Pi_k} \bar{L}_\Phi(p).$$

For large k values, however, the difference

$$\max_{p \in \Pi_k} \bar{L}_\Phi(p) - \min_{p \in \Pi_k} \bar{L}_\Phi(p)$$

is usually not too large, and so p may be selected from Π_k arbitrarily.

If the summation in (6.3.1) is performed over $GF(p)$, then $0 \leqq c_l^{(q)} \leqq p - 1$ for any l, q, and the upper bound on code length for $c_l^{(q)}$ is reduced (see (6.3.5)) from $m + k + 1$ to $k + 1$. There is a corresponding reduction in the capacity of the adder.

For the example of §§6.1 and 6.2, $k = 3$, $\Pi_k = \Pi_3 = \{11, 13\}$, $\bar{L}_\Phi(11) = \bar{L}_\Phi(13) = 0$, and so we put $p = p_{opt} = 11$.

The nonvanishing expansion coefficients of the step function $\Phi(z)$ representing the system of excitation functions in this example, over the field $GF(11)$, are shown in Table 6.3.1. The length of the $c_l^{(q)}$ codes is reduced from 9 to 4 and the number of

TABLE 6.3.1.

$\delta = 2^l + q - 1$	(l, q)	$c_l^{(q)}$	Binary code of $c_l^{(q)}$			
			2^3	2^2	2^1	2^0
0	(0, 0)	10	1	0	1	0
1	(0, 1)	8	1	0	0	0
2	(1, 1)	10	1	0	1	0
3	(1, 2)	2	0	0	1	0
4	(2, 1)	6	0	1	1	0
5	(2, 2)	3	0	0	1	1
6	(2, 3)	6	0	1	1	0
7	(2, 4)	3	0	0	1	1
8	(3, 1)	10	1	0	1	0
10	(3, 3)	5	0	1	0	1
11	(3, 4)	9	1	0	0	1
12	(3, 5)	5	0	1	0	1
13	(3, 6)	7	0	1	1	1
14	(3, 7)	1	0	0	0	1
15	(3, 8)	1	0	0	0	1
20	(4, 5)	10	1	0	1	0
25	(4, 10)	8	1	0	0	0
27	(4, 12)	9	1	0	0	1
28	(4, 13)	7	0	1	1	1
54	(5, 23)	9	1	0	0	1

stages of the adder from 10 to 4, implying a substantial simplification of any realization using the block-diagram of either Figure 6.1.1 or (in particular) Figure 6.2.1.

The block-diagrams of Figures 6.1.1 and 6.2.1 in both serial and parallel versions remain essentially the same, with the sole difference that the adder is replaced by a p-ary adder, $2^k < p < 2^{k+1}$. The complexity estimates for the encoder and adder become, respectively,

$$L_{\text{En}} = \tfrac{1}{2}(k+1)\,L \cdot L_{\text{“1”}} \tag{6.3.6}$$

$$L_{\text{a}\Sigma} \leqq (k+1)\,\tilde{L}^{(1)}_{\text{a}\Sigma} \tag{6.3.7}$$

in the serial case and

$$L_{\text{c}\Sigma} \leqq (m-1)(k+1)\,\tilde{L}^{(1)}_{\text{c}\Sigma} \tag{6.3.8}$$

in the parallel case, where $\tilde{L}^{(1)}_{\text{a}\Sigma}$ and $\tilde{L}^{(1)}_{\text{c}\Sigma}$ denote the complexity of one stage of a mod p adder-accumulator and mod p two-term combinational adder, respectively.

The number L of nonvanishing coefficients does not increase when one goes over to calculations over a finite field. The complexity of all other components of the block-diagrams in Figures 6.1.1 and 6.2.1 (except the encoders and adders) remains unchanged. In addition, there is now no need for mod 2 adders to compute the sign digits. As a result, the upper bound on complexity is reduced by the amount

$$\begin{aligned}\Delta_p L_{\text{Sq}} &= \frac{1}{2}\left(\sum_{l=1}^{m} L_{m-l}(k+l) - L(k+1)\right)L_{\text{“1”}} + L_{\oplus}\\ &\quad + (m+k+1)\,L^{(1)}_{\text{a}\Sigma} - (k+1)\,\tilde{L}^{(1)}_{\text{a}\Sigma}\end{aligned} \tag{6.3.9}$$

for the serial variant and

$$\begin{aligned}\Delta_p L_{\text{Pr}} &= \frac{1}{2}\left(\sum_{l=1}^{m} L_{m-l}(k+l) - L(k+1)\right)L_{\text{“1”}} + mL_{\oplus}\\ &\quad + (m-1)(m+k+1)\,L^{(1)}_{\text{c}\Sigma} - (m-1)(k+1)\,\tilde{L}^{(1)}_{\text{c}\Sigma}\end{aligned} \tag{6.3.10}$$

for the parallel variant.

Setting $L^{(1)}_{\text{a}\Sigma} = \tilde{L}^{(1)}_{\text{a}\Sigma} = L_{\text{ff}}$, $L^{(1)}_{\text{c}\Sigma} = \tilde{L}^{(1)}_{\text{c}} = 9L_{\&}$, $L_{\text{“1”}} = L_{\oplus} = L_{\&}$, we obtain the final estimates for the reduction in complexity:

$$\Delta_p L_{\text{Sq}} = \left(\frac{1}{2}\left(\sum_{l=1}^{m} l \cdot L_{m-l} - L\right) + 1\right)L_{\&} + mL_{\text{ff}}, \tag{6.3.11}$$

$$\Delta_p L_{\text{Pr}} = \left(\frac{1}{2}\left(\sum_{l=1}^{m} l \cdot L_{m-l} - L\right) + 9m(m-1) + m\right)L_{\&}. \tag{6.3.12}$$

An examination of the estimates (6.3.9) through (6.3.12) shows that the passage to expansions over finite fields is justified when $L_{a\Sigma}^{(1)}$ and $\tilde{L}_{a\Sigma}^{(1)}$ or $L_{c\Sigma}^{(1)}$ and $\tilde{L}_{c\Sigma}^{(1)}$ differ from each other only slightly, especially when parallel realization is being employed.

For the example of §4.3, the advantage of computation over $GF(11)$ is

$$\Delta_{11} L_{\text{Sq}} = 11 L_{\&} + 6 L_{\text{ff}}, \qquad \Delta_{11} L_{\text{Pr}} = 286 L_{\&}.$$

Table 6.3.2 summarizes the various complexity estimates obtained for all blocks in the serial and parallel variants, for both ordinary expansions and expansions over finite fields. All data in the table are subject to the assumptions $L_{\text{"1"}} = L_{\oplus} = L_{\&}$, $\tilde{L}_{a\Sigma}^{(1)} = L_{\text{SR}}^{(1)} = L_{\text{Co}}^{(1)} = \tilde{L}_{a\Sigma}^{(1)} = L_{\text{ff}}$, $L_{c\Sigma}^{(1)} = \tilde{L}_{c\Sigma}^{(1)} = 9L_{\&}$. The table also gives complexity estimates for all our realizations of the automaton in the example of §4.3. As regards the estimates for serial and parallel realizations over the field of real numbers ("modulo

		Variant	Modulo	Register z	Counter	Decoder: Single-coordinate PSU	Decoder: Two-coordinate PSU
	1	Serial	∞	$(2m+1)L_{\text{ff}}$	$]\log_2 m[\cdot L_{\text{ff}}$	$2L \cdot L_{\&}$	$(2^{[m/2]+1} + 2^{]m/2[+1} - 8) L_{\&}$
	2	Serial	p	$(2m+1)L_{\text{ff}}$	$]\log_2 m[\cdot L_{\text{ff}}$	$2L \cdot L_{\&}$	$(2^{[m/2]+1} + 2^{]m/2[+1} - 8) L_{\&}$
	3	Parallel	∞	mL_{ff}	0	$\left(\min_l \left(2^{m-l+1} + \sum_{s=1}^{l-1} s \cdot L_{m-l+s} - 4\right)\right) L_{\&}$	
	4	Parallel	p	mL_{ff}	0	$\left(\min_l \left(2^{m-l+1} + \sum_{s=1}^{l-1} s \cdot L_{m-l+s} - 4\right)\right) L_{\&}$	
Example	1	Serial	∞	$13L_{\text{ff}}$	$3L_{\text{ff}}$	$44L_{\&}$	$24L_{\&}$
Example	2	Serial	11	$13L_{\text{ff}}$	$3L_{\text{ff}}$	$44L_{\&}$	$24L_{\&}$
Example	3	Parallel	∞	$6L_{\text{ff}}$	0	$18L_{\&}$	$18L_{\&}$
Example	4	Parallel	11	$6L_{\text{ff}}$	0	$18L_{\&}$	$18L_{\&}$

∞"), the possibility of minimization by approximating the function $\Phi(z)$ to within an error $<\frac{1}{2}$ (see §6.1.4) is not taken into consideration.

It follows from our estimates that, given m and k, the complexity of the realization depends only on the numbers L_{m-l} $(l = 1, 2, \ldots, m-1)$ of nonvanishing series coefficients $c_{m-l}^{(q)}$ $(q = 1, 2, \ldots, 2^{m-l})$.

Thus, the methods of completion and linearization of systems of BF described in §§3.2, 3.3, as well as the methods of state-input assignment, completion of next-state functions and optimal ordering of arguments in the excitation functions, all of which successively minimize the numbers $L_{m-1}, L_{m-2}, \ldots, L_0$, are effective means for minimizing the complexity of network realizations.

The various methods described in this chapter may also be used in the synthesis of reliable (fault-tolerant) digital devices (see §5.4) by spectral methods. If this is done, the block-diagrams of Figures 6.1.1 and 6.1.2 are left unchanged, and one need

TABLE 6.3.2.

Encoder	Adder	Gate block	Sign-production network	Total complexity: Single-coordinate PSU	Total complexity: Two-coordinate PSU
$\frac{1}{2}\sum_{l=1}^{m} L_{m-l}\cdot(k+l)L$	$(m+k+1)L_{\text{ff}}$	$mL_{\&}$	$1\cdot L_{\&}$	$(3m+k+]\log_2 m[+2)L_{\text{ff}} + \left((\frac{1}{2}k+2)L+m+\frac{1}{2}\sum_{l=1}^{m} L_{m-l}\cdot l\right)L_{\&}$	$(3m+k+]\log_2 m[+2)L_{\text{ff}} + \left(\frac{1}{2}k\cdot L+\frac{1}{2}\sum_{l=1}^{m} L_{m-l}+m+2^{[m/2]+1}+2^{]m/2[+1}-8\right)L_{\&}$
$\frac{1}{2}L(k+1)L_{\&}$	$(k+1)L_{\text{ff}}$	$mL_{\&}$	0	$(2m+k+]\log_2 m[+2)L_{\text{ff}} + (\frac{1}{2}k\cdot L+2.5L+m-1)L_{\&}$	$(2m+k+]\log_2 m[+2)L_{\text{ff}} + (\frac{1}{2}k\cdot L+\frac{1}{2}L+m+2^{[m/2]}+2^{]m/2[+1}-9)L_{\&}$
$\frac{1}{2}\sum_{l=1}^{m} L_{m-l}(k+l)L_{\&}$	$9(m-1)(k+m+1)L_{\&}$	0	$m\cdot L_{\&}$	$mL_{\text{ff}} + \left(2^{m-1}+9m^2+m(9k+1)+L_{m-1}+k(\frac{1}{2}L-9)+\frac{1}{2}\sum_{l=1}^{m} L_{m-l}\cdot l-13\right)L_{\&}$	
$\frac{1}{2}L(k+1)L_{\&}$	$9(m-1)(k+1)L_{\&}$	0	0	$mL_{\text{ff}} + (2^{m-1}+9(m-1)(k+1)+L_{m-1}+\frac{1}{2}k\cdot L+\frac{1}{2}L-4)L_{\&}$	
$46L_{\&}$	$10L_{\text{ff}}$	$6L_{\&}$	$1L_{\&}$	$26L_{\text{ff}}+96L_{\&}$	$26L_{\text{ff}}+76L_{\&}$
$35L_{\&}$	$4L_{\text{ff}}$	$6L_{\&}$	0	$20L_{\text{ff}}+85L_{\&}$	$20L_{\text{ff}}+65L_{\&}$
$45L_{\&}$	$450L_{\&}$	0	$6L_{\&}$	$6L_{\text{ff}}+519L_{\&}$	
$35L_{\&}$	$180L_{\&}$	0	0	$6L_{\text{ff}}+233L_{\&}$	

only introduce additional decoders for the error-correcting (or error-detecting) codes. The realization of decoders of this type is considered, e.g., in [81].

With slight modifications, the methods are also applicable when the basis system is the set of Walsh functions, or indeed any other orthogonal system of those considered in Chapter I.

6.4. COMPARISON OF SPECTRAL AND CLASSICAL METHODS FOR SYNTHESIS OF DIGITAL DEVICES

6.4.1. In this, the final section of the book, we shall try to compare our spectral methods for synthesis of combinational networks and finite automata with the classical methods. By "classical" methods we mean those based on the use of Boolean algebra or many-valued logic and finite automata theory. A detailed account of the classical methods of synthesis may be found in, e.g., [62, 70, 92]. Characteristic of these methods is that, generally speaking, they lead to optimal solutions, in the sense, say, of the complexity of the network being designed. However, they all require an exhaustive search of all alternatives in order to tackle such important problems as minimization of systems of BF, optimal state and input assignment, optimal completion of incompletely specified BF and automata, and so on. Moreover, this limitation is apparently intrinsic [75, 143] to these problems. The amount of computational work involved in exhaustive search increases at an enormous rate with increase in the number of arguments of the BF or in the number of states and input signals of the automaton; as a result, exact solution of the above-mentioned problems, even for a relatively small number of arguments, usually presents even the most sophisticated computer with almost insurmountable difficulties. And when the classical methods do yield an exact solution, the cost involved may completely outweigh the advantage of optimality. Moreover, in the classical framework it is frequently difficult to estimate the minimal complexity of the synthesized network before the end of an extremely laborious synthesis process.

On the other hand, the classical methods usually depend essentially on the basis system of elements from which the device is being designed. However, new and technologically economical basis systems are now being rapidly developed, integrated and homogeneous circuitry is extensively used. There is an urgent need for the synthesis of networks with time-variable structure and networks capable of adapting themselves to the external medium, and this situation will in all probability become even more acute in the future. What we need are synthesis methods which depend only weakly on the specific features of the basis system of elements, suitable for application to integrated and homogeneous circuitry, and methods for the design of networks of variable structure and networks adaptable to changing environments.

An additional factor is the increasing use made today of multistable elements, networks built from elements with different numbers of stable states, and digital-analog networks. This requires sufficiently general methods, applicable to design of networks with any number of stable states.

Finally, the design of devices equipped with efficient error-detecting and error-correcting abilities is a no less important and pressing task of synthesis methods.

Spectral methods meet all these requirements to a considerable degree. The main shortcoming of spectral methods is that they are approximate—the network realizations that they produce are not absolutely minimal. Nevertheless, they possess several advantages.

6.4.2. A major advantage is that spectral methods of synthesis are convenient for computer implementation and yield solutions to problems of quite high dimensions. Underlying this circumstance is the fact that in the spectral approach the structure of the device is rigidly fixed (see Figures 3.1.1, 6.1.1 and 6.2.1). Consequently, one can tackle the important problems involved in minimizing the complexity of networks realizing systems of LF (§3.2), optimal state-input assignment for automata (§4.2), optimal completion of incompletely specified LF (§3.3) and automata (§4.3) without having to resort to brute-force techniques.

In addition, the spectral methods provide an easy estimate of the complexity of the network being designed, based on the number of nonvanishing expansion coefficients; this in turn makes it possible to estimate the expected complexity of the network in terms of the number of components required for a physical realization (see §§6.1, 6.2, 6.3).

6.4.3. Another essential merit of spectral methods is their very weak dependence on the basis system of components. Indeed, the block-diagram of a device synthesized by spectral methods contains familiar and standard components of computer technology: registers, adders, counters, decoders, etc. Consequently, realization is simplified, technological feasibility is increased and, in particular, there are improved prospects for microminiaturization and implementation by integrated circuitry or homogeneous structures.

Since the specific features of the functions realized by a network (given the number of arguments or of states and input signals) have an effect only on the content of the memory (encoders in Figure 6.1.1 or 6.2.1), this makes for easy readjustment of the network to the realization of any system of functions. Readjustment may be effected by erasing the information held in the memory and replacing it by the expansion coefficients of the step function representing the new system of LF or excitation functions. A similar principle may be employed to adapt the network to changes in its environment.

6.4.4. Yet another advantage of spectral methods, especially for the synthesis of combinational networks, is their weak dependence on the number of stable states of the basic components. This universal property makes spectral methods particularly suitable for the design of devices with elements having different numbers of stable states, and for digital-analog devices.

Indeed, in many cases the orthogonal expansions of step functions representing systems of logical functions or excitation functions in terms of the basis functions described in §§1.2, 1.3 yield quite simple analog realizations of these step functions. The block-diagram of the device will then contain, apart from a basis-function generator, only an analog adder to compute the sum of the series and L resistors (where L is the number of nonvanishing expansion coefficients) with resistances proportional to the expansion coefficients. This approach may be extremely convenient for small L and small numbers of arguments of the original function or states and inputs of the automaton, especially in the synthesis of digital-analog devices.

In addition, as we saw in §5.4, the spectral approach provides for highly simple techniques of error detection and/or correction using systems of arithmetic codes, and it is also possible to organize simultaneous checking in any system of devices containing arithmetic units.

6.4.5. All the spectral methods of synthesis described in this book, as regards both systems of logical functions and automata, admit not only network interpretations but also program implementation. In other words, they may be utilized to simulate systems of LF and automata on a computer. In so doing, one is again presented with the problem of minimizing the number of nonvanishing expansion coefficients to be stored in the memory of the computer, and this may be done with the aid of methods set forth in this book.

BIBLIOGRAPHY

1. Ahmed, N. and K. R. Rao. Discrete Fourier and Hadamard Transforms.—*Electronics Letters* **6**, No. 7 (1970), 221–224.
2. Ahmed, N. and K. R. Rao. Spectral analysis of linear digital systems using BIFORE.—*Electronics Letters* **6**, No. 2 (January 22 1970), 43–44.
3. Ahmed, N., K. R. Rao, and A. L. Abdusattar.—BIFORE or Hadamard Transform.—*IEEE Trans. Audio Electroacoust.* **AU–19** (September, 1971), 225–234.
4. Ahmed, N., K. R. Rao, and A. L. Abdusattar. On cyclic autocorrelation and Walsh–Hadamard transform.—*IEEE Trans. Electromagnetic Compatibility* **EMC-15** (August 1973), 141–146.
5. Ahmed, N., K. R. Rao, and R. B. Schultz. The generalized transform.—*Proc. 1971 Symp. Application Walsh Functions AD–727–000,* pp. 60–67. Washington, D.C., 1971.
6. Ahmed, N., U. U. Schreiber, and P. V. Lopresti. On notation and definition of terms related to a class of complete orthogonal functions.—*IEEE Trans. Electromagnetic Compatibility* **EMC–15** (May, 1973), 75–80.
7. Aizenberg, I. I., Yu. L. Ivaskiv, D. A. Pospelov, and U. A. Khaidakov. Many-valued threshold functions.—*Kibernetika,* No. 4 (1974). (Russian).
8. Aizenberg, I. I. and Z. L. Rabinovich. Some classes of functional systems of operations and canonical forms of many-valued logical functions.—*Kibernetika,* No. 2 (1965), pp. 37–45. (Russian).
9. Allen, C. M. and D. D. Givone. A minimization technique for multiple valued logic systems.—*IEEE Trans. Computers* **C–17** (February, 1968), 182–184.
10. Andrews, H. C. Digital image transform processing.—*Proc. 1971 Symp. Applications of Walsh Functions, AD–727–000,* p. 26, Washington, D.C., 1971.
11. Andrews, H. C. and K. L. Caspari. A generalized technique for spectral analysis.—*IEEE Trans. Computers* **C–19**, No. 1 (January, 1969).
12. Apple, G. G. and P. A. Wintz. Calculation of Fourier transforms on finite abelian groups.—*IEEE Trans. Information Theory* **IT–16** (March 1970), 233–234.
13. Bartee, D. S. and G. F. Schneider. Computation with finite fields.—*Information and Control* **6** (June, 1963), 79–99.
14. Berlekamp, E. R. *Algebraic Coding Theory.*—New York, McGraw-Hill, 1968.
15. Booth, T. L. An analytical representation of signals in sequential networks.—*Proc. Symp. Mathematical Theory of Automata* (Microwave Research Inst. Symposia Series, Vol. XII), pp. 301–340, New York, Polytechnic Press, 1963.
16. Böbwetter, C. The mutual spectral representation of trigonometric functions and Walsh functions. —*Proc. 1971 Symp. Applications Walsh Functions AD–727–000,* pp. 43–46, Washington, D.C., 1971.
17. Bursenov, N. R. A ternary arithmetic machine.—*Vestnik Moskov. Univ.* **2**, 39–48. (Russian).
18. Butson, A. I. Generalized Hadamard matrices.—*Proc. Amer. Math. Soc.* **13** (1962), 894–898.
19. Butzer, P. D. and H. J. Wagner. Approximation of Walsh polynomials and the concept of a derivative.—*Proc. 1972 Symp. Walsh Functions,* pp. 388–392.

20. Cairns, J. W. On the fast Fourier transform on finite abelian groups.—*IEEE Trans. Computers* **C-20**, No. 5 (May, 1971), 569–572.
21. Carl, J. W. and R. V. Swartwood. A hybrid Walsh transform computer.—*IEEE Trans. Computers* **C-22**, No. 7 (July, 1973).
22. Chang, S. H. and J. Joseph. On ordering of a class of generalized Walsh functions.—*Proc. 1972 Symp. Walsh Functions*, pp. 337–343.
23. Chow, C. K. On the characterization of threshold functions.—*Proc. Ann. Symp. Switching Circuit Theory and Logical Design* (September, 1961), pp. 34–38.
24. Chrestenson, H. E. A class of generalized Walsh functions.—*Pacific J. Math.* **5** (1955), 17–31.
25. Yuen, C. K. Remarks on the ordering of Walsh functions.—*IEEE Trans. Computers* **C-21** (December, 1972) (Correspondence), 1452–1453.
26. Coleman, R. P. Orthogonal functions for the logical design of switching circuits.—*IRE Trans. Electronic Computers* **EC-10**, No. 3 (1961), 379–383.
27. Cooley, J. W. and I. W. Tukey. An algorithm for machine computation of complex Fourier series. —*Math. Comp.* **19** (1965), 297–301.
28. Davies, A. C. On the definition and generation of Walsh functions.—*IEEE Trans. Computers* **C-21**, No. 2 (February, 1972), 187–189.
29. Dertouzos, M. L. *Threshold Logic: A Synthesis Approach.*—Cambridge, Mass., M.I.T. Press, 1965.
30. Dvortsin, V. I. and V. I. Ivanenko. Structural synthesis of controllers in automatic control systems on the basis of threshold element networks.—*Kibernetika.* No. 5 (1966), 45–50. (Russian)
31. Doetsch, G. *Anleitung zum praktischen Gebrauch der Laplace-Transformation und der Z-Transformation.*—Oldenbourg, München, Wien, 1967.
32. Edmonds, J. Paths, trees and flowers.—*Canad. J. Math.* **17**, No. 3 (1965), 449–468.
33. Edmonds, J. Maximum matching and a polyhedron with 0, 1 vertices.—*J. Res. Nat. Bur. Standards* **69B** (1965), 125–130.
34. Edwards, C. R. The application of Rademacher-Walsh transforms to digital circuit synthesis.—*Symp. Theory and Applications of Walsh and Other Nonsinusoidal Functions*, Hatfield Polytechnic Inst., United Kingdom, 1973.
35. Edwards, C. R. The application of Rademacher Walsh transform to Boolean function classification and theshold logic.—*IEEE Trans. Computers* **C-24**, No. 1 (January, 1975).
36. Erosh, I. L. and M. G. Karpovsky. Error correction in arithmetic units based on multistable elements.—*Teoriya Avtomatov*, No. 2 (1970), 78–79. (Russian)
37. Faradzhev, R. G. Analytical methods of computation in linear sequential machines.—*Tekhnicheskaya Kibernetika*, No. 5 (1965), 93–102. (Russian)
38. Faradzhev, R. G. On the spectral approach to problems of sequential machines.—*Avtomat. i Telemekh.*, No. 5 (1966), 50–55. (Russian)
39. Faradzhev, R. G. Methods of sampled-data control systems in the theory of finite automata (sequential machines).—*Izv. Akad. Nauk. Azerb. SSR. Ser. Fiz.-Tekhn. i Mat. Nauk.* No. 1 (1969). (Russian)
40. Ferrel, S. Some Walsh function analysis techniques.—*Proc. 1971 Symp. Application Walsh Functions, AD-727-000*, pp. 151–154, Washington, D.C., 1971.
41. Fine, N. J. The generalized Walsh functions.—*Trans. Amer. Math. Soc.* **69** (1950), 66–77.
42. Frank, J. Implementation on dyadic correlation.—*Proc. 1971 Symp. Application Walsh Functions, AD-727-000*, pp. 111–117, Washington, D.C., 1971.
43. Frankel, H. D. Applications of Walsh functions.—*Proc. 1971 Symp. Application Walsh Functions, AD-727-000*, pp. 134–136. Washington, D.C., 1971.
44. Garaev, M. U. and R. G. Faradzhev. On an analog of Fourier expansion over Galois fields and its applications to problems of generalized sequential machines.—*Izv. Akad. Nauk Azerb. SSR, Ser. Fiz.-Techn. i Mat. Nauk*, No. 6 (1968), 69–75. (Russian)

45. Gelbaum, B. R. On the functions of Haar.—*Ann. of Math.* **51** (1950), 26–36.

46. Gentleman, W. M. Matrix multiplication and fast Fourier transform.—*Bell System Tech. J.* **47** (1968), 1099–1103.

47. Gentleman, W. M. and G. Sande. Fast Fourier transform for fun and profit.—*AFIPS Proc.* **29** (1966), 563–578.

48. Gethöffer, H. Mutual mapping of generalized convolution systems.—*Proc. 1972 Symp. Walsh Functions*, pp. 310–316.

49. Gibbs, J. E. and M. J. Millard. *Walsh functions as solutions of logical differential equations* (Report No. 1, 1969); *Some methods of solution of linear ordinary logical differential equations* (Report No. 2, 1969).—National Physical Laboratory, Division of Electrical Science, Teddington, Middlesex, England.

50. Gibbs, J. E. and F. R. Pichler. Comments on transformation of Fourier power spectra into Walsh power spectra.—*Proc. 1971 Symp. Applications Walsh Functions*, pp. 51–54, Washington, D.C., April, 1971.

51. Givone, D. D. and R. W. Snelsire. *Final report on the design of multiple-valued logic systems.*—Digital Lab. Dep. Elec. Engng., State Univ. of New York, Buffalo, June, 1968.

52. Glukhov, Yu. I. and A. P. Boikov. Spectral representations of Boolean functions.—*Izv. Leningrad. Elektrotekhn. inst. im. Lenina*, No. 91 (1969), 57–60. (Russian)

53. Golomb, S. W. and L. D. Baumert. The search for Hadamard matrices.—*Amer. Math. Monthly* **70** (1963), 12–17.

54. Good, I. J. *The relationship between two fast Fourier transforms.*—*IEEE Trans. Computers* **C–20** (March, 1971), 310–317.

55. Gorbunov, G. V. and E. S. Moskalev. Application of spectral methods in minimization of combinational networks.—*Avtomat. i Vychisl. Tekh.*, No. 3 (1971). 1–8. (Russian)

56. Gorbunov, G. V. and E. S. Moskalev. On a certain method for constructing digital generators for functions of three variables.—In: *Vychislitel'naya Tekhnika*, No. 1 (edited by V. B. Smolov), pp. 195–206, Energiya, Moscow, 1970. (Russian)

57. Gorbunov, G. V. and E. S. Moskalev. Design of many-input digital function generators by the method of approximating functional lattices.—*Kibernetika*, No. 2 (1971), 62–66. (Russian)

58. Gorbunov, G. V. and E. S. Moskalev. Application of Rademacher-Walsh polynomials to approximation of functions of several variables.—*Kibernetika*, No. 3 (1971), 19–22. (Russian)

59. Haar, A. Zur Theorie der orthogonalen Funktionensysteme.—*Math. Ann.* **69** (1910), 331–371.

60. Haas, G. *Fundamentals and Components of Electronic Digital Computers.*—Philips Technical Library, 1963.

61. Hammond, J. L. and L. B. Jonson. Orthogonal square wave functions.—*J. Franklin Inst.* **273** (1962), 211–225.

62. Haring, D. R. *Sequential-Circuit Synthesis: State Assignment Aspects.*—Cambridge, Mass., M.I.T. Press, 1966.

63. Harmuth, H. F. *Transmission of Information by Orthogonal Functions.*—Berlin, Springer, 1970.

64. Harmuth, H. F. A generalized concept of frequency and some applications.—*IEEE Trans. Information Theory* **IT–14**, No. 3 (1968), 375–382.

65. Hawkins, J. K. *Circuit Design of Digital Computers.*—New York, Wiley, 1968.

66. Helm, H. A. Group codes and Walsh functions.—*Proc. 1971 Symp. Application Walsh Functions, AD–727–000*, Washington, D.C., 1971.

67. Henderson, K. W. Some notes on the Walsh functions.—*IEEE Trans. Electronic Computers* **EC–13** (February, 1962), 50–53.

68. Henderson, K. W. Comment on the fast Fourier transform.—*IEEE Trans. Computers* **C–19** (September, 1970), 850.

69. Hewitt, E. and K. Ross. *Abstract Harmonic Analysis.*—Berlin, Springer, 1963.

70. HILL, F. and G. R. PETERSON. *Introduction to Switching Theory and Logical Design.*—New York, Wiley, 1968.
71. HOPCRAFT, J. E. and R. M. KARP. an $n^{5/2}$ algorithm for maximum matching in bipartite graphs. —*SIAM J. Computing* **2**, No. 3 (1973), 225–232.
72. HWA, H. R. and C. L. SHENG. An approach for the realization of threshold functions of order *r*. —*IEEE Trans. Electronic Computers* **EC-18**, No. 10 (October, 1969), 923–939.
73. HURST, S. L. The application of Chow parameters and Rademacher-Walsh matrices in the synthesis of binary functions.—*Computer J.* **16**, No. 2 (1973), 165–173.
74. YABLONSKII, S. W. Functional constructions in *k*-valued logic.—*Trudy Mat. Inst. Steklov* **51** (1958), 5–142. (Russian)
75. YABLONSKII, S. W. On algorithmic obstacles to the synthesis of minimal contact networks.—*Problemy Kibernetiki,* No. 2 (1959), 75–121. (Russian)
76. ITÔ, T. The Applications of the Walsh Functions to Pattern Recognition and Switching Theory. —*Proc. Symp. Application of Walsh Functions,* Naval Res. Lab., Cambridge, Mass, 1970.
77. KACZMARZ, S. and H. STEINHAUS. *Theorie der Orthogonalreihen.*—New York, Chelsea, 1951.
78. KARPOV, YU. G. and M. G. KARPOVSKY. On optimal methods of code redundancy implementation for error correction in finite automata.—*Kibernetika,* No. 1 (1971), 62–67. (Russian) (English translation: *Cybernetics* (1971), pp. 72–89.)
79. KARPOV, YU. G. and M. G. KARPOVSKY. Decomposition of algebras and synthesis of reliable discrete devices by integral moduli.—*Kibernetika,* No. 3 (1973). (Russian) (English translation: *Cybernetics* (1975).)
80. KARPOVSKY, M. G. and E. S. MOSKALEV. Realization of a system of logical functions by means of expansion in orthogonal series.—*Avtomat. i Telemekh.,* No. 12 (1967), 119–129. (Russian). (English translation: *Automation and Remote Control* **28** (1967), 1921–1932.)
81. KARPOVSKY, M. G. Finite automata with error detection and correction.—*Vychislitel'naya Tekhnika i Voprosy Kibernetiki,* No. 5 (1968), 71–74. (Russian)
82. KARPOVSKY, M. G. and E. S. MOSKALEV. On a method for state assignment in automata.—In: *Proc. 3rd Conference on Theory of Transmission and Coding of Information, USSR, sect. 3,* pp. 32–42. Moscow, Nauka, 1967. (Russian)
83. KARPOVSKY, M. G. and E. S. MOSKALEV. Utilization of autocorrelation characteristics for the realization of systems of logical functions.—*Avtomat. i Telemekh.,* No. 2 (1970), 83–90. (Russian). (English translation: *Automation and Remote Control* **31** (1970), 342–350.)
84. KARPOVSKY, M. G. and E. S. MOSKALEV. Realization of partially-defined functions of algebra of logic by means of expanding into orthogonal series.—*Avtomat. i Telemekh.,* No. 8 (1970), 89–99. (Russian). (English translation: *Automation and Remote Control* **31** (1970), 1278–1288.)
85. KARPOVSKY, M. G. Error correction in automata whose combinational parts are realized by expansion in orthogonal series.—*Avtomat. i Telemekh.,* No. 9 (1971), 204–209. (Russian). (English translation: *Automation and Remote Control* **32** (1971), 1524–1527.)
86. KARPOVSKY, M. G., E. S. MOSKALEV, and B. K. PODKLETNOV. On a method for realization of an (*m*, *k*)-terminal logical network.—*Voprosy Radioelektroniki, ser. 12,* No. 28 (1966), 73–80. (Russian)
87. KARPOVSKY, M. G. and A. A. TROYANOVSKII. Analysis of error-correcting capability of logical functions.—In: *Proc. 5th Symp. on Use of Redundancy in Information Systems,* Leningrad, 1972. (Russian)
88. KARPOVSKY, M. G. and A. A. TROYANOVSKII. Methods for analyzing the correcting power of automata.—*Avtomatika i Vychisletel'naya Tekhnika* **8**, No. 1 (1974). (Russian) (English translation: *Automatic Control and Computer Science* **8**, No. 1 (1974), 22–27.)
89. KARPOVSKY, M. G. and N. S. SHCHERBAKOV. Automata with transition self-correction.—*Kibernetika* **7**, No. 6 (1971), 63–67. (Russian) (English translation: *Cybernetics* (1971), 1011–1015.)

90. KARPOVSKY, M. G. Harmonic analysis over finite commutative groups in linearization problems for systems of logical functions.—to be published in *Information and Control*.

91. KITAHASHI, T. and K. TANAKA. Orthogonal expansion of many-valued logical functions and its application to the realization with a single threshold element.—*IEEE Trans. Computers* **C–21**, No. 2, Short Notes (February 1972), 211–218.

92. KOHAVI, Z. *Switching and Finite Automata Theory*.—New York, McGraw-Hill, 1972.

93. KOHONEN, T. *Digital Circuits and Devices*.—New Jersey, Prentice-Hall, 1972.

94. KREMER, H. Algorithmen zur Berechnung der Haar-Funktionen und der schnellen Haar-Transformation.—*ANT-Forschungsbericht*, No. 12, Inst. für allg. Nachrichtentechnik, Darmstadt, Dezember, 1970.

95. KUHN, H. W. The Hungarian method for the assignment problem.—*Naval Res. Logist. Quart.* 2 (1955), 83–97.

96. KUHN, H. W. Variants of the Hungarian method for assignment problems.—*Naval Res. Logist. Quart.* **3** (1956), 253–258.

97. LANG, S. *Algebra*.—Reading, Mass., Addison–Wesley, 1965.

98. LANGE, F. H. *Correlation Techniques*.—Princeton, N.J., Van Nostrand, 1967.

99. LECHNER, R. J. A transform approach to logic design.—In: *Proc. 9th Symp. Switching and Automata Theory* (1968), pp. 213–214. (Also: *IEEE Trans. Computers*, **C-19**, (1970), 627–640.)

100. LECHNER, R. J. Harmonic analysis of switching functions.—In: *Recent Developments in Switching Theory* (edited by A. MUKHOPADHYAY), New York, Academic Press, 1971.

101. LE DINH, C. T., P. LE, and R. GOULET. Sampling expansion in discrete and finite Walsh-Fourier analysis.—*Proc. 1972 Symp. Application Walsh Functions*, pp. 265–271.

102. LE DINH, C. T. and R. GOULET. Time-sequency-limited signals in finite Walsh transforms.—*IEEE Trans. Information Theory* **IT–20**, No. 2 (March, 1974), 274–277.

103. LEE, J. D. Review of recent work on applications of Walsh functions in communications.—*Proc. 1970 Symp. Application Walsh Functions*, pp. 26–35.

104. LUPANOV, O. B. On a method of network synthesis.—*Izv. Vuzov, Radiofizika*, No. 1 (1958), 43–45. (Russian)

105. LUPANOV, O. B. On the possibilities of synthesis of networks from different types of elements. —*Dokl. Akad. Nauk SSSR* **103** (1955), 561–563. (Russian)

106. MAR, H. Y. L. and C. L. SHENG. Fast Hadamard transform using the *H*-Diagram.—*IEEE Trans. Computers (Correspondence)*, 957–960.

107. MENGER, K. S. A transform for logic networks.—*IEEE Trans. Electronic Computers* **C–18**, No. 3 (1969), 241–250.

108. MOTZKIN, T. S. The assignment problem.—In: *Sixth Symp. Applied Mathematics*, pp. 109–125. New York, Mc Graw-Hill, 1956.

109. MULLER, D. E. Boolean algebras in electric circuit design.—*Amer. Math. Monthly* **61**, No. 7, Part II (1954), pp. 27–28.

109a. NECHIPORUK, E. I. On the synthesis of gate networks.—*Problemy Kibernetiki*, No. 9 (1963). (Russian)

110. NICHOLSON, P. J. M. *Algebraic Theory of the Finite Fourier Transform*.—Ph.D. Thesis, Stanford University, Stanford, Calif. 1969.

111. NINOMIYA, I. A theory of the coordinate representation of switching functions.—*Mem. Fac. Engrg. Nagoya Univ.* **10** (1958), 175–190 (Reviewed in *IEEE Trans. Electronic Computers* **EC–12**, No. 2 (1963), 152).

112. NINOMIYA, I. *A Study of the Structures of Boolean Functions and its Applications to the Synthesis of Switching Circuits*.—Ph.D. Thesis, Univ. Tokyo, 1961.

113. OSTAPKO, D. L. and S. S. YAN. Realization of an arbitrary switching function with two-level network of threshold and parity elements.—*IEEE Trans. Electronic Computers* **C–19**, No. 3 (1970), 262–269.

114. Oran Brigham, E. *The Fast Fourier Transform.*—Englewood Cliffs, N. J., Prentice Hall, 1974.

115. Paley, R. E. A. C. A remarkable system of orthogonal functions.—*Proc. London Math. Soc.* **34** (1932), 241–279.

116. Peterson, W. W. *Error-Correcting Codes.*—Cambridge, Mass., M.I.T. Press, 1961.

117. Pichler, F. *Some Aspects of a Theory of Correlation with Respect to Walsh Harmonic Analysis.* —Technical Research Report R–70–11, Department of Electrical Engineering, Univ. of Maryland, Baltimore, Md., 1970.

118. Pichler, F. On discrete dyadic systems.—*Proc. 1971 Symp. Theory and Applications of Walsh Functions,* Hatfield Polytechnic, Hatfield, Herts, U.K., 1971.

119. Pichler, F. Walsh functions and linear system theory.—*Proc. 1970 Symp. Application of Walsh functions,* pp. 175–182, Washington, D.C., 1970.

120. Pichler, F. On state space description of linear dyadic invariant systems.—*IEEE Trans. Electromagnetic Compatibility* **EMC–13**, No. 3 (1971), 166–170.

121. Pierce, W. H. *Failure-Tolerant Computer Design.*—New York–London, Academic Press, 1965.

122. Pitassi, D. A. Fast convolution using the Walsh transform.—*Proc. 1971 Symp. Applications of Walsh Functions, AD–727–000,* pp. 130–133, Washington, D.C., 1971.

123. Polyak, B. and Yu. Shreider. Application of Walsh Functions in Approximate Computations.— In: *Voprosy Teorii Matematicheskikh Mashin,* Vol. 2 (edited by Bazilevskii), pp. 174–190. Moscow, Fizmatgiz, 1962.

124. Posner, E. C. Combinatorial structures for planetary reconnaissance.—*Proc. 1968 Symp. Error Correcting Codes* (edited by H. B. Mann), New York, Wiley, 1968.

125. Pospelov, Yu. D. *Logical Methods for Network Analysis and Synthesis.*—Moscow, Energiya, 1968. (Russian)

126. Pratt, W. K., J. Kane, and H. C. Andrews. Hadamard Transform Image Coding.—*Proc. IEEE* **57**, No. 1 (January, 1969), 58–68.

127. Rademacher, H. Einige Sätze über Reihen von allgemeinen Orthogonalfunktionen.—*Math. Ann.* **87** (1922), 112–138.

128. Redinbo, G. R. An implementation technique for Walsh functions.—*IEEE Trans. Computers* **C–20** *(Corresp.)* (June, 1971).

129. Rejchart, V. J. Signal flow graph and Fortran program for Haar-Fourier transform.—*IEEE Trans. Computers* **C–21**, No. 9 (September, 1972), 1026–1027.

130. Revuluri, K., K. R. Rao, M. A. Narishan, and N. Ahmed. Cyclic and dyadic shifts, Walsh-Hadamard transform and the *H*-diagram.—*IEEE Trans. Computers* **C–23**, No. 12 (December, 1974), 1303–1306.

131. Robinson, G. Discrete Walsh and Fourier power spectra.—*Proc. 1972 Symp. Walsh Functions,* pp. 298–309.

132. Ross, I. and J. J. Kelly. A new method for representing Walsh functions.—*Proc. 1972 Symp. Walsh Functions,* pp. 359–361.

133. Rudin, W. *Fourier Analysis on Groups.*—New York, Wiley (Interscience), 1967.

134. Rushforth, C. K. Fast Fourier-Hadamard decoding of orthogonal codes.—*Information and Control* **15** (1969), 37–39.

135. Schauder, J. Eine Eigenschaft des Haarschen Orthogonalsystems.—*Math. Z.* **28** (1928), 317–320.

136. Schmidt. R. O. A collection of Walsh analysis programs.—*Proc. 1971 Symp. Application Walsh Functions, AD–727–000,* pp. 88–94, Washington, D.C., 1971.

137. Searle, N. H. A logical Walsh-Fourier transform.—*Conf. Applications Walsh Functions,* April, 1970.

138. Selfridge, R. E. Generalized Walsh transforms.—*Pacific J. Math.* **5** (1955), 451–480.

139. Sellers, F. Jr., M. Y. Hsia, and L. W. Bearnson. Analyzing errors with the Boolean difference. —*IEEE Trans. Electronic Computers* **C–17**, No. 7 (July, 1968), 676–683.

140. Shanks, J. R. Computation of the fast Fourier transform.—*IEEE Trans. Electronic Computers* **C-18**, No. 5 (1969), 457–459.

141. Shannon, C. E. The synthesis of two-terminal switching circuits.—*Bell System Tech. J.* **28**, No. 1 (1949).

142. Sherman, H. Some optimal signals for time measurement.—*Trans. IRE* **IT-2**, No. 1 (1956).

143. Sholomov, L. A. Complexity criteria for Boolean functions.—*Problemy Kibernetiki*, No. 17 (1966). (Russian)

144. Shore, J. E. On the application of Haar functions.—*IEEE Trans. Communications* **COM-21**, No. 3 (March 1973), 209–216.

145. Siemens, K. H. and R. Kitai. Walsh series to Fourier series conversion.—*Proc. 1972 Symp. Walsh Functions*, pp. 295–297.

146. Sobol', I. M. *Multidimensional Quadrature Formulas and Haar Functions*.—Moscow, Nauka, 1969. (Russian)

147. Su, S. V. H. and P. T. Cheung. A computer-oriented algorithm for minimizing multi-valued switching functions.—*Conf. Rec. 1971 Symp. Theory and Applications of Multiple-Valued Logic Design*, pp. 140–152.

148. Swick, D. A. Walsh function generation.—*IEEE Trans. Information Theory* **IT-15**, No. 1 (January, 1969), 67.

149. Thause, A. and M. Davio. Boolean difference calculus and its applications to switching theory. —*IEEE Trans. Computers* **C-22**, No. 4 (April, 1973).

150. Trakhtman, A. M. *Introduction to Generalized Spectral Theory of Signals*.—Moscow, Sovetskoe Radio, 1972. (Russian)

151. Trakhtman, V. A. Factorization of Walsh function matrix ordered according to Paley and by sequence frequencies.—*Radiotekhnika i Elektronika*, No. 12, 1973. (Russian)

152. Trakhtman, A. M. Fundamentals of the linear theory of signals defined on a finite set of points. —*Avtomat. i Telemekh.*, No. 4, 1974. (Russian). (English translation: *Automation and Remote Control* 35 (1974), Part 1, pp. 521–612.) September, 1974.

153. Ulman, L. J. Computation of the Hadamard transform and *R*-transform in ordered form.—*IEEE Trans. Computers* **C-19**, No. 4 (April, 1970), 359–360.

154. Vilenkin, N. I. Concerning a class of complete orthogonal systems.—*Izv. Akad. Nauk SSSR Ser. Mat.*, No. 11 (1947). (Russian)

155. Von Neumann, J. A certain zero-sum two-person game equivalent to the optimal assignment problem.—In: *Contributions to the Theory of Games*, Vol. II (Annals of Mathematics Studies 28), pp. 5–12, Princeton Univ. Press, 1953.

156. Vranesic, Z. G., E. S. Lee, and K. C. Smith. A many-valued algebra for switching systems.—*IEEE Trans. Computers* **C-19** (October 1970), 964–971.

157. Walsh, J. L. A closed set of orthogonal functions.—*Amer. J. Math.* **45** (1923), 5–24.

158. Walsh, J. L. A property of Haar's system of orthonormal functions.—*Math. Ann.* **90** (1923), 38–45.

159. Watari, C. A generalization of Haar functions.—*Tôhoku Math. J.*, No. 8 (1965), 286–290.

160. Whelch, J. E., Jr., and D. F. Guinn. The fast Fourier-Hadamard transform and its use in signal representation and classification.—*IEEE Trans. Aerospace and Electronic Systems* (1968), 561–573.

161. Winder, R. O. The status of threshold logic.—*First Ann. Conf. Inform. Sci. Systems, Princeton, 1967*, pp. 59–67. Princeton, N.J., Princeton Univ. Press. 1967.

162. Yuen, C. K. Walsh functions and Gray code.—*Proc. 1971 Symp. Applications Walsh Functions, AD-727-000*, pp. 68–73. Washington, D.C., 1971.

163. Yuen, C. K. Approximation errors of a Walsh series.—*Proc. 1972 Symp. Walsh functions*, pp. 289–292.

LIST OF MAIN ABBREVIATIONS AND SYMBOLS

BF	Boolean function(s)
LF	logical function(s)
DD	digital device
TE	threshold element(s)
DFG	digital function generator(s)
SG	sequence generator(s)
FA	finite automaton (automata)
PSU	permanent storage unit
CP	clock pulse(s)
$\{R_i(z)\}$	system of Rademacher functions
$\{W_\omega(z)\}$	system of Walsh functions
$\{K_{r,i}^{(p)}(z)\}$	system of generalized Rademacher functions
$\{\chi_\omega^{(p)}(z)\}$	system of Chrestenson functions
$\{H_l^{(q)}(z)\}$	system of Haar functions
$\{M_{r,l}^{(p,q)}(z)\}$	system of generalized Haar functions
$z^{(i)}$	i-th component of binary (p-ary) expansion of number z
$S_f(\omega)$	spectrum of function f
$B^{(f_1,f_2)}(\tau)$	cross-correlation function of functions f_1, f_2
$[z]$	greatest integer $\leqq z$
$]z[$	smallest integer $\geqq z$
$\text{res}_a m$	remainder upon division of m by a (where m and a are integers)
$[a, b), [a, b]$	half-open and closed intervals
$GF(p^k)$	finite (Galois) field with p^k elements
$\lvert\sigma\rvert_p$	determinant of matrix σ over $GF(p)$
Ξ	class of nonsingular matrices over $GF(p)$

$$\text{sign } z = \begin{cases} 1 & \text{if } z > 0 \\ 0 & \text{if } z \leqq 0 \end{cases} \qquad \text{sgn } z = \begin{cases} 1 & \text{if } z > 0 \\ -1 & \text{if } z \leqq 0. \end{cases}$$

$$H(t_0, \ldots, t_{k-1}) = -\sum_{i=1}^{k-1} t_i \log_2 t_i \qquad \text{entropy function}$$

SUBJECT INDEX